# AIR CONDITIONING AND REFRIGERATION REPAIR MADE EASY

# AIR CONDITIONING AND REFRIGERATION REPAIR MADE EASY

**A Complete Step-by-step Repair Guide for Commercial and Residential Air-Conditioning and Refrigeration Units**

2009 EDITION
REPAIRS AND INSTALLATION OF: AIR CONDITIONERS AND HEAT PUMPS,
WALK-IN COOLERS/FREEZERS, REFRIGERATORS, WATER COOLERS
SALAD BARS, OPEN AND CLOSED REFRIGERATED CASES, ICE MACHINES
STEP-BY-STEP RETROFIT INSTRUCTIONS FOR SUVA REFRIGERANT

Copyright © 2010 by Hooman Gohari.

Library of Congress Control Number:     2009904992
ISBN:                                   9781075627040

All rights reserved. No part of this book may be reproduced or transmitted in any form or by any means, electronic or mechanical, including photocopying, recording, or by any information storage and retrieval system, without permission in writing from the copyright owner.

SINCE CONDITIONS OF USE OF THE INFORMATION IN THIS BOOK ARE OUTSIDE OF OUR CONTROL, NEITHER AIR ZONE NOR THE AUTHOR ASSUME ANY RESPONSIBILITY IN CONNECTION WITH ITS USE.

PUBLISHED BY AIR ZONE
503 GRABO DR., SAN ANTONIO, TEXAS 78216

**To order additional copies of this book, contact:**
Xlibris Corporation
1-888-795-4274
www.Xlibris.com
Orders@Xlibris.com

# CONTENTS

Acknowledgment ........................................................................... vii
Introduction.................................................................................... 1
Rules for Personal Safety ............................................................... 3
Good Tools Are Half the Job .......................................................... 5
Basic Refrigeration......................................................................... 7
Residential Refrigerators and Freezers ....................................... 13
The Brazing and Flaring Methods of Connecting Tubing ............ 37
Testing Residential and Commercial Refrigeration Units........... 47
Procedures for the Evacuation and Charging of Commercial and Residential Refrigeration Units ......................................... 107
Additional Controls for Commercial Units and Compressor Service Valves........................................................ 131
Ice Machines .............................................................................. 185
Water Coolers and Fountains ..................................................... 203
Refrigerated Display Cases and Walk-In Coolers ........................ 209
Repair Techniques in Commercial and Residential Air-Conditioning Units............................................ 225
Basic Electricity........................................................................... 289
Troubleshooting Refrigerant Flow Controls............................... 357
Checking Out the Solid-State TEV (Thermal Electric Valve) .......... 383
Servicing a Cooling (Water) Tower ............................................. 393
Substituting SUVA Refrigerants for CFCs .................................... 399
Glossary of Terms ....................................................................... 419
Index............................................................................................ 433

# ACKNOWLEDGMENT

The author and AIR ZONE Refrigeration Company wish to express their deep appreciation to the following companies for their many excellent contributions of illustrations and information, without which the completion of this unique book would not have been possible.

ALCO Controls
Bally Engineered Structures, Incorporated
Beckman Industrial Corporation
Beverage-Air
Buchbinder, Chicago
Coari and Associates
Coleman Heating and Air-Conditioning
Copeland, Division of Emerson Electric Company
DuPont Fluoroproducts
Eaton Corporation, Controls Division
EBCO Manufacturing Company
Gates Rubber Company
Gibson
Henry Valve Company
Honeywell (Controls)
Ice-O-Matic, A Welbilt Company
Johnson Controls
MALCO Products, Incorporated
Marvel Industries, Division of Northland Corporation
OMRON Electronics, Incorporated
Paragon Electric Company, Incorporated
RAM Freezers and Coolers Manufacturing, Incorporated
Robinair Division, SPX Corporation
Tecumseh Product Company
TIF Instruments Inc.
Wagner Products Corporation
White-Rodgers, Division of Emerson Electric Company
Wilshire Corporation, Schaumburg, Illinois

# INTRODUCTION

This comprehensive book has been developed to put an average person into the vast commercial and residential refrigeration and air-conditioning market within a short period of time. It provides all the technical knowledge needed to start a successful refrigeration and air-conditioning business anywhere in the world.

As opposed to the existing publications in this field, this unique book has been written neither at the third-grade level nor does it require a PhD to understand. It is the essence of several years of experience containing the most up-to-date information and methods of troubleshooting and repairing commercial and residential refrigeration and air-conditioning units.

By successfully studying this book and applying its techniques, even those already working in this field can substantially increase their hands on repairing knowledge and technique. For those who would like to pursue this profession as a sideline, just to supplement their incomes, they will find that as they begin to be known and become more and more in demand, they find less and less time for their "regular" jobs; and in a short time, they will get into this profession full time and really generate more business they ever thought possible. Considering the tremendous size of this market, this is, after all, a profession that puts its practitioners in great demand worldwide.

Providing these services to the consumer in his home or business is the most rewarding. Frequently, a unit fails when the customer needs it the most. Therefore, a service call must be prompt and efficient.

This book is written in plain language coupled with hundreds of pictures and illustrations to make it easily understood. The step-by-step troubleshooting, repair techniques and charts in this book make the most complicated jobs very simple.

Considerable effort has been made to keep the text clear and concise. Study each page until it is clear before going on to the next. By studying carefully and paying attention to all the details, the technician will be able to provide his customers with high-quality service.

This is a very profitable business, and the market is as big as the world. Information gained and used from this course is probably the best gift anyone can give his or her loved ones; truly, a gift to last a lifetime.

The average person can develop this into a very lucrative business during his first year of operation.

This book is dedicated to those who are serious about succeeding in this profession, but cannot afford the time or expense of a full-time formal training course.

# RULES FOR PERSONAL SAFETY

WARNING! DISCONNECT POWER SUPPLY BEFORE BEGINNING ANY TYPE OF SERVICING TO AVOID INJURY OR POSSIBLE DEATH FROM ELECTRICAL SHOCK.

1. Always wear safety goggles or glasses when working on refrigeration or air-conditioning units where there is danger of flying particles from compressed gases.
2. Never breathe refrigerant fumes of any kind. Always wear a gas mask when working in a refrigerant-laden atmosphere or near brazing fumes.
3. Do not expose yourself to electrical shocks. Do not work on electrical circuits in moist or wet areas, and always keep open electrical terminals covered with an insulating material.
4. Make sure there is an adequate and proper fire extinguisher available, especially when brazing or using a torch.
5. Prevent severe shocks by discharging capacitors before touching them. Use a capacitor discharger by following the instructions in this book.
6. Never use any gas other than Refrigerant printed on the name plate for developing pressure in residential refrigeration systems.
7. Avoid contact with liquid refrigerant. It can severely burn your skin. When working outside, and plants or shrubs are downwind, do not allow the refrigerant to be released into the air as it kills grass and other small plant life.
8. Always unplug the unit before beginning any electrical or mechanical work.
9. Always use a three-wire extension cord or power supply cord. This is the common grounded three-prong plug found on most new appliances. It is dangerous to personnel and the equipment to disregard proper grounding techniques.
10. When transferring liquid refrigerant from a storage cylinder to a service cylinder, never fill it completely full; the pressure buildup can cause the cylinder to explode.
11. Always keep refrigerant containers and acetylene and oxygen cylinders away from any heat source or flame.

12. Compressor oil mixed with refrigerant in a system may become acidic. Avoid burns by using rubber gloves and safety glasses.
13. Always remove the refrigerant from a system before brazing. This can be done by installing a service valve in the system. When the valve is opened, the gas in the system can be recovered.
14. Never touch liquid refrigerant placed in an open vessel. It can create severe injuries when coming in contact with skin.
15. Always use leg muscles (instead of back muscles) to lift heavy objects.
16. Be cautious of sharp edges or corners when performing any service on refrigerating units.
17. Get into the habit of keeping one hand in a pocket when troubleshooting any equipment with high-voltage circuitry.
18. Do not forget that even a small shock can be dangerous. A reflex reaction to it, could cause a fall against a higher voltage source.
19. Never drop refrigerant cylinders. When transporting two or more, be sure that they are blocked or tied so they cannot strike one another.
20. Never allow cylinders to be moved with a lifting magnet.
21. Always replace the valve protector caps on cylinders after each use.
22. Make sure that the test leads on all meters have the proper insulation. Never use those with frayed or missing pieces of insulation.
23. Always be cautious when vapor refrigerant is released. Vaporizing refrigerant causes a severe "freeze-burn" when it comes in contact with skin. Also, do not touch a valve with bare hands during or immediately after releasing the gas.
24. Never touch any uninsulated wire or terminal while the supply power is connected to the unit.
25. Make sure the unit is properly grounded. In a properly grounded unit, should a short occur, the electrical charge will be carried harmlessly to ground instead of flowing through the person who may be touching it at that time. Electricity flows through the path of least resistance. Since the human body has far more resistance than a grounded circuit, stray current flows directly to ground. Conversely, if someone touches a nongrounded circuit and a short-caused electrical charge decides that his/her body has the least amount of resistance to its path, that person may become the ground.

NOTE: When transporting a refrigerator, avoid laying it down because the compressor oil leaves the compressor. If it is restarted too soon, the compressor will operate without proper lubrication and be destroyed. If it must be laid down for any length of time over five minutes, it must be placed in its upright position for at least twelve hours before a start-up. This gives the oil a chance to settle back into the compressor housing.

# GOOD TOOLS ARE HALF THE JOB

Good hand tools play a very important role in doing high-quality, professional work. Good service technicians need good hand tools. Buy top quality tools and keep them clean and neatly arranged in a good toolbox. Some of the best brands available are Proto, MAC, Stanley, and Craftsman (Sears). The better ones are guaranteed forever against breakage or excessive wear. The best tools give pride in ownership, make the job easier, last longer, and consequently become the least expensive. So buy the best.

It is important to use wrenches that fit the nut or bolt, screwdrivers that fit the slot in the screw head, etc. The wrong-size wrench does not fit or it is slightly too large causes the edges of the nut or bolt to become rounded. Too small of a screwdriver causes the blade to twist or "shave" the edges of the screw slot. The proper-size screwdriver allows the application of more torque with less effort.

When turning a nut or bolt, draw the wrench toward you rather than pushing it away. While pushing, if the wrench slips, it may cause a hand injury. Never try to drive a wrench with a hammer or use a "cheater bar" (a short length of pipe) on the shaft, as it may damage the tool, the nut, or bolt, or cause an injury. If more torque is necessary, use a longer wrench or a loosening agent such as "Screw Loose."

Make it a habit to remove watches and rings and secure any loose clothing before beginning to work. Wearing safety-toe shoes (shoes with a hard steel cap in the toe) is also a good idea.

# BASIC REFRIGERATION

## MOST IMPORTANT: KEEP THE INFORMATION ON THIS PAGE IN MIND AS YOU PROCEED WITH THIS COURSE.

### DISPLACING HEAT BY VAPORIZING REFRIGERANT

Different liquids boil at different temperatures. Some need a great deal of heat to reach their boiling point to vaporize while others require very little.

When speaking of boiling a liquid, we immediately think of heating water in a pan on a stove and watching it bubble. The fact is that the boiling points of some liquids are so low that they vaporize far below room temperature.

In an imaginary experiment, if R-134A liquid refrigerant were placed in a tea cup, it would immediately vaporize, leaving a thick, heavy layer of frost on the cup. This would happen because the relatively high temperatures in the room would cause the liquid refrigerant, with a boiling point of -14.9°F (-26.1°C), to instantly vaporize. The refrigerant absorbs the heat required to boil from the closest object to it, the cup. When the cup gives up its heat to the liquid R-134A, it becomes extremely cold, causing the moisture in the surrounding air to condense on the cup surface and form a thick layer of frost.

### THE FOUR ESSENTIALS IN REFRIGERATION

1. Liquid refrigerant tends to vaporize as pressure imposed upon it is reduced. (It is already known that water boils at a lower temperature at high altitudes where the atmospheric pressure is lower.) This is what happens to the refrigerant as it is drawn into evaporator coil of a refrigeration unit and the reason the evaporator feels cold.
2. Vapor refrigerant tends to return to its liquid state (condensate) as pressure imposed upon it is increased. (This is what happens as refrigerant is forced into the condenser of unit.)
3. Vaporizing refrigerant absorbs heat from its immediate surrounding environment. (This is what happens in the evaporator.)
4. When vapor refrigerant gives up a certain amount of heat, it changes back to its liquid state. (This is what happens as hot refrigerant vapor is forced into the condenser coil and that is why the condenser feels warm.)

# THE PRINCIPLE OF REFRIGERATION

Heat is absorbed from the inside of a refrigerated area through the evaporator coil. It is then delivered to the condenser where it is dissipated in the air outside the refrigeration unit.

Look at figures 1 and 2 to see the route the refrigerant takes, and how it vaporizes or changes back into its liquid state as it is circulated through areas under different pressures.

The compressor is the heart of a refrigerator, freezer, air conditioner, ice machine, etc., all of which operate on the same principle. The compressor circulates refrigerant throughout the system by creating a pressure difference.

The compressor suction power from its suction port reduces pressure in the evaporator, causing the refrigerant to vaporize upon reaching there, absorbing the heat in the freezer compartment (essentials 1 and 3). It is then drawn into the compressor and compressed and forced into the condenser through the compressor discharge port. This action creates high pressures that speed up the refrigerant molecular motion causing the cold vapor to change to hot vapor. In the condenser coil, this added heat, and the heat absorbed in the evaporator, is radiated into the surrounding air causing the vapor refrigerant to cool down and return to its liquid state (essentials 2 and 4).

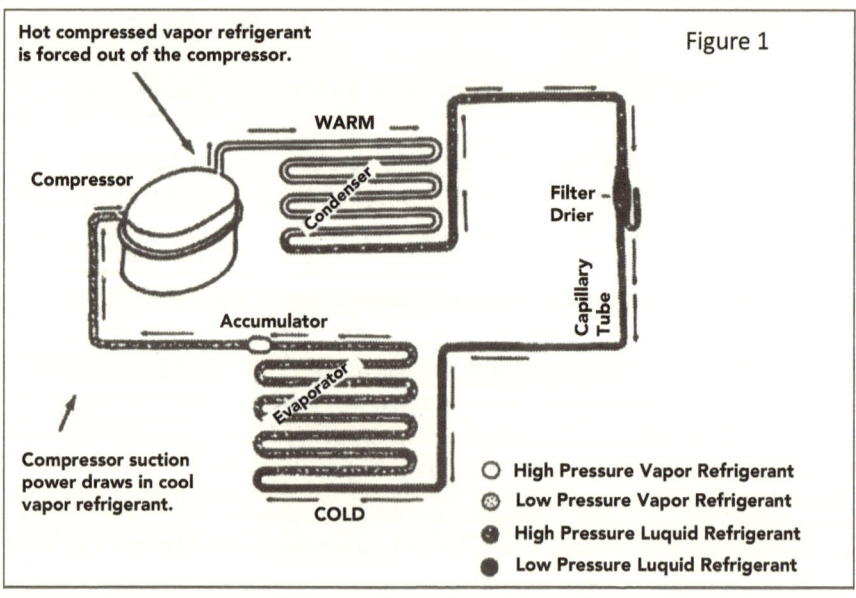

Figure 1

The liquid refrigerant is then circulated through the filter-drier and capillary tube on its route to the evaporator coil connected to the suction side of the compressor.

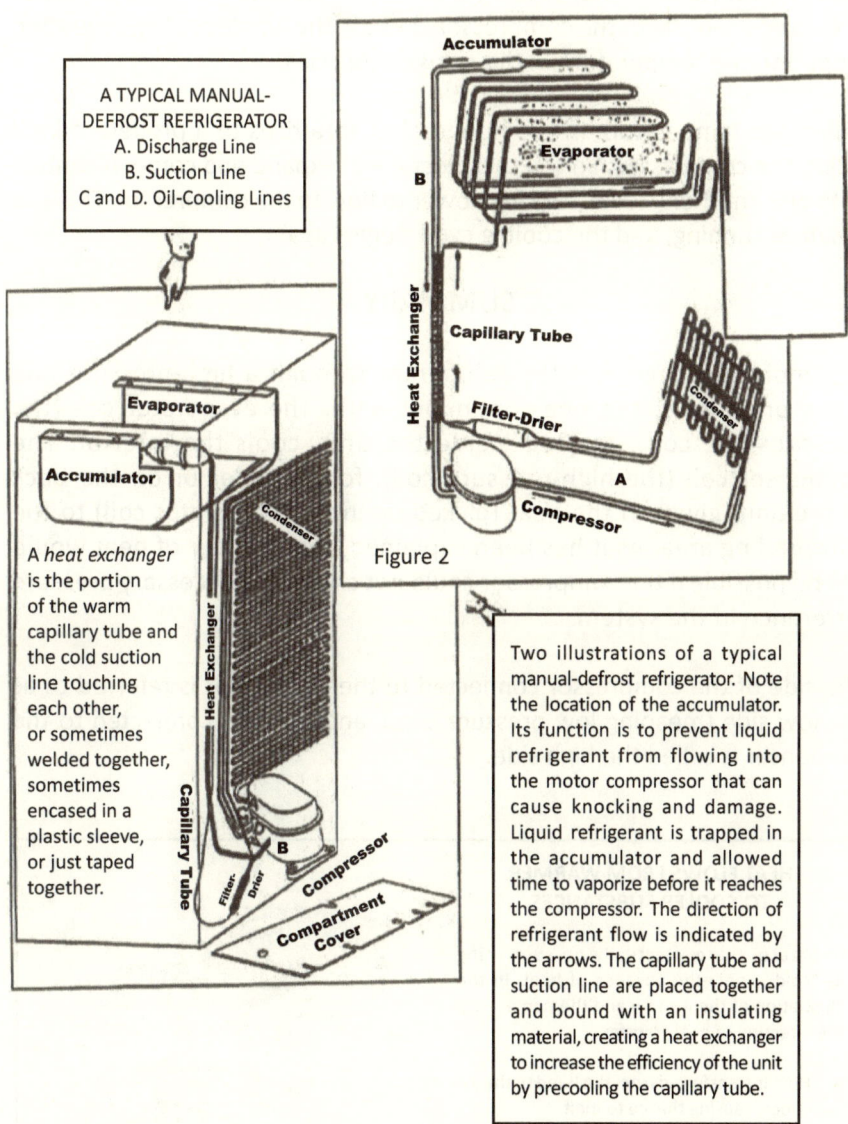

A TYPICAL MANUAL-DEFROST REFRIGERATOR
A. Discharge Line
B. Suction Line
C and D. Oil-Cooling Lines

Figure 2

A *heat exchanger* is the portion of the warm capillary tube and the cold suction line touching each other, or sometimes welded together, sometimes encased in a plastic sleeve, or just taped together.

Two illustrations of a typical manual-defrost refrigerator. Note the location of the accumulator. Its function is to prevent liquid refrigerant from flowing into the motor compressor that can cause knocking and damage. Liquid refrigerant is trapped in the accumulator and allowed time to vaporize before it reaches the compressor. The direction of refrigerant flow is indicated by the arrows. The capillary tube and suction line are placed together and bound with an insulating material, creating a heat exchanger to increase the efficiency of the unit by precooling the capillary tube.

The suction power of the compressor creates a very-low-pressure environment in the evaporator, low enough to vaporize the refrigerant. Once the refrigerant reaches the evaporator low-pressure environment, it is immediately vaporized, and heat from its surrounding area is absorbed, causing the unit to cool.

The cooling cycle continues until a preset temperature is sensed by the cold control (thermostat) causing it to disconnect the electrical circuit to the compressor. With the compressor shut off, the whole cooling operation stops and the temperature in the refrigerator rises.

When the temperature in the refrigerated area rises to a predetermined point, the contacts within the thermostat will expand and come in contact with one another (close), causing power to flow to the compressor. The unit resumes running, and the cooling cycle begins again.

## SUMMARY

A compressor circulates the refrigerant through a high-pressure and a low-pressure coil in one continuous path. The evaporator coil (the low-pressure coil), located inside the unit, cools the interior. The condenser coil (the high-pressure coil), found under or on the back of the unit, gives off the heat (picked up in the evaporator coil) to the surrounding area. As it has been explained, this transfer of heat would not be possible if the compressor could not create the necessary pressure difference in the system.

The side of the compressor connected to the evaporator is referred to as the low side (meaning low pressure side), and the side connected to the condenser is called the high side.

---

**HEAT FLOWS FROM WARMER TO COOLER SUBSTANCES**

In nature, there is actually no such thing as "cold"—only the absence of heat. In the illustration of the ice tray and the flame, note the direction of heat transfer.

A. Heat is transferred from the hand to the ice cubes, causing the ice to melt.
B. Heat is transferred from the flame to the hand, causing the hand to get warm.

Just as water seeks its own level, heat tends to equalize.

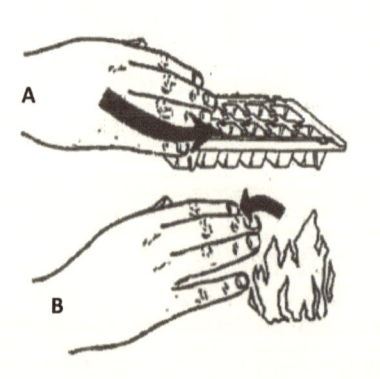

## A TEST OF KNOWLEDGE ON BASIC REFRIGERATION

1. What are the other names for the suction and discharge sides of a compressor? (p. 10)
2. Which part of the unit radiates the heat absorbed by the refrigerant? (p. 8)
3. What causes the liquid refrigerant in the evaporator to vaporize? (pp. 9,10)
4. Why does vaporizing refrigerant in the evaporator make the freezing compartment cold? (pp. 10,8)
5. Explain the four essentials of refrigeration. (p. 7)
6. It is already known that water boils at 212°F at sea level. At what temperature does refrigerant R-134A vaporize at sea level? (p. 7)
7. Never _____ while the power supply is connected to the unit. (p. 3)
8. The side of the compressor connected to the evaporator is referred to as the _____ side. (pp. 10,9)
9. The side of the compressor connected to the condenser is referred to as the _____ side. (pp. 9,10)
10. What component de-energizes the compressor upon reaching the desired cabinet temperature? (p. 10)
11. What causes the unit to start running again after a compressor shut off? (p. 10)
12. Why can refrigerant R-134A not be kept in an open container? (p. 7)
13. When turning a nut or bolt, how should force be applied to the wrench? (p. 5)
14. Why is caution advised when handling refrigerant? (pp. 3,4)
15. What is the purpose of an accumulator? (p. 9)
16. Under normal operating conditions, what is the state of refrigerant flowing through the filter-drier? (p. 9)
17. Why should transporting a refrigerator on its back be avoided? (p. 4)
18. What should be done before removing a capacitor? (p. 3)
19. Where is the filter-drier located? (figs. 1,2)
20. How are a refrigerator, an ice machine, an air conditioner, and a freezer alike? (p. 8)

# RESIDENTIAL REFRIGERATORS AND FREEZERS

This section explains how residential refrigeration units work. It illustrates and describes the function of the various components used in these refrigerators and freezers. It includes manual-defrost, cycle-defrost, and automatic-defrost units.

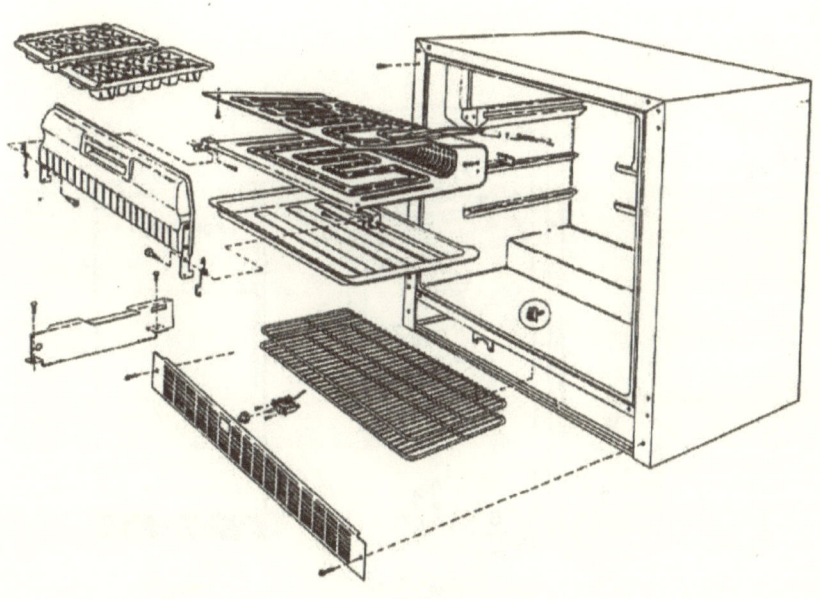

Courtesy of Marvel Industries
Division of Northland Corporation

## THE BASIC REFRIGERATOR

The sealed system consists of a compressor, evaporator, condenser, tubing, filter-drier, and an accumulator. The illustration below is a rear view of a frost-free refrigerator. The evaporator is accessible from the front, and the compressor, condenser, and the condenser fan are accessible when the rear cover panel is removed (it is retained by a few screws). The upper part of the drain system has a small heating element (not shown) to prevent restriction by an ice buildup during the defrost cycle. The front grille is retained by screws or spring clips.

1. Condensation tray
2. Condensation hose
3. Defrost heater
4. Condenser fan motor
5. Compressor
6. Grille
7. Rear cover panel
8. Evaporator fan motor
9. Defrost thermostat

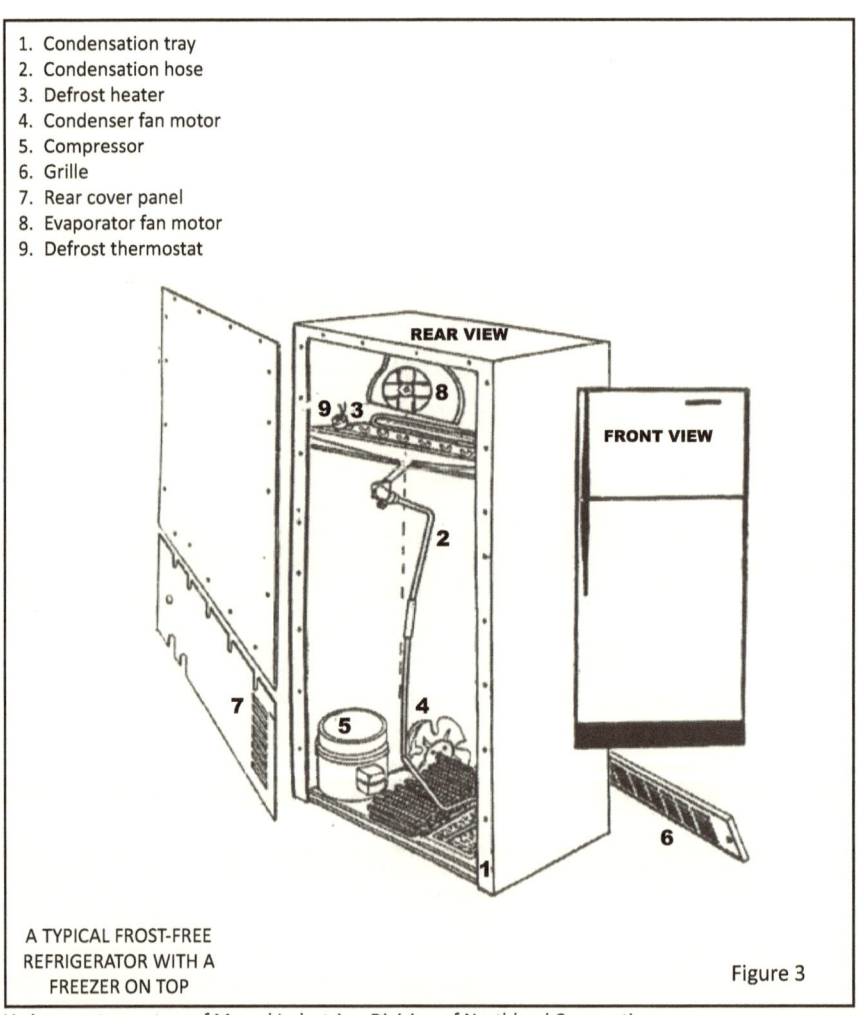

A TYPICAL FROST-FREE REFRIGERATOR WITH A FREEZER ON TOP

Figure 3

Various parts courtesy of Marvel Industries, Division of Northland Corporation

The cabinet, or body of the refrigerator, contains and supports the evaporator, condenser, and the compressor. The evaporator is located inside the freezer compartment where it absorbs heat from the food. This heat is transferred to the condenser where it is radiated to the outside air. The condenser is usually mounted at the back of the unit or in base of the cabinet next to the compressor (figs. 3 and 4). The air in a refrigeration unit is very dry because the moisture in the cabinet tends to collect and condense on the evaporator surface where the temperature is very low. That is why food should be covered with a moisture-proof cover to be kept from drying out.

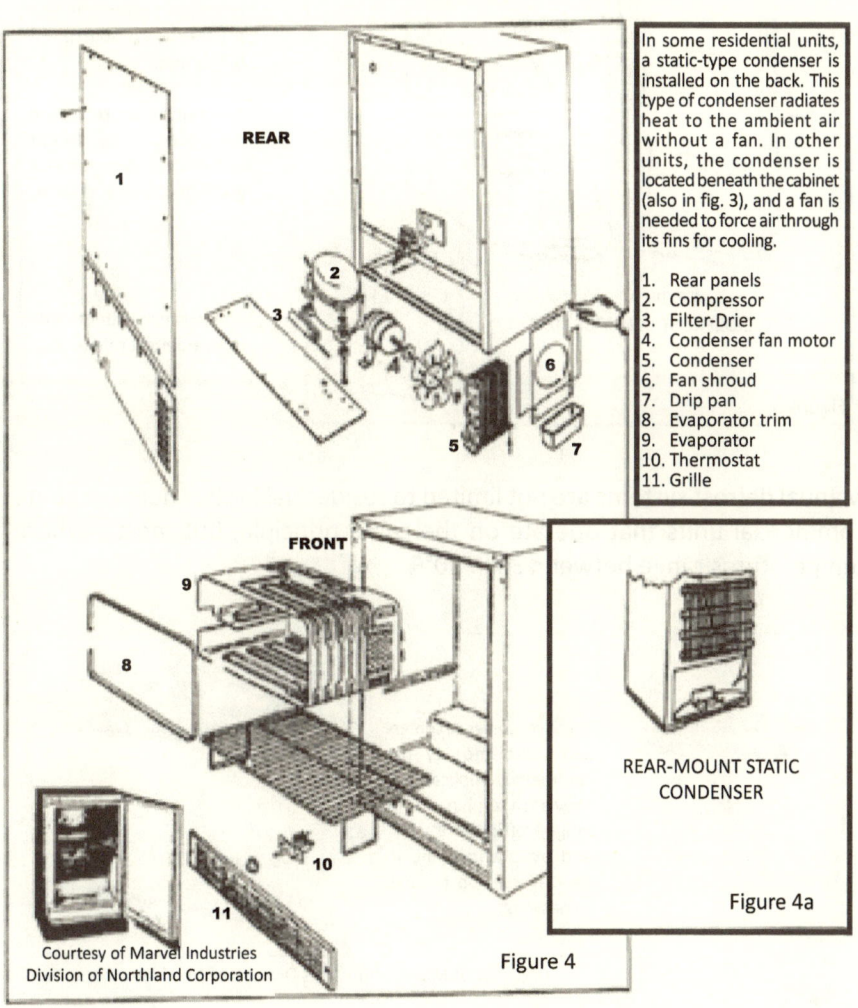

In some residential units, a static-type condenser is installed on the back. This type of condenser radiates heat to the ambient air without a fan. In other units, the condenser is located beneath the cabinet (also in fig. 3), and a fan is needed to force air through its fins for cooling.

1. Rear panels
2. Compressor
3. Filter-Drier
4. Condenser fan motor
5. Condenser
6. Fan shroud
7. Drip pan
8. Evaporator trim
9. Evaporator
10. Thermostat
11. Grille

REAR-MOUNT STATIC CONDENSER

Figure 4a

Courtesy of Marvel Industries
Division of Northland Corporation

Figure 4

The temperature inside the fresh-food cabinet ranges between 33°F and 42°F, while the temperature in the freezer section is between 0°F and 10°F.

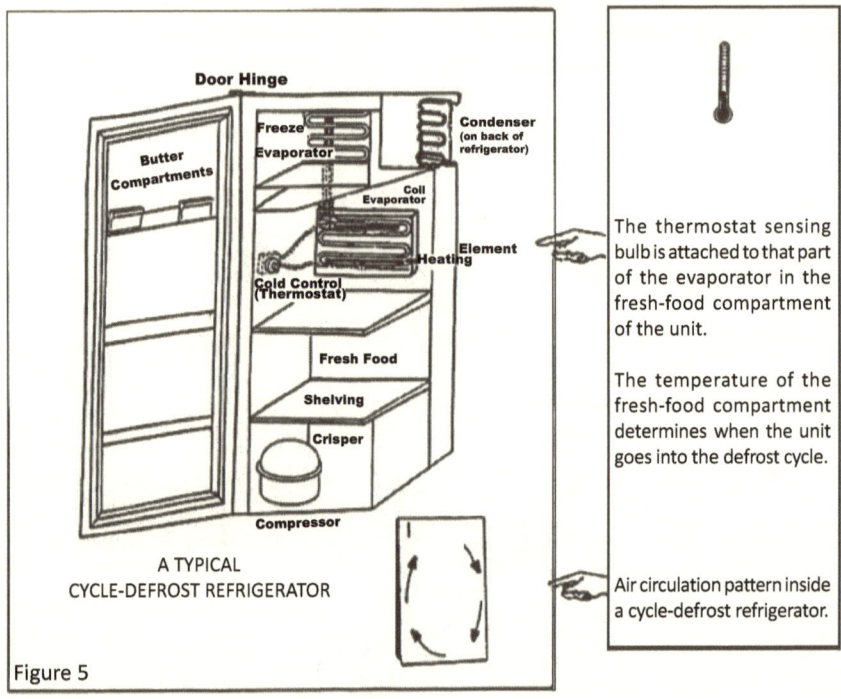

The thermostat sensing bulb is attached to that part of the evaporator in the fresh-food compartment of the unit.

The temperature of the fresh-food compartment determines when the unit goes into the defrost cycle.

Air circulation pattern inside a cycle-defrost refrigerator.

Figure 5

Manual defrost systems are not limited to residential units. There are some commercial units that operate on the same principle, but mostly where temperatures range between 35°F-50°F.

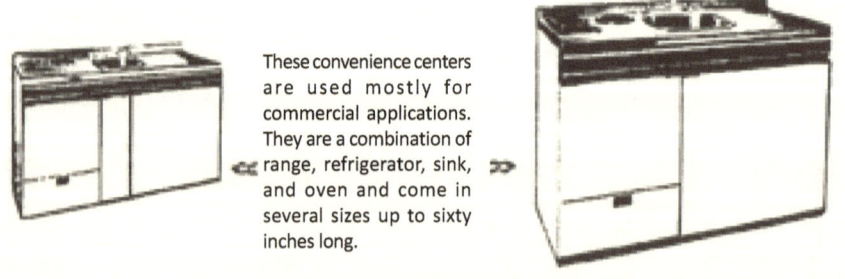

These convenience centers are used mostly for commercial applications. They are a combination of range, refrigerator, sink, and oven and come in several sizes up to sixty inches long.

Courtesy of Marvel Industries Division of Northland Corporation

Generally, the lower the temperature, the longer the food can be preserved.

The walls of freezers and refrigerators contain an insulating material to prevent the loss of cold air and to keep heat from penetrating. Either urethane foam or fiberglass is used as insulation in almost all units. In some commercial and industrial units, other materials may be used as well. The walls of refrigeration units are made of cold rolled steel and are welded together.

In a simple fresh-food refrigerator, the evaporator is located in the top of the cabinet and cold air is distributed automatically because cold air is heavy and flows down from the evaporator surface and settles on the bottom of the cabinet (see fig. 5). In frost-free refrigerators, cold air from the evaporator is forced into the fresh-food compartment by means of a fan because the evaporator and fresh-food compartment are separated by a wall. More about this later. (See fig. 6).

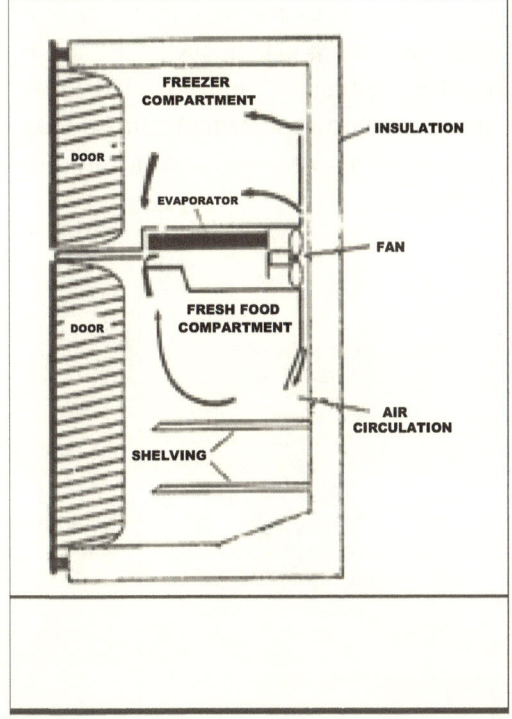

Figure 6

## MANUAL DEFROST AND CYCLE DEFROST

Manual defrost units are the least expensive and should be defrosted every four weeks in the summer, and every eight to ten weeks in the winter. Usually, the refrigerator is shut off and permitted to stay off overnight to become completely defrosted. (Or for faster results, a pan of hot water is placed in the freezer while the unit is turned off for a few hours). The freezer temperature ranges between 5°F and 15°F, and the temperature in the fresh-food compartment ranges between 38°F to 45°F. Normal low-side pressure in manual defrost units varies between 10-15 psi. Cycle-defrost refrigerators are less troublesome than the manually defrosted units, but they are not as sophisticated as the fully automatic units. In the cycle-defrost type, defrosting occurs each time the thermostat becomes satisfied. An electric heating element attached to the extension of the evaporator in the fresh-food compartment (fresh-food evaporator) becomes energized to ensure thorough de-icing of the fresh-food evaporator in the off cycle (see fig. 5).

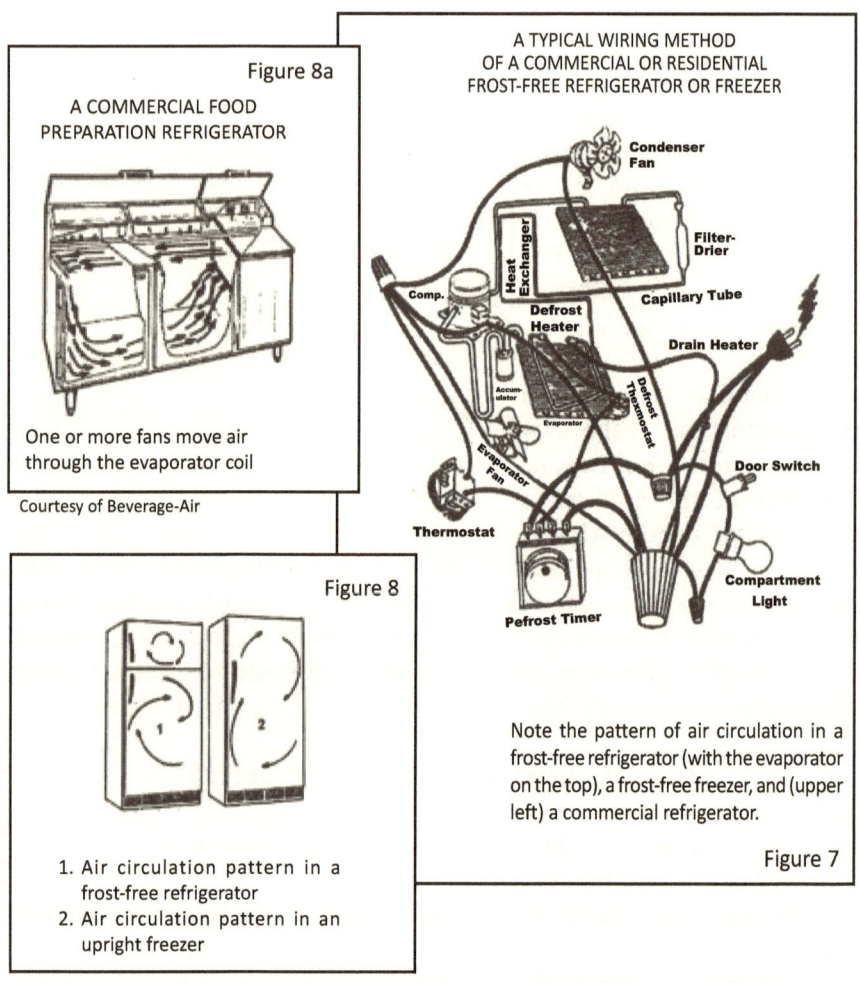

Figure 8a
A COMMERCIAL FOOD PREPARATION REFRIGERATOR

One or more fans move air through the evaporator coil

Courtesy of Beverage-Air

Figure 8
1. Air circulation pattern in a frost-free refrigerator
2. Air circulation pattern in an upright freezer

A TYPICAL WIRING METHOD OF A COMMERCIAL OR RESIDENTIAL FROST-FREE REFRIGERATOR OR FREEZER

Note the pattern of air circulation in a frost-free refrigerator (with the evaporator on the top), a frost-free freezer, and (upper left) a commercial refrigerator.

Figure 7

Cycle-defrost refrigerators should be manually defrosted two to four times a year. In cycle-defrost units, the temperature ranges between 0°F and 10°F in the freezer compartment, and 38°F to 45°F in the fresh-food compartment. The normal low-side pressure ranges between five and ten pounds per square inch (5-10 psi).

## FROST-FREE REFRIGERATORS AND FREEZERS

In these units, the evaporator coil is automatically defrosted in regular intervals every day. A fan is mounted on the evaporator to draw air from one side, move it over the evaporator coil, and force it into the freezer and fresh-food compartments through a damper (see figs. 6, 8, and 9). A thermostat controls the operation of the compressor to maintain food at desired temperatures. It is regulated by a manually adjustable knob. A baffle control knob inside the cabinet of the unit manually controls the temperature of the fresh-food compartment by increasing or reducing the flow of cold air into that compartment.

There are two defrosting systems used in residential frost-free refrigerators and freezers:

1. Electric heater system
2. Hot refrigerant vapor system (hot gas system)

*Electric-defrost system.* A timer (see figs. 7 and 11a-11e) automatically de-energizes the compressor and the evaporator fan motor, and at the same time it energizes the evaporator defrost heater every six, eight, or twelve hours (depending on the type of timer being used) to de-ice the evaporator plate. The power from the timer to the defrost heater flows through a defrost thermostat (also referred to as a "defrost termination switch"), which is clipped to the evaporator coil. When the ice buildup on the evaporator coil is melted, the temperature of the evaporator coil rises to about 50°F ±6°F (10°C ±3°C). The rise in temperature causes the bimetal within the termination switch to open its contacts, thus, de-energizing the defrost electric heating element to end the process of defrosting even with the timer still in its defrost cycle. Contacts within defrost thermostats close at 20°F (-7°C). The whole defrosting process takes between eight and thirty minutes. Over 90% of the frost-free residential refrigeration units of today are equipped with this type of defrosting system.

*Hot gas, automatic defrost system.* In a hot gas defrost system, a bypass line connects the compressor discharge line to the evaporator inlet (between the capillary tube and the evaporator inlet [see figs. 16 and 16a]). When the unit is taken into the defrost mode, a solenoid defrost valve becomes energized, allowing discharged hot refrigerant to circulate directly through the evaporator and quickly defrost it. It is then drawn back into the compressor through the suction line. In this way, the condenser and the capillary tube are bypassed. In the normal cooling cycle, the valve is de-energized (its defrost port is closed) allowing refrigerant to flow through the condenser coil, capillary tube, evaporator, and then back into the compressor suction side. Figure 129 shows a different defrost system in which hot vapor refrigerant is circulated through a second coil adjacent to the evaporator.

Normally, a frost-free unit is equipped with a fan-forced condenser, which is mounted at the bottom rear of the unit.

A fan is mounted next to the condenser and moves ambient air through the condenser to help it dissipate heat.

NOTE: The most common problem in this type of unit is related to a dirty or lint-filled condenser. A dirty condenser loses its ability to transfer heat causing the unit to run continuously but not cool sufficiently. The condenser in a frost-free refrigerator or freezer requires annual cleaning.

In side-by-side refrigerators, the evaporator is mounted at the back of the freezer compartment. A damper regulated by the consumer controls flow of cold air from the freezer to the fresh-food compartment. This cold air forces the warmer air back toward the evaporator cold coil through the bottom of the fresh-food compartment. (See fig. 9)

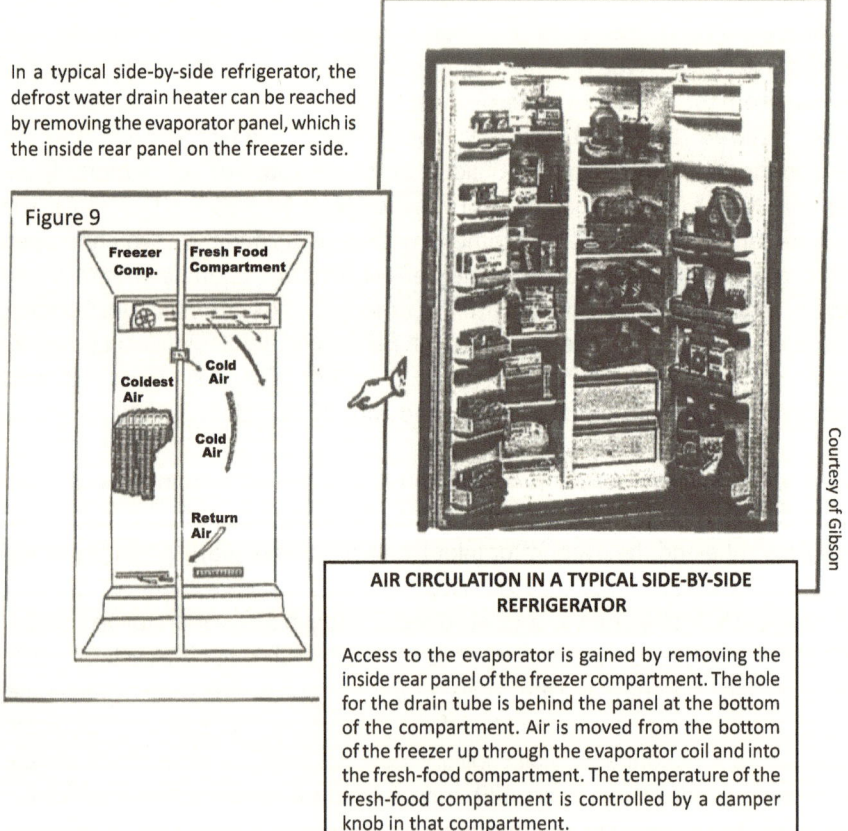

In a typical side-by-side refrigerator, the defrost water drain heater can be reached by removing the evaporator panel, which is the inside rear panel on the freezer side.

Figure 9

**AIR CIRCULATION IN A TYPICAL SIDE-BY-SIDE REFRIGERATOR**

Access to the evaporator is gained by removing the inside rear panel of the freezer compartment. The hole for the drain tube is behind the panel at the bottom of the compartment. Air is moved from the bottom of the freezer up through the evaporator coil and into the fresh-food compartment. The temperature of the fresh-food compartment is controlled by a damper knob in that compartment.

In units with the freezer compartment on top, the evaporator is placed either in the back or on the bottom of the freezer. An evaporator fan forces cold air to the fresh-food compartment through a damper. In frost-free refrigerators, the freezer air temperature runs between 0°F and 10°F, and the fresh-food compartment temperature ranges between 38°F and 45°F (see fig. 3, 6 and 9).

## CHEST-TYPE FREEZERS

Chest-type freezers have an important advantage. Since cold air is heavier than warm air, it has more of a tendency to stay in the bottom of the freezer and not "spill out" each time the door (lid) is opened. Consequently, a lot of moisture does not accumulate in the cabinet. (See fig. 10)

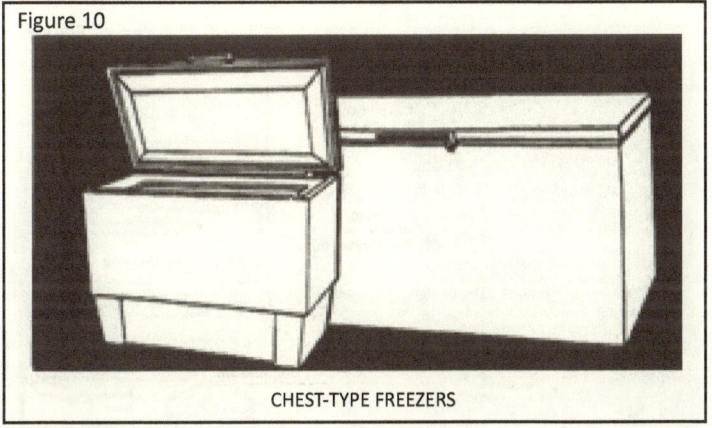

Figure 10

CHEST-TYPE FREEZERS

Due to this fact, chest-type units are manually defrosted as defrosting is usually not needed more than once or twice a year. To defrost the freezer, remove the food from the freezer and wrap it in newspaper to keep it from thawing. Unplug the unit or engage the defrost switch (usually located in the lid or rim of the cabinet). The water from defrosting runs out through a drain in the bottom of the cabinet. (With a garden hose attached, water could drain directly to the outside of the house.)

Most test lights are capable of indicating up to 600 VAC.

Refrigeration and Air-Conditioning Service Technician's Thermometers.

Figure 10a

Courtesy of MALCO Products

Figures 14 and 15 show different types of evaporators used in frost-free, cycle-defrost, or manual-defrost refrigerators. Usually, frost-free refrigerators and residential freezers have fan-forced evaporators and condensers while manual-defrost and cycle-defrost units use static-type (non-fan-forced) evaporators and condensers.

## UPRIGHT FREEZERS

The principle is the same as in the regular refrigerator with the exception that they come with heavier insulation and a different cold-control range. Normal operating temperatures in freezers range between -14°F and +7°F (see fig. 11).

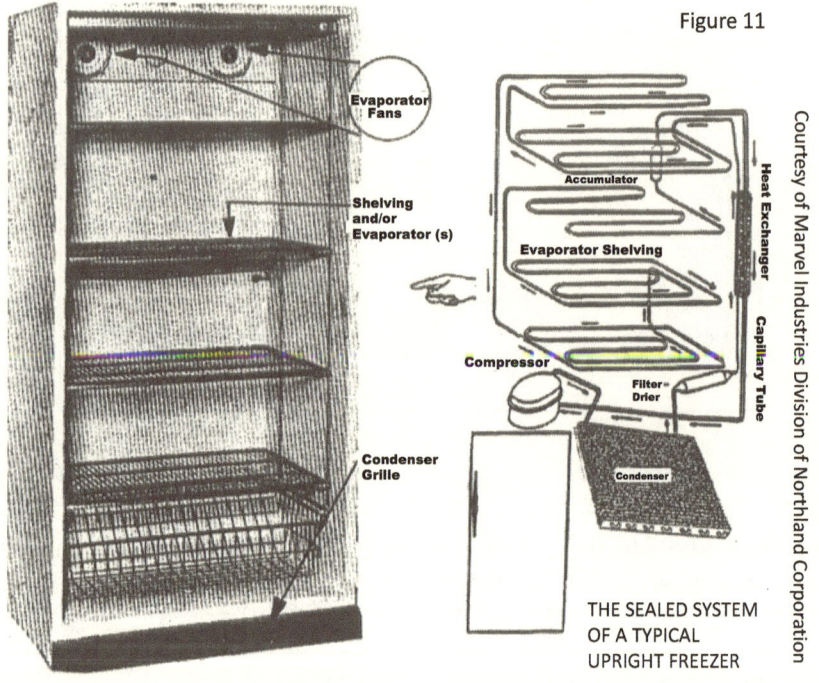

Figure 11

THE SEALED SYSTEM OF A TYPICAL UPRIGHT FREEZER

Courtesy of Marvel Industries Division of Northland Corporation

In freezers with double fans, one draws air from the inside of the cabinet to be circulated over the evaporator while the other one moves the chilled air from the evaporator to the cabinet. Larger commercial units may have three or more fans. They all move air in the same pattern. Depending on the model of freezer, the evaporator coil could be located in the walls, in the back, or even in one or more of the shelves.

## REFRIGERATION UNIT COMPONENTS

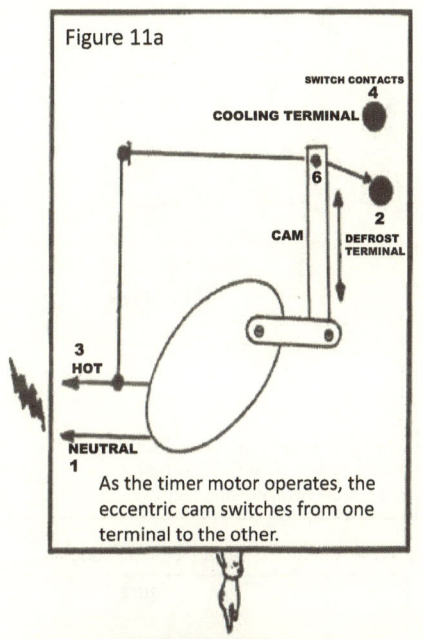

Figure 11a

As the timer motor operates, the eccentric cam switches from one terminal to the other.

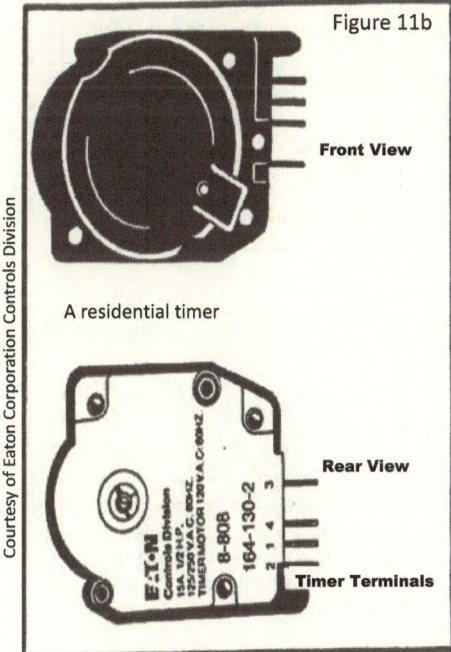

Figure 11b

A residential timer

Courtesy of Eaton Corporation Controls Division

*Automatic defrost timer.* It is a device used in automatic-defrost refrigerators and freezers to take the system into the defrost and then back to the cooling mode in regular intervals.

It consists of an electric motor and a switching mechanism (see fig. 11a). The electric motor is geared to move a cam in such a way that after every six, eight, or twelve hours of compressor operation (depending on the design of the timer), the cam rotates from one contact to another for no longer than thirty minutes. This causes a temporary change in the direction of current flow from the cooling terminal to the defrosting terminal and then back to the cooling terminal again. The cooling terminal of the timer is wired to the compressor and the evaporator fan motor circuits. The defrost terminal is wired to the defrost heater (or a defrost solenoid valve) circuit. (See figs. 54C through 54L for more detail.)

When the timer takes the system into the defrost cycle, the compressor and fan motors remain de-energized, and the defrost heater becomes energized. When it takes the system into the cooling mode, the compressor and evaporator fan motors become energized while the defrost heater remains de-energized.

The four timer terminals are numbered on the back.

When air (which contains moisture) comes in contact with a cold evaporator coil, the moisture condensates and immediately changes to a layer of ice. If the evaporator coil is not heated from time to time, the accumulation of ice will insulate the coil, preventing it from absorbing the heat from its surrounding air inside the unit, and the unit will lose its ability to cool.

When the unit is in the defrost mode, the defrost heaters are energized and the evaporator fan(s) de-energized. The water produced by the melted ice flows by gravity through a plastic pipe and collects in a tray on or beside the compressor where it is heated and evaporated by the condenser and the heat from the compressor.

In many models, an electric heating element is placed around the inlet of the drain tube. It is wired into the defrost heater circuit. Every time the timer takes the unit into the defrost mode, this drain heater is also energized to prevent drain water from freezing in the drain line and clogging it.

In side-by-side units, this drain heater can be seen at the bottom of the evaporator around the drain opening when the evaporator panel is removed (see fig. 9). In the case of units with the freezer on top, it is on the bottom of the freezer compartment around the drain opening.

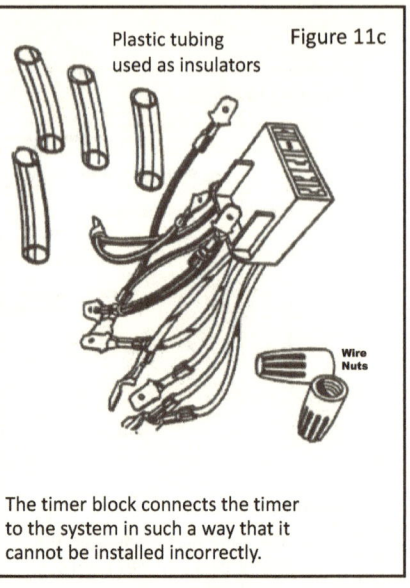

Figure 11c

Plastic tubing used as insulators

Wire Nuts

The timer block connects the timer to the system in such a way that it cannot be installed incorrectly.

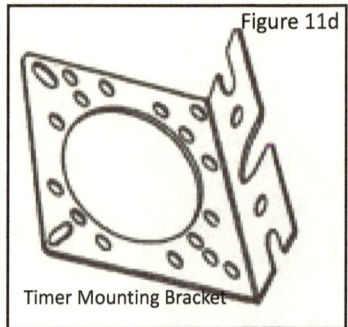

Figure 11d

Timer Mounting Bracket

Courtesy of Eaton Corporation, Controls Division

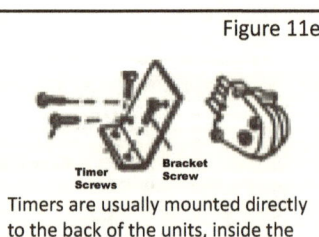

Figure 11e

Timer Screws   Bracket Screw

Timers are usually mounted directly to the back of the units, inside the cabinet, or fastened to a bracket, which is mounted behind the grille.

Courtesy of Marvel Industries, Division of Northland Corporation

Different evaporator designs have different defrosting requirements for optimum efficiency. Some evaporators must be defrosted every six or eight hours, some require defrosting every twelve hours, and some in commercial systems are designed to be defrosted after every cooling cycle in addition to the periodic defrosting at regular intervals. Timers are usually mounted inside the cabinet next to the cold control or behind the toe plate in the front.

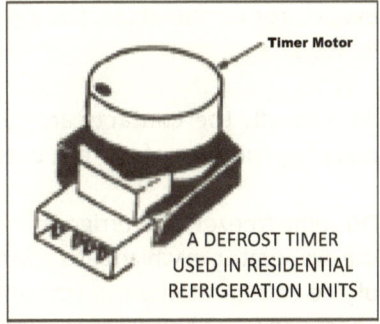

A DEFROST TIMER USED IN RESIDENTIAL REFRIGERATION UNITS

One of the most common problems leading to timer replacement occurs when the timer motor freezes in one position. When it gets stuck in the defrost cycle, the refrigeration unit remains in the defrost cycle, the compressor no longer runs, and the evaporator panel in the freezer compartment feels warm to the touch. When it gets stuck in the cooling cycle, temperature in the freezer and fresh-food compartments goes abnormally high as the compressor runs continuously, causing a thick layer of ice to form on the evaporator coil.

Timer terminals rarely get fused together. This causes the defrost heater to remain on while in the cooling cycle. When this happens, the compressor runs continuously and the unit no longer cools. In this case, the timer must be replaced.

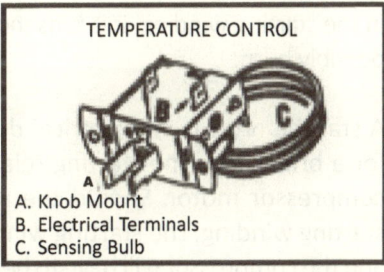

TEMPERATURE CONTROL
A. Knob Mount
B. Electrical Terminals
C. Sensing Bulb

*Thermostat (Temperature Control or Cold Control).* A device for automatically regulating the temperature inside a refrigeration unit by controlling the operation of the compressor. The manually adjustable control knob attached to its operating mechanism is mounted inside the cabinet, usually with two screws. It consists of a gas-charged capillary tube (called a sensing bulb) connected to the operating mechanism. The sensing bulb is attached to the evaporator. The operating mechanism of the thermostat responds to the pressures exerted from the trapped gas within the sensing bulb. As the temperature of the evaporator coil drops, the volume of gas decreases (due to contraction), and thus the amount of pressure exerted upon the accordion-type bellows within the thermostat operating mechanism is reduced, causing the bellows to contract and open a set of contacts, disrupting the flow of current to the compressor motor. With the rising evaporator temperature, the gas expands and exerts more pressure on the bellows.

As a result, the contacts are forced to close by the expanded bellows, restoring the flow of current to the compressor motor.

On some freezer and refrigerator models, the sensing bulb is not attached to the evaporator, in which case, the temperature of the refrigerated air is transmitted to the operating mechanism.

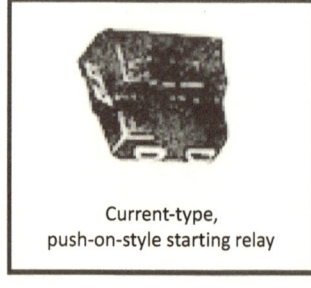

Current-type, push-on-style starting relay

*Starting relay.* Because starting a compressor requires more torque and draws at least three times the amperage of its normal running speed, compressor motors are equipped with a starting winding and a running winding and both compose the stator. At the instant of starting, current flows through both windings. When motor speed increases to about 75% of its normal running speed, a starting relay disconnects power to the starting winding, and the running winding continues operating the motor (see fig. 13). Since heavy wire is used in the starting winding, if it runs more than a few seconds, it will overheat and possibly burn.

A starting relay is an electrical device that energizes the starting winding for a brief time. The starting relay plays an important role in the life of a compressor motor. Should the relay not disengage from energizing the starting winding, the starting winding (in the compressor motor) will burn, and the compressor will have to be replaced. In household refrigeration units, they are installed under the compressor terminal cover and connected to the compressor start and run terminals. Some relays come with three openings that connect the starting relay to the three terminals of the compressor. Many GE units are of this type. (See figs. 46 and 46a.)

*Overload protector.* This is overheat/overload protection for the compressor in the form of a bimetal, reusable safety fuse. If, for any reason, the temperature of the compressor motor goes too high, the bimetal strip within

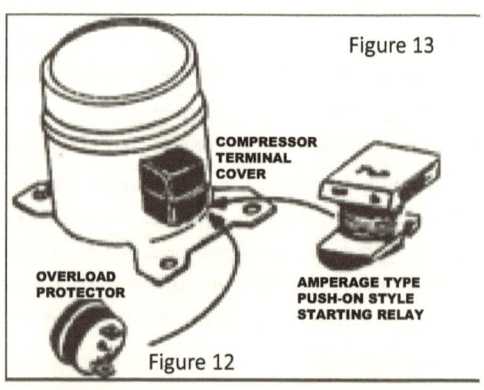

Figure 13

Figure 12

the overload protector will shut off the power to the compressor until it cools sufficiently for safe operation. It is installed under the compressor terminal cover. (See figs. 12, and 55.)

Under this cover, there are three compressor terminals: a run winding terminal marked R, a start winding terminal marked S, and a common terminal marked C. The overload protector is connected in series with the common terminal of the compressor.

**A HERMETIC COMPRESSOR**

1. Plastic cover
2. Discharge line
3. Suction line

*Courtesy of Marvel Industries Division of Northland Corporation*

*Compressor.* The compressor is the heart of the refrigeration unit. It is a motor-operated device that circulates refrigerant much as a pump would in a sealed system. All household and many commercial refrigeration units employ hermetic compressors. (This means that the compressor with its motor are sealed in an airtight canister as opposed to belt-driven compressors.)

When energized, it creates enough pressure difference to circulate the confined refrigerant in the entire sealed system. (See fig. 1.) Through the compressor suction side, vapor refrigerant in the evaporator is drawn in and changed to hot vapor by compression. It is then forced into the condenser (through the compressor discharge tube) where it is cooled to its liquid state again before reaching the evaporator. An efficient compressor must be able to remove the refrigerant vapor at the same rate that liquid refrigerant enters the evaporator and vaporizes. The low-pressure side of the compressor is connected to a tubing having a larger diameter than that of the high-pressure side. Generally, in a regular residential unit, a good compressor should create a pressure between 15 inch of vacuum and 22 psi on the low side, and between 80 psi and 160 psi on the high side at ambient air temperature of 70°F. These pressures are checked by installing access valves (such as piercing valves) on the copper tubing on the suction and discharge lines of the compressor (see fig. 79). The installation of piercing valves mainly applies to the residential units as compressors used in most commercial systems are equipped with service valves (see figs. 32, 118, and 120). Many refrigeration problems can be diagnosed simply by checking the compressor discharge and suction pressures. The recommended high- and low-side pressures for every model of residential unit

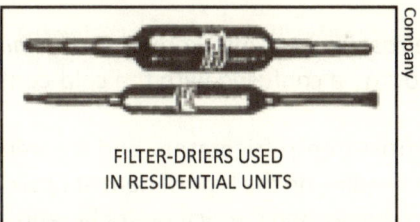

FILTER-DRIERS USED IN RESIDENTIAL UNITS

*Courtesy of Henry Valve Company*

manufactured in the United States can be looked up in a reference book called *Tech Master* published by Master Publications.

*Filter-drier.* A cylinder made of copper or brass (of varying sizes depending on the capacity of the unit) filled with alumina or silica gel. Its function is to remove moisture that penetrates the sealed system by absorption. It is installed at the inlet of the capillary tube. The ends are either silver brazed or coupled by flare connections.

In the sealed system, only pure refrigerant and a small amount of clean refrigerant oil should be circulated. Any moisture penetrating the system will cause a lot of trouble. The air in the atmosphere can penetrate the sealed system through very tiny pinhole leaks in the tubing or at the joints. Air contains moisture. When it enters the system, the excessive low temperatures cause the moisture to freeze and block the circulation. Hence, the reason for the installation of filter-driers. Since the narrowest passage through which refrigerant moves is the capillary tube, this is the most likely place for the restriction. When a restriction occurs, the unit will no longer cool. When a sealed system is opened, moisture gets in and the filter-drier must be replaced prior to evacuating the system with a vacuum pump and recharging the unit with refrigerant.

The drier should be installed as close to the capillary tube as possible. Always keep the drier sealed until installed. (If left open, it will absorb moisture from the air and very soon become saturated.) In a properly operating system, the filter-drier should feel slightly warmer to the touch than the ambient temperature.

*Tubing.* These are the pipes through which refrigerant is circulated throughout the system. The tubing used in household units is made of copper. Aluminum and steel are seldom used. Copper is relatively soft, flexible, and easy to bend and flare. The nine most common sizes in use are 3/16", 1/4", 5/16", 3/8", 7/16", 1/2", 9/16", 5/8", and 3/4" outside diameter (OD). Handle the tubing with care to prevent damage.

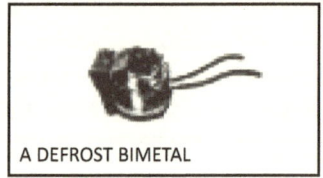
A DEFROST BIMETAL

*Defrost bimetal.* Also known as a *termination switch* or *defrost thermostat*. (Not to be confused with the cold-control thermostat discussed earlier.)

In frost-free refrigerators and freezers, every six, eight, or twelve hours (depending on the type of defrost timer being used), a timer de-energizes the compressor and the evaporator fan and, at the same time, energizes an electric resistance heater which is clipped to the underside of the evaporator fin area to defrost the evaporator for no longer than thirty minutes. A defrost thermostat is wired in series with the electric heater. It is clipped to the evaporator. When the temperature of the evaporator rises to 50°F (10°C), the defrost thermostat opens the electrical circuit to the heater to end defrosting. Contacts within the defrost bimetal close at 20°F (-7°C). No later than thirty minutes after the beginning of the defrost cycle, the timer takes the unit into the cooling cycle—at which time the operation of the compressor and the evaporator fan is restored—and the electrical circuit to the heater is opened.

Defrost bimetals play a very important role as the second component to control the defrost heater. If the heater is not de-energized, the excessive heat can cause damage to the unit. If it is not energized, the unit will no longer defrost. Often a bad defrost bimetal is mistaken for a bad defrost heater.

*Condenser.* The compressor draws in low-pressure, cool refrigerant vapor from the evaporator. This cool vapor is compressed (squeezed) and changed to hot vapor within the compressor, and then forced into the condenser. In the condenser, heat from the refrigerant is radiated into the surrounding air, causing the refrigerant to return to liquid. At 70°F, pressure in the condenser of operating residential refrigerators and freezers ranges between 80 and 160 psi. While a residential refrigeration unit is running, the temperature of the condenser should feel well above room temperature when touched. For optimum efficiency, the condenser should be cleaned every year.

There are two types of condensers commonly employed in residential units: static and fan-forced convection. The static type is mounted on the back of the freezers and refrigerators. It radiates heat through natural convection without the use of a fan. As air in contact with the condenser tubing or fins absorbs heat from the hot refrigerant and becomes heated, it expands and rises, and cooler air occupies its space (see fig. 4a). The fan-forced type is mounted beneath the unit. When the compressor operates, a fan moves air through the condenser tubing fins (see fig. 3). Linted condensers should be cleaned regularly to prevent any restriction of air circulation. When air circulation is blocked by a linted condenser, heat cannot be removed from the vapor refrigerant fast enough to bring it back into its liquid state as it leaves the condenser. Consequently, the evaporator will no longer cool, the unit runs continually, the temperature never drops to a point low enough to satisfy the cold control, and the high-side pressure rises higher than normal, causing the compressor to burn out.

Be sure that enough clearance is always provided for proper air circulation.

*Evaporator.* When liquid refrigerant reaches the evaporator, it vaporizes (or boils) and absorbs heat from the freezer. Sometimes the evaporator is referred to as a cooling coil. Due to the compressor suction power on the outlet of the evaporator and the fact that the capillary tube (with its very small inside diameter) is placed on the inlet side of the evaporator, the pressure in the evaporator is reduced to an average of 3.5 psi in an ambient temperature of 70°F (see fig. 1). It is these low pressures that cause the refrigerant to boil and absorb the surrounding heat during its change of state (essentials 1 and 3).

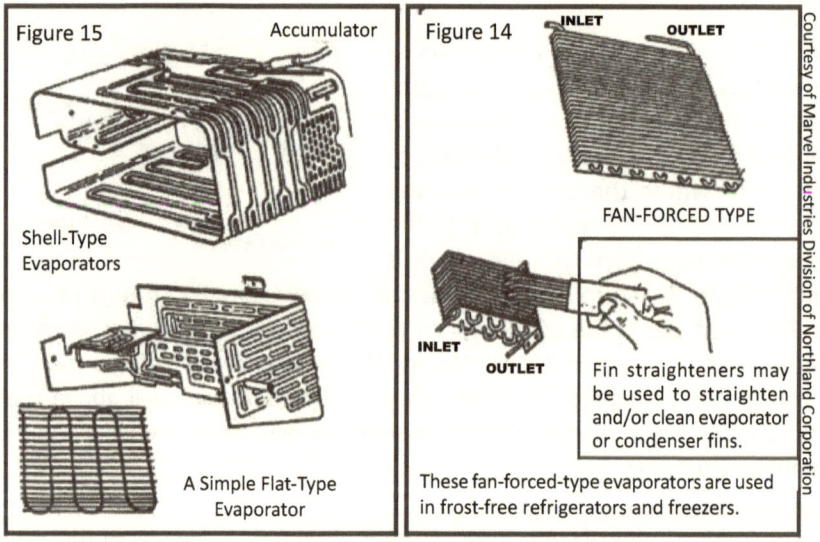

Figure 15 — Shell-Type Evaporators / Accumulator

A Simple Flat-Type Evaporator

Figure 14 — FAN-FORCED TYPE

Fin straighteners may be used to straighten and/or clean evaporator or condenser fins.

These fan-forced-type evaporators are used in frost-free refrigerators and freezers.

Courtesy of Marvel Industries Division of Northland Corporation

Four of the most common types of evaporators are illustrated above and in figures 3, 4, and 5. Figure 5 shows the cycle-defrost type of evaporator. This is basically a flat aluminum plate with a cooling coil in it. About three-fourth of this plate is in the freezer compartment while the remainder extends into the fresh-food compartment. Since fewer loops of the coil are in the fresh-food compartment, only a small portion of the cold air is produced there but adequate to maintain the proper temperature.

Frost-free refrigeration units are equipped with fan-forced circulation evaporators in which a fan increases the airflow through the coil. This type of evaporator cools the refrigerator cabinet or freezer very rapidly.

Manual-defrost refrigerators have a shell-type evaporator, which is located in the top of the cabinet. The shell type is similar to the evaporators used in the cycle-defrost type, which is a flat plate, but bent into a boxlike configuration (see fig. 15). Frozen food and ice trays are kept inside the shell while the fresh food in the rest of the cabinet is kept at the proper temperature by the cold air emitted from it. Shell-type evaporators are almost always found in small office-type or inexpensive refrigerators.

Normally, as a safety measure, an accumulator is installed at the outlet of all evaporators (the small cylinder in fig. 15) to trap any unvaporized refrigerant. It prevents liquid refrigerant from getting into the compressor, causing serious damage.

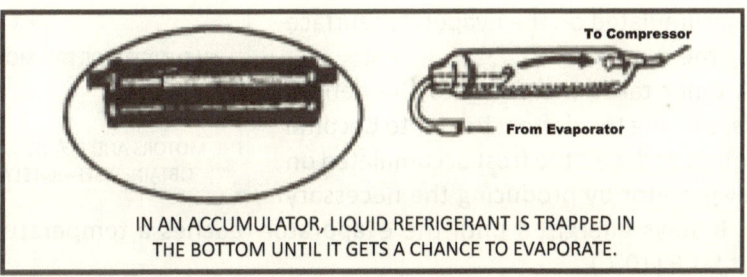

IN AN ACCUMULATOR, LIQUID REFRIGERANT IS TRAPPED IN THE BOTTOM UNTIL IT GETS A CHANCE TO EVAPORATE.

(This can be evidenced by a loud knocking when the compressor runs.) The liquid refrigerant trapped there will get a chance to vaporize before entering the compressor.

*Capillary tube.* A length of thin tubing connected to the high-pressure side (liquid line) from its inlet side and to the low-pressure side of the system (the evaporator) from its outlet side. Liquid refrigerant is forced to flow through the capillary tube by these two forces. Because of the small inside diameter of the capillary tube, a constriction in the flow of refrigerant is created in the sealed system. This constriction maintains the pressure difference between the high and low side. Without continually maintaining this pressure difference, the vaporization and liquification of the refrigerant would not be possible. When refrigerant reaches the larger space of the evaporator (by the suction power of the compressor), the low pressures in this environment immediately cause the refrigerant to vaporize and absorb the heat from the evaporator. Commonly used in commercial and household refrigerators, freezers, and window or rooftop-type air conditioners, capillary tubes are installed between the filter-drier on the liquid line and the evaporator. If a capillary tube must be replaced for any reason (such as a restriction that cannot be cleared with a tube cleaner), it is most important to replace it with one of exactly the same length and inside diameter. The diameter is measured by a capillary tube sizing kit. Capillary tubes are used in different sizes and lengths according to the capacity of the unit (see fig. 45b, 45c, 45d, 45e).

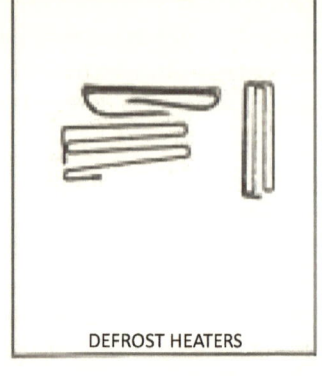
DEFROST HEATERS

*Defrost heater.* It is an electric resistant heating element clipped to the evaporator in frost-free refrigerators and freezers. The purpose of the defrost heater is to melt the frost accumulated on the evaporator surface during the run cycle.

A timer takes the unit into the defrost mode, causing the defrost heater to become energized and melt the frost accumulated on the evaporator by producing the necessary heat. It stays energized until the evaporator reaches a temperature of about 50°F (10°C).

AN EVAPORATOR FAN MOTOR

THE MOTORS AND BLADES MAY BE OBTAINED SEPARATELY

At this point, a termination bimetal senses the rise in temperature and disrupts the flow of current to the defrost heater, even before the defrost cycle ends. In many models, a length of electrical resistance heating element is also installed on the inlet of the drainpipe. It is energized during the defrost cycle to prevent condensation from the evaporator from freezing and blocking the flow down the drain tube.

In cycle-defrost refrigerators, an electric resistant heating element (wired in series with the compressor) is clipped to the inlet and outlet tubing of the fresh-food evaporator to defrost the frozen food evaporator during the off cycles (see fig. 5). The heater is energized when the temperature control is satisfied.

*Evaporator fan.* In frost-free refrigerators and freezers, a fan is installed over the evaporator to move air through the evaporator and circulate it in the freezer and the fresh-food compartments. If this fan fails to operate, there will not be enough air velocity for an effective heat exchange to prevent a frost buildup on the evaporator coil. Consequently, frost will build up on the coil. The temperature in the freezer compartment will rise from -10°F to 0°F or higher. As a result, the temperature in the fresh-food compartment will rise to room temperature. The same thing happens in air conditioners. If the evaporator blower fails, air will no longer be circulated through the cold evaporator fins, causing an accumulation of ice on the evaporator plate, and a sudden temperature rise in the air-conditioned area.

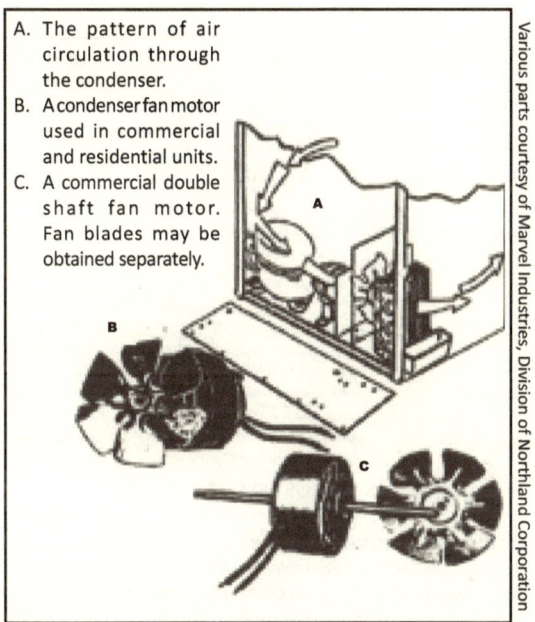

A. The pattern of air circulation through the condenser.
B. A condenser fan motor used in commercial and residential units.
C. A commercial double shaft fan motor. Fan blades may be obtained separately.

Various parts courtesy of Marvel Industries, Division of Northland Corporation

*Condenser fan.* In automatic-defrost freezers and refrigerators (and in larger-capacity units) that require a more rapid cooling of the condenser, additional air movement through the condenser is provided by a fan (called forced draft cooling). Air is drawn into the compressor compartment from one side of the front grille, circulated through the condenser, and expelled through the other side of the front grille.

In some side-by-side models, air is expelled at the rear of the compressor compartment. In addition to the rapid cooling of the condenser, the condenser fan also causes rapid evaporation of water in the condensation tray.

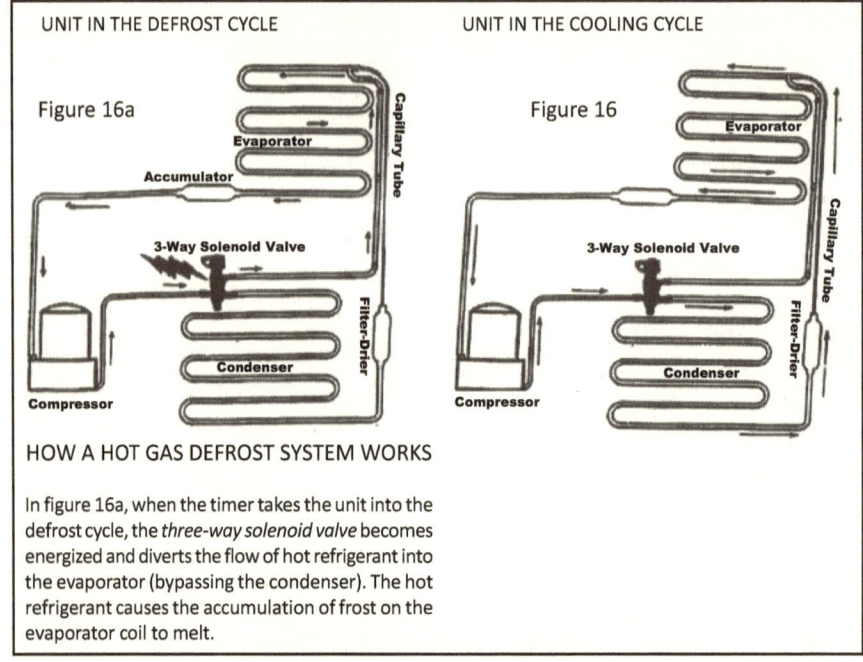

**HOW A HOT GAS DEFROST SYSTEM WORKS**

In figure 16a, when the timer takes the unit into the defrost cycle, the *three-way solenoid valve* becomes energized and diverts the flow of hot refrigerant into the evaporator (bypassing the condenser). The hot refrigerant causes the accumulation of frost on the evaporator coil to melt.

## TEST YOUR KNOWLEDGE ON BASIC REFRIGERATORS AND FREEZERS

1. What are the primary parts of a refrigeration unit? (fig. 4a)
2. Where is the evaporator located? (figs. 3,4,5,6,8a,9,11 and 5)
3. What is a sealed system? (p. 14, fig. 11)
4. What is the function of an evaporator? (pp. 5,30)
5. What is the primary function of a defrost timer? (pp. 23,24)
6. What are the most common troubles with defrost timers? (p. 25)
7. What happens when loose connections on a defrost timer fuse together? (p. 25)
8. What is the primary function of a thermostat? (p. 25)
9. What is a thermostat sensing bulb, and where is it attached? (pp. 25,26)
10. What purpose does an overload protector serve? (p. 26)
11. Where is an overload protector located? (pp. 26, 27, fig. 12)
12. What is the role of a starting relay? (p. 26)
13. Which compressor winding(s) is (are) are energized at the instant a compressor starts? (p. 26)
14. How many windings are there in a compressor? (p. 26)
15. What is the suction port of a compressor connected to? (p. 28)

16. How are the pressures in a sealed system checked? (p. 27)
17. What is the primary purpose of a filter-drier? (pp. 27,28)
18. What are other names for a defrost bimetal? (p. 28)
19. What is the primary purpose of a termination switch? (p. 28)
20. What is the defrost bimetal clipped to? (p. 28)
21. How many types of condensers are used in residential refrigeration? (p. 29)
22. How many types of evaporators are there? (figs. 14,15)
23. Where is an accumulator installed? (p. 31)
24. What is the primary function of a capillary tube? (p. 31)
25. What is the most important thing to consider before replacing a capillary tube? (p. 31)
26. How many types of defrost systems are there? (figs. 16,18,17,19, pp. 30,31,32)
27. What are the most common types of condensers used in residential refrigerators and freezers? (pp. 29,30)
28. What is the purpose of insulating materials in refrigerator walls? (fig. 6, p. 17)
29. How is airflow regulated in a side-by-side refrigerator? (p. 20, fig. 9)
30. What is the principal difference between a refrigerator and a freezer? (p. 22)
31. What components are activated in opening the flow of hot gas in a hot gas defrost system? (fig. 16,16a, p.19)
32. Why shouldn't ice build up on the evaporator coil? (p. 32)
33. Why is fan-forced circulation used in frost-free refrigeration? (p. 30)
34. What is the capillary tube connected to? (p. 31)
35. What is the function of an accumulator? (p. 31)

# THE BRAZING AND FLARING METHODS OF CONNECTING TUBING

This section covers the most practical methods and techniques of connecting tubing. It includes flaring, brazing, swaging, bending, and cutting tubing.

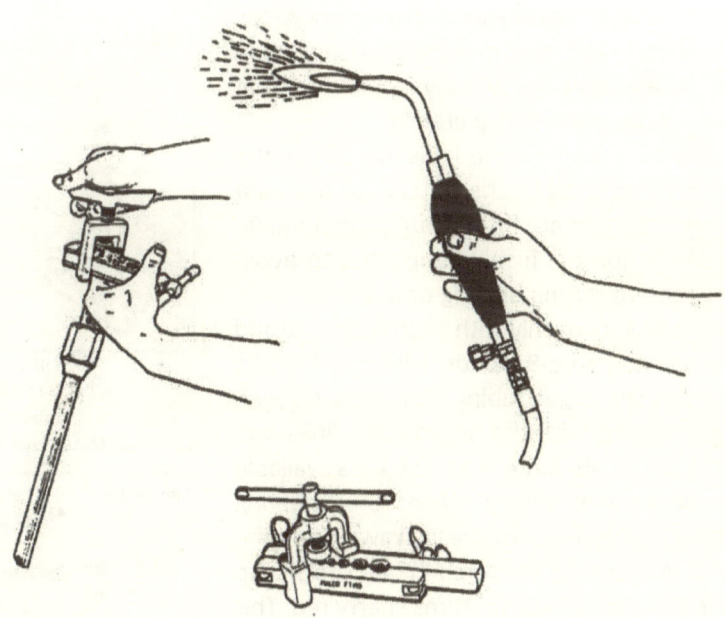

## THE BRAZING AND FLARING METHODS OF CONNECTING TUBING

There are two common methods of connecting tubing: silver brazing (or silver soldering, or just brazing) and flaring.

Silver brazing: Brazed joints are very strong and considered to be the best method of making leakproof connections. The correct procedure is as follows:

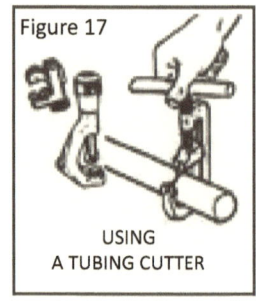

Figure 17

USING A TUBING CUTTER

CAUTION: Before any brazing is begun, all of the refrigerant *must* be evacuated from the system as instructed in the section on repairing leaks in the sealed system.

1. Clean and burnish the joints with fine sandpaper. The parts to be brazed must be fitted snugly and accurately, clean, and securely supported so that no part can move during brazing.
2. Apply the recommended flux (fig. 19) for the alloy being used to the outside of the joint after the tubes are fitted snugly together. Be sure the joint is firmly supported to avoid movement during brazing or cooling.
3. Heat the joint evenly with a torch (figs. 20 and 21) using a figure-8 motion. More heat will be needed for larger tubing. Acetylene/oxygen tanks (along with the appropriate lines and tips) can create more heat. The kit is available at refrigeration supply houses. Start heating about one-half inch to one inch away from the joint. Never hold the torch in one spot. Keep it moving until the joint turns cherry red. This color indicates that the joint has reached the temperature at which silver alloy flows, 1,145°F.
4. Apply the brazing alloy at the top and allow it to seep into the heated joint. Since alloy always flows toward heat, hold the torch at the back of the joint to let the alloy flow into the joint and fill it up.
5. Cool the joint with a piece of wet rag, then use hot water and a brush to clean it. This is

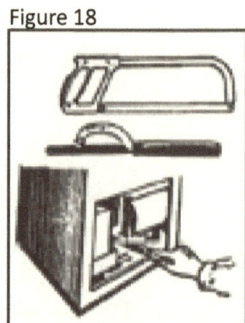

Figure 18

Use either style hacksaw to cut tubing.
Courtesy of MALCO Products Inc.

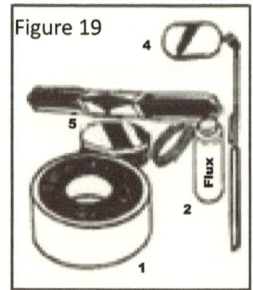

Figure 19

1. Teflon Tape
2. Flux
3. Brazing Rods
4. Inspection Mirror

Courtesy of Wagner Products Corporation, Miami, Florida

important because any remaining flux will tend to corrode the tubing and block a leak that might show up.

CAUTION: Always buy alloy that does not contain cadmium (Cd) as cadmium fumes are highly poisonous! If the brazing alloy contains any amount of cadmium, do not inhale the fumes or allow them to come in contact with your eyes or skin.

*Swaging copper tubing.* Swaging is a process by which the end of one tube is enlarged to allow the end of another tube of the same diameter to fit inside for brazing. This method of joining two lines of the same diameter eliminates the use of fittings. As a general rule, the length of the swaging, or overlap, should equal the outside diameter of the tubing being joined. For instance, when joining one-half-inch tubing, the swaged overlap should be one-half inch; for three-fourth-inch tubing, the length of the swage should be three-fourth-inch, etc. Figure 22 illustrates two pieces of tubing where one has been swaged to create a connection with the other. There are two types of swaging tools available: the lever type and the more popular punch type (fig. 23).

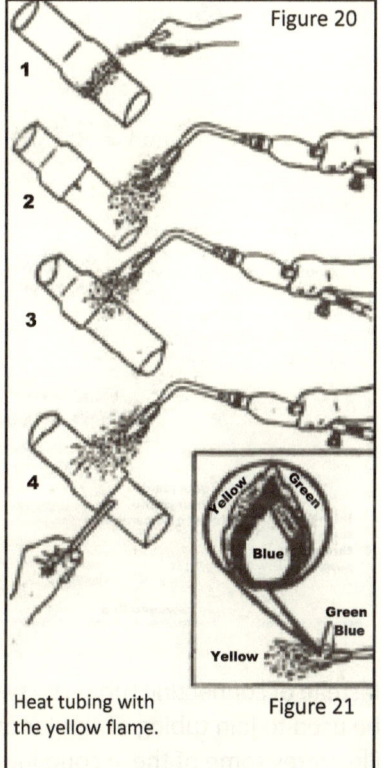

Figure 20

Heat tubing with the yellow flame.

Figure 21

Correct Brazing Technique for Joining Copper Tubing

1. Use fine sandpaper to clean the joint before applying flux.
2. Heat tubing starting one-half inch to one inch from the joint.
3. Move torch toward joint and heat that area briefly.
4. Concentrate heat a little above the joint until joint turns cherry red and apply flux around joint.

Different sized collars and punches are used to swage the various sizes of tubing. To use the punch type, insert the tubing in the correct size hole in the anvil block. Select the proper size punch and insert it in the tubing. Hammer it down into the tubing until it reaches the proper overlap distance.

## DISASSEMBLING A BRAZED JOINT

It is easy to take brazed joints apart using the same method by which they were joined.

Heat the joint with a torch until it becomes cherry red in color, and then grab the tubing near the joint with a pair of pliers and pull it apart.

It will be necessary to disassemble the tubing for replacement of the compressor or any other part in the sealed system. Before removing the old part, you can use this method to disconnect the tubing.

CAUTION: Before applying heat to any joint, it is imperative to evacuate all the gas (Freon) in the system. To do this, use an access valve.

(Refer to figs. 45f and 79.)

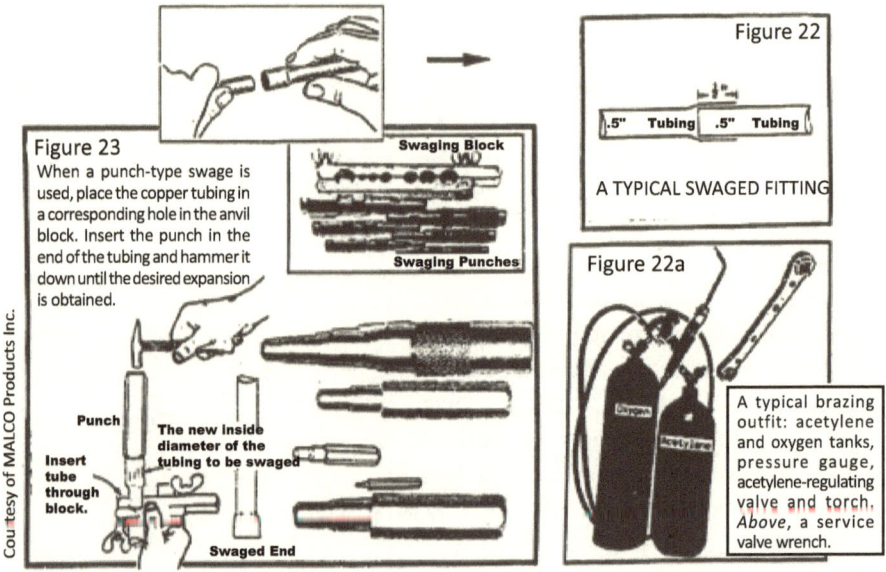

Figure 23
When a punch-type swage is used, place the copper tubing in a corresponding hole in the anvil block. Insert the punch in the end of the tubing and hammer it down until the desired expansion is obtained.

Figure 22 — A TYPICAL SWAGED FITTING

Figure 22a — A typical brazing outfit: acetylene and oxygen tanks, pressure gauge, acetylene-regulating valve and torch. Above, a service valve wrench.

Instead of connecting tubing by swaging before brazing, special couplings can be used to join tubing of similar or different sizes by silver brazing. Figure 24 illustrates some of these couplings. They are available in most tubing sizes.

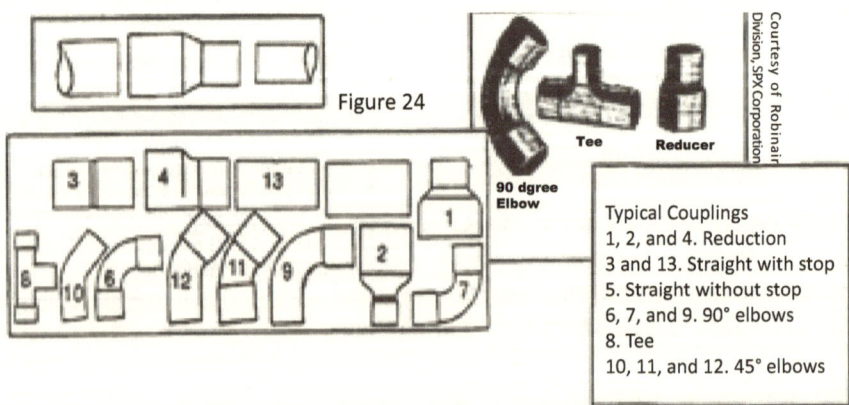

Figure 24

Typical Couplings
1, 2, and 4. Reduction
3 and 13. Straight with stop
5. Straight without stop
6, 7, and 9. 90° elbows
8. Tee
10, 11, and 12. 45° elbows

*Flared connection.* This is a metal-to-metal connection without the use of a solder. To create a flared connection, the ends of the tubing to be joined should be cut straight and square with a tubing cutter (fig. 17) to prevent an off center and, consequently, leaky joint. To do this, the tubing must be held securely while cutting. Use a small vice, C-clamps or Vice-Grip pliers. Then ream the inside of the cut to make it smooth. Most tubing cutters have a reamer attached. The tubing may also be cut with a fine-tooth hacksaw (thirty-two teeth/inch) (see fig. 18). When using either method, make sure that no chips or shavings remain inside the tubing.

*Flaring couplings.* The flared fitting relies on the airtight connections of the fittings rather than brazing. The fittings are a flared half union coupling and two female flare nuts. They are retained on the tubing by a small flaring on each of the ends to be joined. For this type of connection, a different set of tools is needed from those used in making a brazed connection. They are inexpensive and available almost everywhere in hardware and refrigeration supply stores. (See figs. 24a and 24b.)

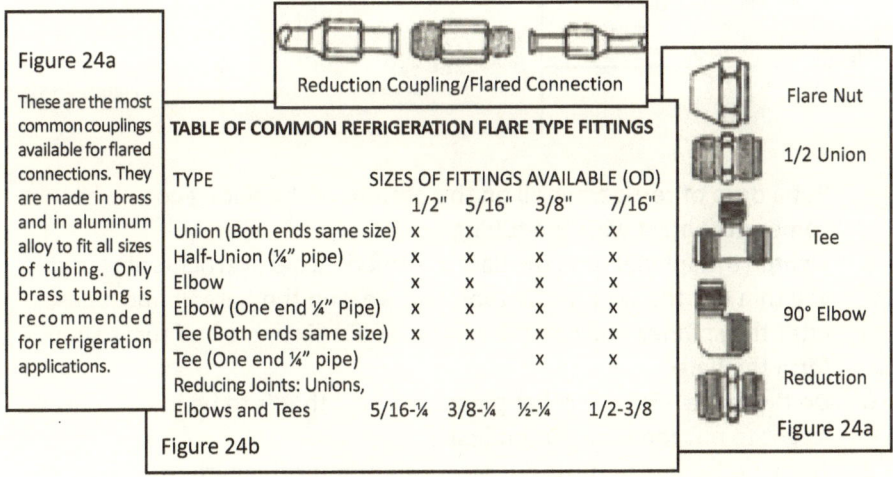

Figure 24a

These are the most common couplings available for flared connections. They are made in brass and in aluminum alloy to fit all sizes of tubing. Only brass tubing is recommended for refrigeration applications.

Reduction Coupling/Flared Connection

**TABLE OF COMMON REFRIGERATION FLARE TYPE FITTINGS**

| TYPE | SIZES OF FITTINGS AVAILABLE (OD) | | | |
| --- | --- | --- | --- | --- |
|  | 1/2" | 5/16" | 3/8" | 7/16" |
| Union (Both ends same size) | x | x | x | x |
| Half-Union (¼" pipe) | x | x | x | x |
| Elbow | x | x | x | x |
| Elbow (One end ¼" Pipe) | x | x | x | x |
| Tee (Both ends same size) | x | x | x | x |
| Tee (One end ¼" pipe) |  |  | x | x |
| Reducing Joints: Unions, Elbows and Tees | 5/16-¼ | 3/8-¼ | ½-¼ | 1/2-3/8 |

Figure 24b

Flare Nut

1/2 Union

Tee

90° Elbow

Reduction

Figure 24a

## HOW TO MAKE A FLARED CONNECTION

1. Cut and clean the ends of the tubes as described earlier. Use a fine-toothed file to smooth the ends or to make a slight correction if the tubing is not cut perfectly square.
2. Place a female flare nut on each of the tubes with the larger end facing the cuts. (Do this prior to making the flare because the nut will not slide over the end after the flare is made.)

3. Slide the nuts back on the tubes to provide enough working room. Then put the end of the first tube into the flaring block hole with the same diameter as the tube. The end of the tube should extend slightly above the chamfered end of the hole to allow enough metal to form a satisfactory flare. The tools have directions with them for guidance in this step. As a rule of thumb, the extension above the block should be about one-third of the height of the flaring. (See fig. 25a and 26.)

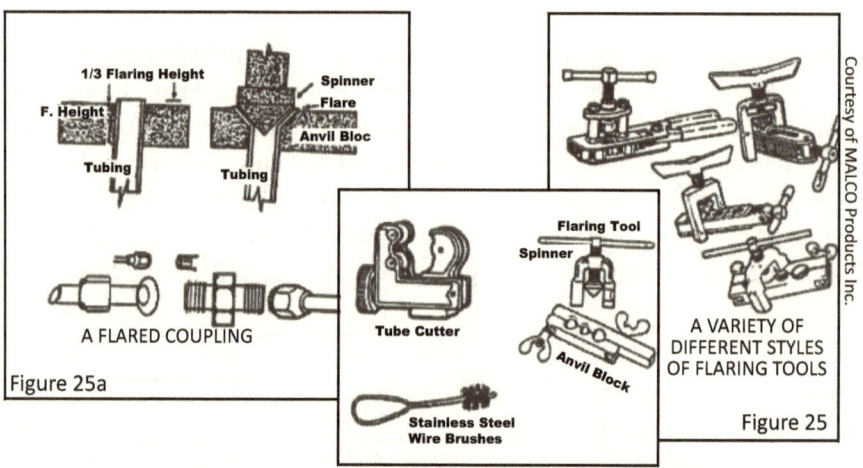

Figure 25a

Figure 25 — A VARIETY OF DIFFERENT STYLES OF FLARING TOOLS

Courtesy of MALCO Products Inc.

4. Put a drop of refrigerant oil on the bottom of the flaring cone where it comes in contact with the tubing.
5. Tighten the spinner until the flare is formed. Avoid overtightening as this will thin out the wall of the tube and weaken the flare. In most cases, after the spinner touches the tubing, about six and a half turns should form the flare.
6. Do the same with the other piece of tubing that is to be joined. (Don't forget to put the flare nut on first!)
7. Use Pipetite (a pipe-fitting compound) or a short length of Teflon tape around the male threads to establish an airtight seal. Teflon is the better of the two. If the compound is used, be sure to apply it sparingly to prevent it from getting into the lines when the flare nuts are connected. (NOTE: Connections made in plastic tubing [such as a water supply to an ice maker] use compression-type fittings since plastic cannot be flared.)

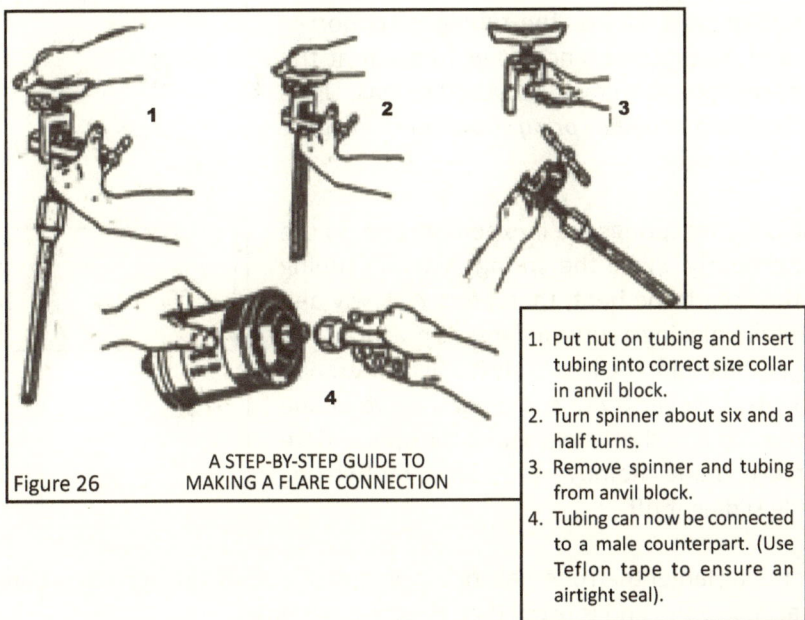

Figure 26 — A STEP-BY-STEP GUIDE TO MAKING A FLARE CONNECTION

1. Put nut on tubing and insert tubing into correct size collar in anvil block.
2. Turn spinner about six and a half turns.
3. Remove spinner and tubing from anvil block.
4. Tubing can now be connected to a male counterpart. (Use Teflon tape to ensure an airtight seal).

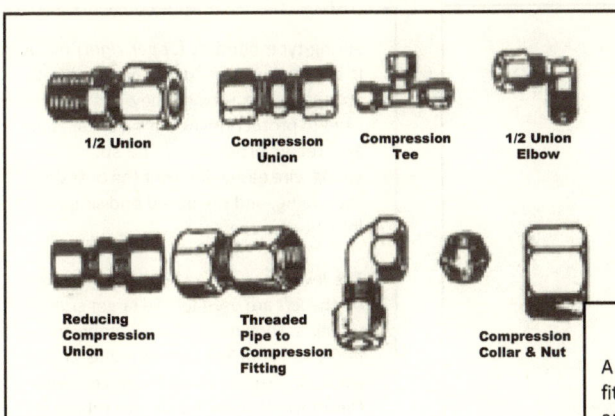

Figure 26a

1/2 Union
Compression Union
Compression Tee
1/2 Union Elbow
Reducing Compression Union
Threaded Pipe to Compression Fitting
Compression Collar & Nut

A variety of compression fittings used in refrigeration and air-conditioning. To prevent possible leaks, do not overtighten nuts and couplings.

## BENDING TUBING

Do not attempt to bend tubing by hand or around a pipe or pole. Unless the tubing is supported around its circumference, it will flatten at the bend and possibly crack, causing it to leak. Use a tool called a *bending spring* shown here in figure 27 on the right.

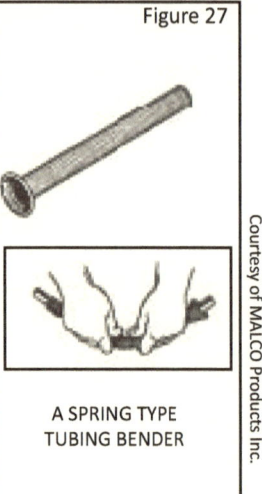

Figure 27

A SPRING TYPE TUBING BENDER

Courtesy of MALCO Products Inc.

The bending springs are inexpensive and do the job properly. Place the spring over the tubing at the area to be bent, then bend it slowly and carefully, making a curve as large as practical. The larger the bend, the less likely it will be flattened or kinked. Never try to complete a curve in one stroke. Do it in short stages until the appropriate curve is made. Remove the spring by twisting while sliding it off.

For tubing larger than one-half inch in diameter, a bending lever with a flange attached to the end is used. (See illustration below.)

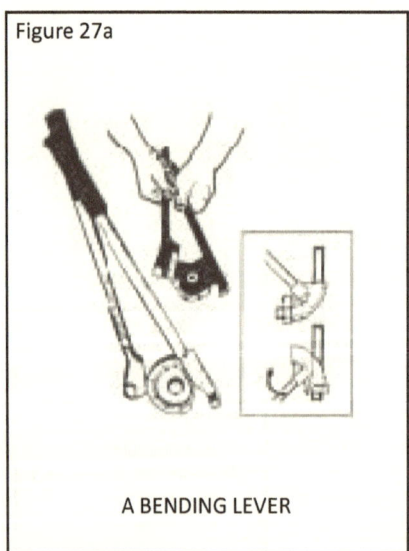

Figure 27a

A BENDING LEVER

Courtesy of Robinair Division, SPX Corporation

Spring-type benders (*upper right*) make it easy to form bends by hand without collapsing the tubing. They can also be used to protect tubing during installation and repair procedures. The special coil spring wire easily slips over the outside of the tubing, and the belled end simplifies insertion and removal.

The lever-type tubing benders illustrated on the *left* are used for the larger sizes of tubing as they require considerably more effort to bend. The curved radius on the stationary bar has a scale in degrees. When the movable lever bends the tubing, its index mark indicates the degree of the curve. The lever type shown at the lower left has no scale, and the bend must be judged by eye or separate measurement.

## TEST YOUR KNOWLEDGE ON CONNECTING TUBING

1. What method is used to bend larger diameter tubing? (fig. 27a)
2. Generally, how many half turns of the spinner should be sufficient to form a flare? (p. 42)
3. Name 4 different types of flare unions. (fig. 24a)
4. What are the most commonly practiced ways of joining tubing? (p. 38)
5. How should the torch be positioned while silver brazing? (p. 38, fig 21)
6. When applying heat to a joint, how is it determined when to apply the silver solder? (p. 38)
7. What steps should be taken after a joint is silver-brazed? (p. 38,39)
8. When should flux be applied to the joint? (p. 38)
9. At what temperature does silver alloy melt? (p. 38)
10. How is refrigeration tubing normally cut? (p. 41, figs. 17, 18)
11. What kind of coupling is used to braze tubes of the same diameter? (fig. 24)
12. How and why should a silver-brazed joint be cleaned? (p. 38)
13. What is the general rule to determine the overlap length when swaging copper tubing? (p. 39)
14. What is the purpose of punches with different sized diameters? (p. 39, fig. 23)
15. How is tubing bent with a spring-type bender? (p. 44, fig. 27)
16. What should be done before applying heat to the tubing? (p. 38)
17. What is used when more heat is required for brazing larger-diameter tubing? (p. 38)
18. How is a brazed joint disassembled? (pp. 39, 40)
19. In what direction does alloy flow while it is being heated? (p. 38)

# TESTING RESIDENTIAL AND COMMERCIAL REFRIGERATION UNITS

This section covers the methods of quickly and easily checking each major component in commercial and residential refrigeration units.

This knowledge, coupled with the information on the troubleshooting charts, provides a quick diagnosis of the most common problems encountered in this career field.

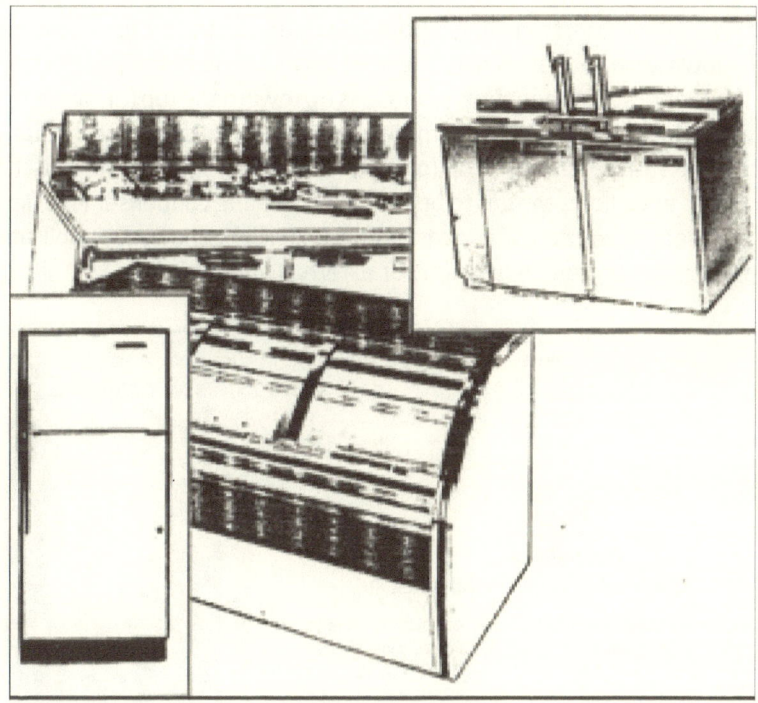

Courtesy of Beverage-Air

# THE COMPRESSOR

There are generally four types of compressors in use today: reciprocating, rotary, centrifugal, and the screw type.

Some of these compressors, used commercially, could be belt-driven or hermetic. The hermetic type has its motor and compressor coupled directly together in a sealed (airtight) metal housing, whereas the others are powered with a separately mounted motor and driven with a V belt(s) and pulleys.

All residential and many light commercial refrigeration units are operated with hermetically sealed motors and compressors because they are compact and require little space. In heavy commercial and industrial use, where large capacity units are required and space is not a problem, the separately mounted motors with v-belts and pulleys to drive the compressors work well.

A *reciprocating compressor* is similar to an automotive engine with a piston moving up and down in a cylinder. Instead of relying on exploding and expanding gas to drive the piston, the piston is powered by an electric motor. As the piston moves downward (or backward), the inlet valve opens and vapor is drawn in from the evaporator. When the piston starts up toward the top (or forward), the inlet valve closes and the gas is compressed (thereby raising its temperature). Before the piston reaches the top (or its most forward position), the discharge valve opens and allows the gas to be propelled into the condenser (see fig. 28) They are used in commercial and residential units, light and heavy applications. These compressors can be hermetic or externally driven.

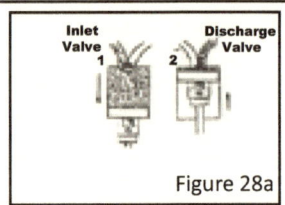

Figure 28

Figure 28a

**THE PRINCIPLE OF THE RECIPROCATING COMPRESSOR**

As the piston moves down, cold vapor is drawn in from the evaporator. As the piston moves upward, the vapor is compressed and converted to a hot gas, which is then forced into the condenser (fig. 28a). In a reciprocating hermetic compressor, the circular motion of the rotor causes a back-and-forth movement of the piston. *Right*: (A) motor stator, (B) motor rotor, (C) compressor cylinder, (D) compressor piston, (E) connecting rod, (F) compressor winding, (G) compressor terminals under a plastic cover, (H) compressor housing weld, (I) compressor suction tube, (J) compressor discharge tube, (K) compressor housing.

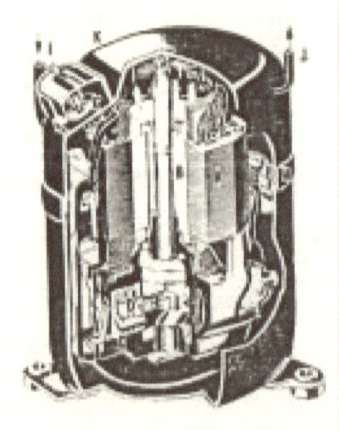

Courtesy of Tecumseh Products Company

A *rotary compressor* also operates on a principle similar to the automotive rotary engine. It can be hermetically sealed or rely on an external conventional electric motor. They are used in commercial and residential units. As the rotor revolves inside the cylinder on an eccentric cam, the spring-loaded vanes pass the intake and discharge ports. As one vane passes the intake port, suction begins and cold vapor is drawn into the cylinder. As the rotor continues its turn, the gas is compressed. When the other vane clears the discharge port, the gas is propelled through the high-pressure line into the condenser. (See fig. 29.)

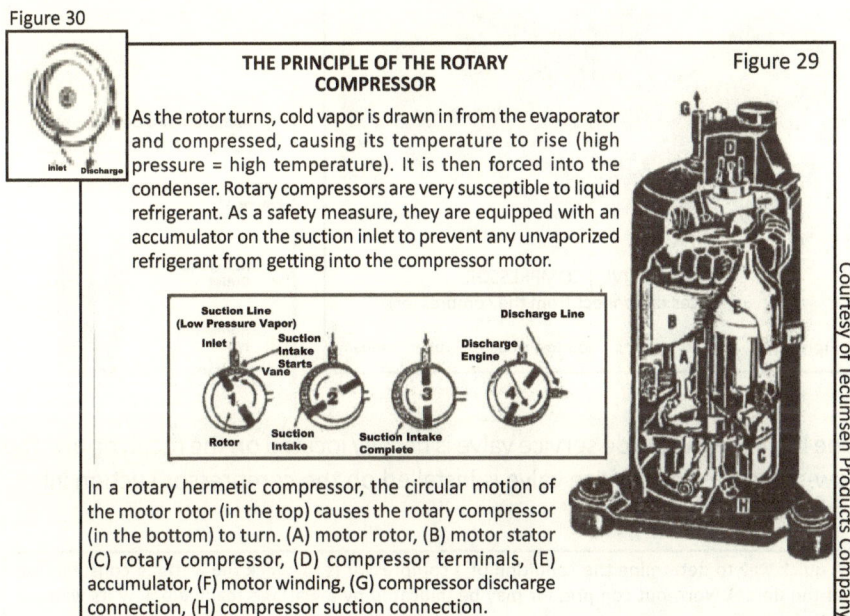

Figure 30

**THE PRINCIPLE OF THE ROTARY COMPRESSOR**

As the rotor turns, cold vapor is drawn in from the evaporator and compressed, causing its temperature to rise (high pressure = high temperature). It is then forced into the condenser. Rotary compressors are very susceptible to liquid refrigerant. As a safety measure, they are equipped with an accumulator on the suction inlet to prevent any unvaporized refrigerant from getting into the compressor motor.

Figure 29

In a rotary hermetic compressor, the circular motion of the motor rotor (in the top) causes the rotary compressor (in the bottom) to turn. (A) motor rotor, (B) motor stator (C) rotary compressor, (D) compressor terminals, (E) accumulator, (F) motor winding, (G) compressor discharge connection, (H) compressor suction connection.

Courtesy of Tecumseh Products Company

A *centrifugal compressor* might be thought of as a "squirrel cage"-like blower motor. As the impeller turns, a vacuum is created at its center, causing cold vapor to be drawn in. It is then compressed and expelled into the discharge port through the sides (caused by high rpm). These compressors can be hermetic or externally driven and are used commercially. (See fig. 30.)

A *screw-type compressor* is similar in operation to a turbocharger. It has two cylindrical vanes with deep, spiraling flutes that mesh together like gear teeth. The extremely high rpm of the vanes meshing into each other creates vacuum on one side and high pressures on the other, causing refrigerant vapor to be drawn in (from the intake port), compressed, and forced out through the compressor discharge port. Screw-type compressors are used in heavy commercial applications.

A liquid receiver is a liquid storage tank used on larger commercial systems and on systems equipped with expansion valve or a low-side float-type refrigerant control explained on pages 140 through 170. (See figs. 33 and 118.)

A BELT-DRIVEN COMPRESSOR
(Service valves disconnect from the compressor).

Figure 31 — Courtesy of Gates Rubber Company

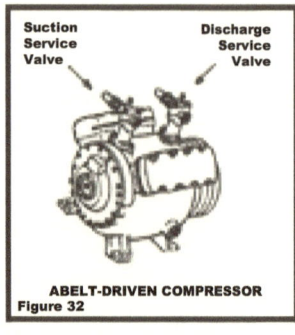

A BELT-DRIVEN COMPRESSOR
Figure 32

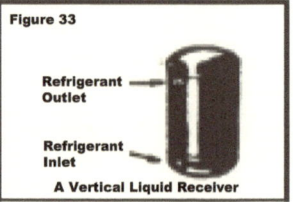

Figure 33

A Vertical Liquid Receiver

The high-pressure-side service valve is usually located on the receiver, and the low-pressure-side service valve is installed on the compressor suction inlet.

A quick way to determine the condition of a compressor is to check the wattage consumption of the unit. A worn-out compressor may be indicated by a wattage reading below its wattage rating.

1. Connect the wattmeter to the compressor circuit to be measured as shown in the diagram. Wires A, B, C, and D are connected to the wattmeter in accordance with the manufacturer's instructions.
2. Depress button E. At first the meter needle will fluctuate to the right; then, immediately, it deflects to the combined reading (the compressor start and run windings both engaged). In a second or two, the needle will deflect to the wattage consumption of the run winding. The wattage readings are then compared with figures published by the manufacturer.

These wattage values, along with a multitude of other useful information, are found in a publication titled *Tech Master for Refrigerators and Freezers*.

A shorted winding is indicated by a high wattage drain. An open circuit is indicated by a low wattage drain.

CHECKING A COMPRESSOR BY A WATTMETER

Figure 33a

*Bolted-type serviceable semihermetic motor compressors* are equipped with high- and low-pressure-side service valves attached to the compressor housing. If the compressor has to be replaced, it can be isolated and removed from the rest of the sealed system without having to discharge the refrigerant from the system (see figs. 119, 32.) The service valves are bolted to the compressor housing with two bolts from one side and connected to the tubing by flared connections from the other side.

## TESTING THE EFFICIENCY OF A COMPRESSOR

### CAUTION
Conduct this test only with the cranking motor. Refrain from running the motor to prevent pressure from climbing to a dangerous level too fast.

When the unit does not cool as well as it is expected to, and the compressor is suspected to be inefficient (an inability to pump), conduct this test as indicated in figure 45h for residential units.

Testing the efficiency of commercial units is an easy task because they come with service valves. This eliminates the necessity of installing piercing valves and pinching off tubing. (See figs. 100, 101, 102.)

## TESTING THE EFFICIENCY OF A COMMERCIAL OR RESIDENTIAL COMPRESSOR

Remove the valve stem caps and gauge connection caps from compressor discharge, and suction valves, and connect your gauges to the valves(compound gage to the suction, and high pressure gauge to the discharge valve). With the help of a service wrench, turn the discharge valve stem all the way clockwise, and crack open the suction valve gauge port by turning the suction valve stem 1/2 a turn clockwise, and turn on the compressor and observe your low pressure gauge. The compound gauge reading should drop to 29" inches of vacuum within 60 seconds or so. At this point turn off the compressor. If within this period, the reading on the compound guage begins to move toward zero, the compressor has an internal lear (Bad Gasket, O' Ring, Bearing, etc.), and must be replaced (only bolted type compressor are repairable (Figs. 119, 120) if the compound guage reading never reaches 28" or 29" of vacuum while running, it has lost its pumping ability, and must be replaced. An efficient compressor must reach about 29" or vacuum, an maintain the level of vacuum after it is turned off.

Residential refrigerators and freezers that do not come with access valves can be tested for efficiency by pinching off the liquid line (see Figure 45g) by using a pinch off tool, and by installing a tap value (see Fig. 45f) on the suction line before you follow the same procedure.

## TESTING THE COMPRESSOR

Compressors become inoperative through either a mechanical or an electrical failure.

Before beginning a compressor test, check the power supply to see if low voltage is the problem. The voltage must be high enough to bring the initial speed up to 75% of its normal revolution per minute (rpm) to allow the run winding to take over. If after start-up this initial speed is never reached, the run winding will never engage, causing the start winding to overheat in a few seconds. When the overload protector senses the high amperage/heat, it shuts off the power to the compressor. When the compressor start winding cools down, the overload protector closes the circuit to the compressor again, and the short-cycling continues. Line voltage may not vary beyond +10% of the compressor specification printed on its nameplate.

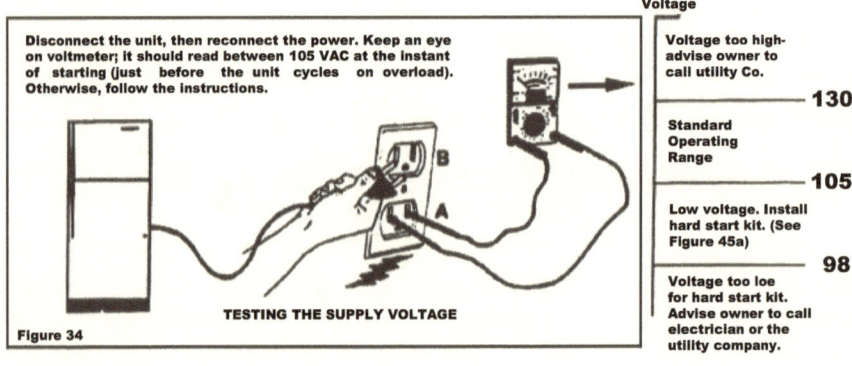

Figure 34 — TESTING THE SUPPLY VOLTAGE

Equipment designed to operate between 108 VAC and 115 VAC will not operate well when the voltage drops below 98 VAC. Should this occur, advise the customer to call the power company to remedy the problem as there may be nothing wrong with the unit. There are transformers on the market today that can remedy the low-voltage problem by increasing the supply voltage. Sometimes there are too many pieces of equipment connected to the same receptacle causing a voltage drop below the acceptable range. Just disconnect some of the load and reconnect it to another wall outlet.

**HOW TO TEST THE SUPPLY VOLTAGE**

1. Set the voltmeter on the proper scale, i.e., on the 110 scale for 110 VAC power supply, etc.
2. Put the voltmeter probes into receptacle A (fig. 34).
3. Plug the unit into socket B while keeping an eye on the voltmeter needle. If at the first instant the unit is plugged in and starts running, the voltage drop registered exceeds (or falls under) the minimum compressor requirement, chances are that there is too much load on that particular circuit. Advise the customer to have an electrician provide the unit with an independent line to meet the requirement of the compressor manufacturer.

## A QUICK CHECK FOR COMPRESSORS THAT FAIL TO START

(CAUTION: These tests have some shock hazard.
Do not touch any wires until the power source is disconnected.)

---

Unit dead. Won't run. No hum. Insert the two check light probes into the power receptacle.
a) Light won't glow.
b) Light glows.

→

a. Check circuit breaker. Flip breaker to ON position if open.
b. Check fuse box for a burnt fuse as described in figure 152f.

---

1. Set cold control at its lowest setting.
2. Remove compressor compartment cover in the back of the unit.
3. Remove the compressor plastic terminal cover with a screwdriver blade.
4. Connect the two test light alligator clips to the two relay terminals that connect it to power.
   a. The light glows.
   b. The light won't glow

→

1. Remove every wire from the compressor terminals.
2. Connect the three test cord clips (see fig. 35) to the three compressor terminals. Wire marked S to compressor terminal S, R to R and C to C.
3. Connect power to the unit.
4. Depress the button on the test cord momentarily and release it.
   a. If compressor starts and continues running, replace starting relay.
   b. If compressor does not start, replace compressor.
   c. If compressor starts, but stops when test cord button is released, replace compressor.

   \* You can use a hard start kit (fig. 45a) instead of a test cord. It can be used on compressors that operate with or without start capacitors to provide sufficient voltage. This is a good tool to use on compressors that no longer run (due to wear and tear) to make them operational again.

---

1. Connect the two test-light probes to the two terminals (behind the compressor compartment cover) where the power cord connects to the unit.
2. Plug the power cord into the power receptacle.
   a. If the light won't glow, replace the power cord.
   b. If light glows, proceed.

---

1. Find the cold-control (thermostat) adjusting knob in the unit cabinet. Remove plastic cover and component out far enough for inspection of charred or disconnected terminals. If it looks OK, and without letting terminals touch anywhere else ...
2. Fasten the two jumper wire clips to the two terminals.
3. Turn on the power.
   a. If the unit starts running, replace the thermostat.
   b. If unit remains dead, proceed.

---

SIGNS OF A FROZEN TIMER
If freezer compartment stays warmer than usual (defrost heaters on) and the unit won't run (including the evaporator fan), but the compartment light turns on when the door is opened, check for a frozen defrost timer. Using a screwdriver, turn the timer cam clockwise until you hear a click. If the unit starts running, replace the timer. (See fig. 54m.)

---

1. Connect the two test wire clips to the two terminals on the overload protector.
2. Plug the unit into the power supply.
   a. If the unit starts running, replace the overload protector.
   b. If it doesn't start, proceed.

△ CAUTION: Never troubleshoot the electrical circuits while power is connected to the unit. Always disconnect the power supply before touching any wire or electrical component.

This quick electrical check for compressors that fail to start does not cover those compressors that operate with capacitors. If you are checking a compressor with a capacitor(s), follow the same procedure after checking the capacitor(s) as outlined below. If the capacitor is bad, it must be replaced with one of the same microfarad (mfd or mf) rating.

IMPORTANT: After disconnecting power from a unit, discharge the capacitor first by shunting between its two poles with well-insulated heavy wire before handling it. If there is no exact capacitor replacement (or a capacitor tester) available, you can make a compressor test cord that can test capacitors too. See figure 34a and follow these instructions:

1. Disconnect the unit from the power supply.
2. Discharge the capacitor and remove it from the unit. (See step b in "Testing Capacitors Using an Ohmmeter.")
3. Put a 60-watt light bulb in socket A (fig. 34a).
4. Insulate the alligator clip marked S and connect the alligator clips marked C and R to the two capacitor poles.
5. Connect plug B to power.

The lamp will,

   a. glow dimly if the capacitor is good,
   b. not glow if the capacitor is open (replace the capacitor),
   c. glow brightly if the capacitor is shorted (replace the capacitor).

6. If the capacitor checks good, replace the 60-watt bulb with a 200-watt bulb and leave the capacitor as it is for no longer than five seconds.
7. Disconnect the cord from the power. Using a heavy insulated wire short across the two capacitor terminals. A spark is the indication of a good capacitor (meaning that the capacitor can load and discharge).

Bigger compressor motors that require more starting or running torque due to heavier loads use capacitors to increase their torque. This is especially true in commercial refrigeration and air-conditioning. They are referred to as *start* or *run capacitors*. Some compressors use only run capacitors, and some use both.

Capacitor testers can be purchased at refrigeration supply houses for very affordable prices.

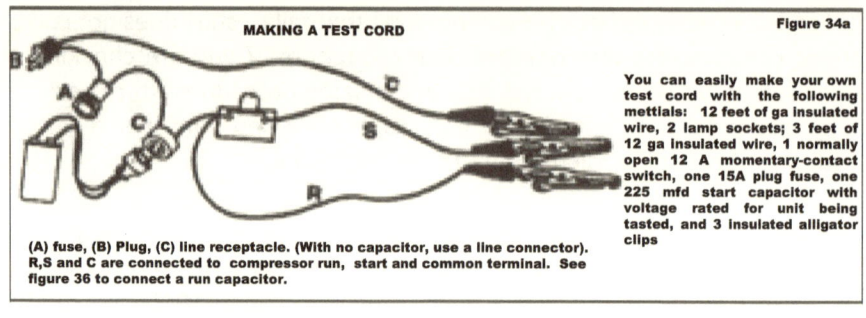

**MAKING A TEST CORD** — Figure 34a

You can easily make your own test cord with the following mettials: 12 feet of ga insulated wire, 2 lamp sockets; 3 feet of 12 ga insulated wire, 1 normally open 12 A momentary-contact switch, one 15A plug fuse, one 225 mfd start capacitor with voltage rated for unit being tasted, and 3 insulated alligator clips

(A) fuse, (B) Plug, (C) line receptacle. (With no capacitor, use a line connector). R,S and C are connected to compressor run, start and common terminal. See figure 36 to connect a run capacitor.

## TESTING COMPRESSORS WITH A TEST CORD

1. Disconnect the power supply.
2. Remove the compartment cover, compressor terminal cover, start relay, and the wire connecting the overload protector to the common terminal.
3. Connect the appropriate test cord wires to the proper compressor terminals. (The three test wires are marked C, S, and R.)
4. Plug the test cord into a proper power supply and depress the switch for no longer than three seconds.

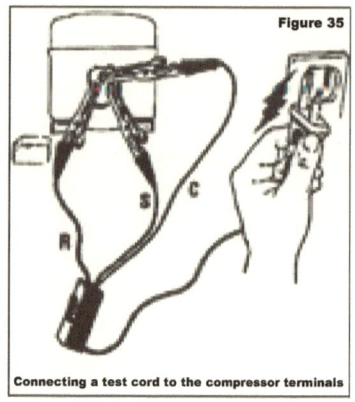

Figure 35

Connecting a test cord to the compressor terminals

If the compressor never starts, or if it stops when the switch is released, it must be replaced. A good compressor will continue to run because the power is still connected to its run and common windings through the test cord. Poor connections also cause compressor failure. Be sure to check these too, prior to replacing a compressor. Tight and clean connections are essential for good current flow. If a compressor must be replaced, all the data for a duplicate replacement can be copied from the compressor nameplate to ensure getting the right one. *Do not remove the nameplate!* The compressor may still be under warranty; if it is, removing the nameplate will void it.

For compressors that require capacitor(s) to operate, figure 36 illustrates a testing cord that can be easily made. (You must use fourteen gauge or heavier wire). It shows wiring to the compressor terminals (a run capacitor, a start capacitor, and a momentary-contact push-button switch). If the compressor being tested does not require a run capacitor or a start capacitor, either one may be left out of the circuit. Consequently, the test cord should be made with quick-disconnect fittings and adapters.

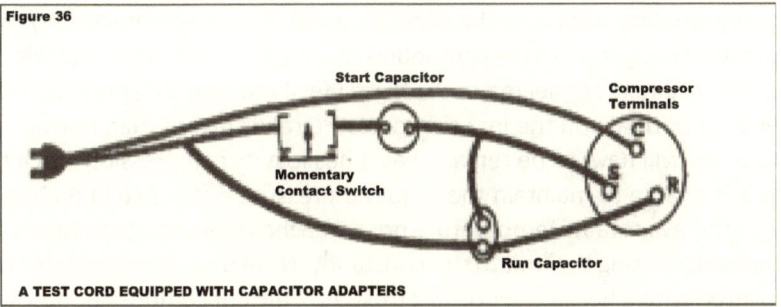

Figure 36

A TEST CORD EQUIPPED WITH CAPACITOR ADAPTERS

## LOCATING A COMPRESSOR INTERNAL ELECTRICAL DEFECT WITH AN OHMMETER

When the two ohmmeter probes touch any two compressor terminals, the meter should register an ohm reading. Otherwise, the compressor is defective and it must be replaced. When the ohmmeter probes touch a single terminal and the housing (ground), the meter should register a no-continuity reading (an open circuit, an infinity reading, a figure 8 on the meter lying on its side); otherwise, the compressor is shorted and it must be replaced. (Scratch or sand the paint off a small spot on the compressor housing to get a good contact.)

**CHECKING A COMPRESSOR MOTOR WITH AN OHMMETER**

1. Disconnect the unit from the power source.
2. Remove the compressor terminal cover.
3. Remove the overload protector and starting relay from the three compressor terminals. (See fig. 12.)
4. Set the ohmmeter on its RX1 scale and zero it.
5. Touch the probes to the compressor C and S terminals. The meter should register a continuity reading. Otherwise, replace the compressor.
6. Touch the probes to the compressor C and R terminals to get a continuity reading. Otherwise, replace the compressor.
7. Touch the probes to the compressor S and R terminals to get a continuity reading. If not, the compressor is defective.
8. Touch one probe to the compressor housing and the other one to each terminal in turn. In each case, the meter should register no-continuity reading. Otherwise, replace the compressor.

## COMPRESSOR MECHANICAL FAILURE

Another problem leading to compressor failure is regular wear and tear. The compressor runs without being able to create the necessary pressure difference in the system simply because the parts are worn. This can be checked by using the pressure gauges. With the compound gauge connected to the low side and the pressure gauge connected to the high side, if the high-side pressure reads lower than normal and the low-side pressure reads higher than normal, the compressor will have to be replaced as it has lost its compression efficiency. Since it is unable to maintain the required pressure difference in the sealed system, the evaporator temperature never reaches low enough to satisfy the thermostat, causing the unit to run constantly. Note that the evaporator may be covered with a heavy layer of soft frost. An efficient compressor produces a layer of hard frost on the evaporator coil. (As more experience is gained, the evaporator frost pattern will become very evident.) When a compressor is turned off, the evaporator frost pattern disappears very quickly. The frost on the accumulator disappears in few seconds when placing a hand around it. An accumulator in a properly operating system is covered with hard frost.

## HOW TO IDENTIFY UNMARKED COMPRESSOR TERMINALS

Occasionally you will encounter some compressor terminals which have no markings (or which may have been obliterated). There is an easy way to determine the compressor C, S, and R terminals. Here is how:

Set the ohmmeter on its lowest scale. In figure 38, imagine the unmarked terminals as 1, 2, and 3. Place the two ohmmeter probes on terminals 1 and 2; make a note of the reading (7 Ω, for instance). Then place the probes on terminals 2 and 3 (assume 8 Ω, and note it).

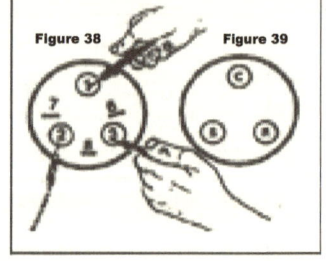

Finally, take a reading between 1 and 3 (assume the meter indicates 6 Ω). The highest reading between any two terminals means that the remaining one is the *common terminal*.

   Since number 2 and number 3 terminals have the highest reading, it can be deduced that number 1 has to be the *common terminal*.

Since it has already been noted that the reading between terminals 1 and 2 is 7 Ω and the reading between terminals 1 and 3 is 6 Ω, the highest of these last two readings determines the *start terminal*. In other words, number 2 is the *start terminal*, so the remaining terminal (number 3) will have to be the *run terminal* (see fig. 39). Some terminal configurations appear as in figure 40. Use the same method to identify these too.

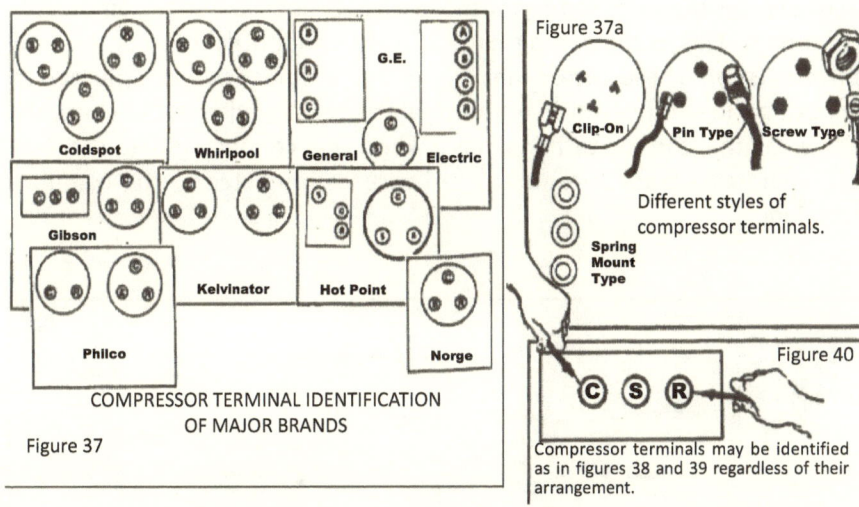

COMPRESSOR TERMINAL IDENTIFICATION OF MAJOR BRANDS
Figure 37

Figure 37a

Different styles of compressor terminals.

Figure 40

Compressor terminals may be identified as in figures 38 and 39 regardless of their arrangement.

### HERMETIC UNIT TERMINAL COLOR CHART

B-Black    G-Green    R-Red    T-Tan    W-White

| BRAND NAME | STARTING | RUNNING | COMMON |
|---|---|---|---|
| Admiral | B | W | R |
| Airtemp | W | R | B |
| Carrier | W | R or T | B |
| Coldspot | W | R | B |
| Copeland | W | R | B |
| Crosley | B | W | R |
| Frigidaire | R | B | W |
| General electrical | W | G | B |
| Gibson | G | W | B |
| Grunow | R | W | B |
| HotPoint | W | G | B |
| Kelvinator | W | R | B |
| Kel-Kold | W | G | B |
| Leonard | W | R | B |
| Norge | R | B | W |
| Philco | W | G or R | B |
| Servel | W | R | B |
| Stewart Warner | W | R | B |
| Tecumseh | W | R | B |
| Universal Cooler | W | R | B |
| Westinghouse | R | B | W |
| Zenith | R | W | B |

Figure 40a

## COMPRESSOR TERMINAL COLOR CODING A QUICK REFERENCE GUIDE FOR TWENTY-TWO POPULAR BRANDS

Each manufacturer has a color code for the wiring in the circuit. The chart above shows the color coding for those wires connected to the compressor terminals. This color coding can be used to identify the compressor terminals when the letter marking is not visible. For example, when checking the compressor in an Admiral freezer, a glance at the chart will show you that the run terminal has a white wire connected to it, the start terminal has a black wire, and a red wire is connected to the common terminal.

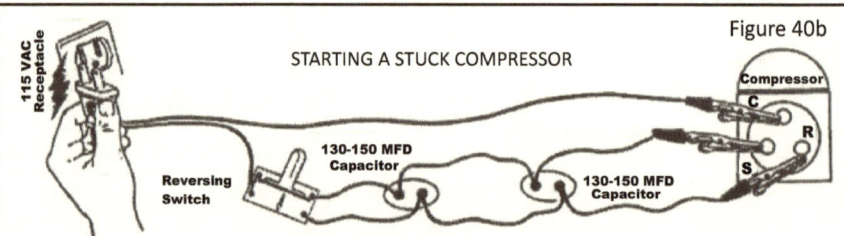

Figure 40b

Occasionally after a long run in high ambient temperatures, a compressor may become stuck and fail to start. Reversing the rotation of the compressor by the following method may break it loose:

A. Prepare a single-pole, double-throw (SPDT) switch, two capacitors rated 130-50 mfd, a line plug, and three alligator clips.
B. Disconnect the unit from the power supply and remove all wires from the compressor terminals.
C. Hook up the reversing circuit as shown and plug the cord into a 115 VAC wall receptacle.
D. Operate the reversing switch by rocking it back and forth to alternately reverse the rotation, causing the compressor to break loose. Add SUPCO 88 additive to the system to prevent it from becoming stuck again.

## TYPES OF HERMETIC COMPRESSOR MOTORS AND HOW THEY ARE WIRED

So far we have talked only about the kind of hermetic motors installed in residential units, the split-phase type. These are called *hermetic* because they are mounted inside an airtight container with the compressor. Basically, there are four types of hermetic motors in commercial use today:

1. *Split-phase hermetic motor* (as it has separate run and start windings) used mainly in residential units with limited use in small commercial units.
2. *Capacitor-start, induction-run motor.*
3. *Capacitor-start, capacitor-run motor.*
4. *Permanent-split, capacitor-run motor.*

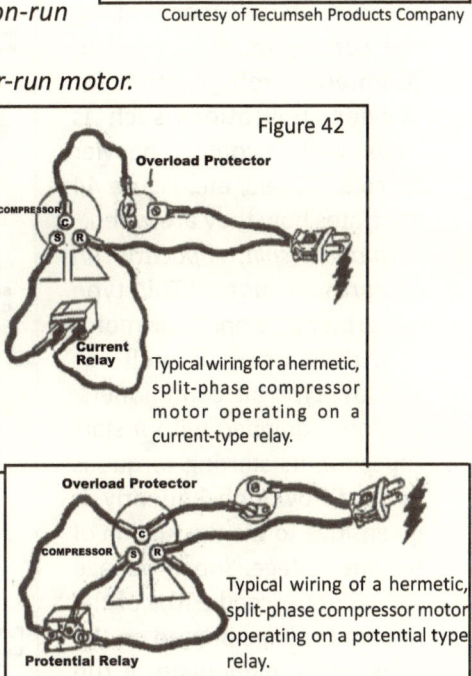

Figure 41
HERMETIC COMPRESSORS

1. Split-phase compressor (generally used in residential and commercial units)
2. A capacitor-start compressor used in commercial equipment
3. A rotary compressor

Courtesy of Tecumseh Products Company

1. *Split-phase hermetic* is the simplest kind, used mostly for household refrigerators since the compressor motors do not require a lot of starting torque. In these units, when the thermostat shuts the system down, the high- and low-side pressures equalize through the capillary tube. In heavy commercial units, the pressures do not equalize because they use TEV (thermostatic electric valve) or automatic valves

Figure 42

Typical wiring for a hermetic, split-phase compressor motor operating on a current-type relay.

Typical wiring of a hermetic, split-phase compressor motor operating on a potential type relay.

(covered later) as well as capillary tubes. These valves isolate the high and low sides, and the pressures in the system do not equalize very easily in the off cycle. In these motors, a relay controls the engagement of the start windings. The starting relay used may be of the thermal type, the current type or the potential type, which will be covered later.

2. *Capacitor-start compressor motor* is a popular type of hermetic motor in refrigeration units. A capacitor is installed in series with the motor start winding to produce more starting torque by providing more initial voltage. This capacitor is isolated from the circuit during the run cycle (see fig. 43). This type of compressor is normally used in walk-in coolers, salad bars, beverage coolers, ice machines, and similar commercial refrigeration units.

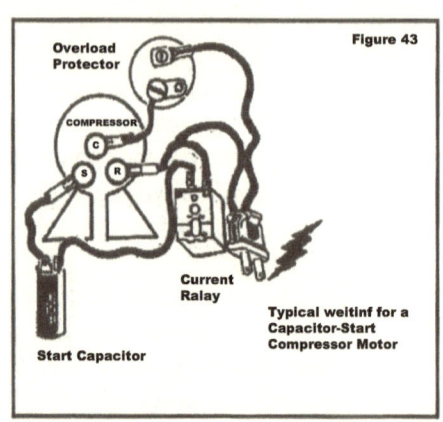

Typical weitinf for a Capacitor-Start Compressor Motor

3. *Capacitor-start, capacitor-run compressor motor* is a very efficient type of motor. A start capacitor increases its starting torque, and a run capacitor increases its efficiency during the run cycle. It is used in commercial refrigeration for heavier applications such as larger walk-in coolers, heavier air conditioners, etc. Figure 44 illustrates how they are wired.

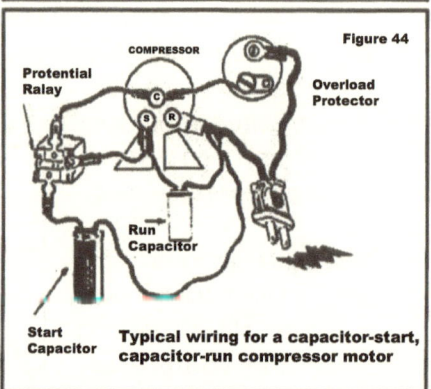

Typical wiring for a capacitor-start, capacitor-run compressor motor

4. *Permanent-split, capacitor-run compressor motor*. This type of hermetic compressor motor is widely used in wall- or window-type air conditioners. It is not equipped with a start capacitor. Its starting torque is (almost) low. Consequently, it is sensitive to the fluctuation of the line voltage. Supply voltage should not exceed ±10% of the required voltage printed on the compressor nameplate. A run capacitor is installed between its start and run windings to provide more efficiency during the run cycle. Figure 45 shows a typical wiring diagram of these motor compressors.

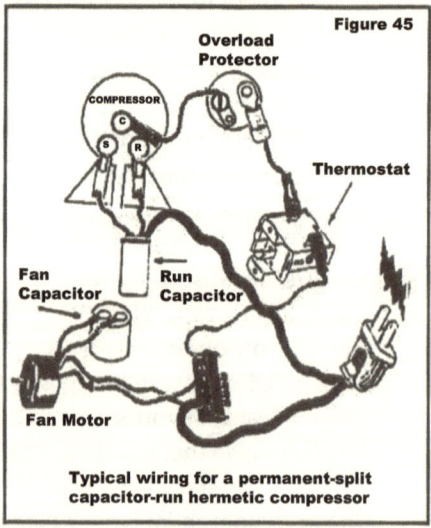

Typical wiring for a permanent-split capacitor-run hermetic compressor

## HARD START KIT

A hard start kit is a combination relay, start capacitor, and overload protector. Primarily, this kit (see fig. 47a) was designed to start older compressors that due to their age and normal wear become hard to start. Very often, a compressor that is thought to be defective can be restored to service by using a hard start kit.

A hard start kit is a great tool for the service technician. It eliminates a lot of work and time spent in testing several components and looking for the defective part when a compressor fails to run or cycles on overload. It is inexpensive and can be used on almost all 115 VAC compressors under 1/2 hp.

Some hard start kits are designed for use on compressors operating on 208-230 VAC and on air conditioners up to several tons. The wires on the kit are either color coded or individually labeled Start, Run, and Common.

---

Figure 45a

### HOW TO USE A HARD START KIT

a. Disconnect the power from the unit.
b. Remove the overload protector and starter relay.
c. Connect the start terminal of the compressor to the wire on the kit marked Start.
d. Connect the run terminal of the compressor to the wire on the kit marked Run.
e. Connect the common terminal of the compressor to the wire on the kit marked Common.
f. Find the lines that supply power to the compressor. (Normally, one goes into the overload protector and one to the start relay.)
g. Connect these lines to the remaining two wires on the hard start kit. Plug the unit back in to the power source, and if the compressor starts and continues to run, then the problem is solved. If it does not, the compressor is defective. In which case, remove the new hard start device and replace the compressor.

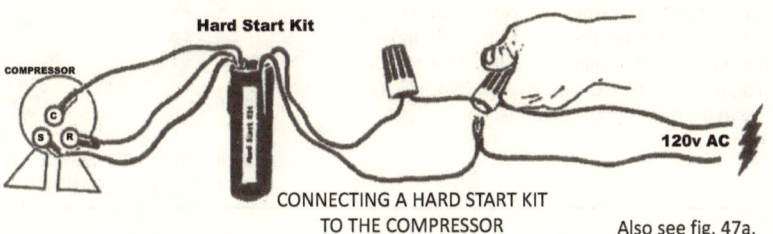

CONNECTING A HARD START KIT
TO THE COMPRESSOR          Also see fig. 47a.

# HOW TO UNCLOG A CAPILLARY TUBE

Sometimes a wax buildup or dirt obstructs the passage of refrigerant through a capillary tube. Due to its length, the capillary tube may run through places difficult to reach. In such cases, it is easier to unclog it rather than replacing it.

A capillary tube cleaner can be purchased from most major refrigeration supply houses.

The way the device works is that it forces wax and dirt out of the capillary tube under high pressure. Some of these devices are capable of producing pressures as high as 3,000 psi.

An obvious sign of a clogged capillary tube is that the back pressure reads lower than normal (or even vacuum), the head pressure reads higher than normal, the unit no longer cools while running constantly, and the condenser feels cooler than normal.

To use a capillary tube cleaner, disconnect the capillary tube at both ends. (Flux and apply heat to the brazed joint to remove it.) Connect the tube cleaner to one open end of the capillary tube by using an adapter fitting; then turn the handle to create the pressure necessary to clear the tube. In these devices, either oil or R-11 is used as a pressure fluid. (See fig. 45b.)

After removing the obstruction from the tube, install a new filter-drier and silver-braze the tube back into the system before evacuating and charging the unit.

---

Figure 45b

Capillary tubes are not expensive, but sometimes they are hard to reach (particularly in residential units). The capillary tube begins in the compressor compartment in residential refrigerators and freezers, runs through the body of the unit and it is silver-brazed to the evaporator inlet. A lot of work and replacement time can be saved if a blockage can be cleared with a capillary tube cleaner.

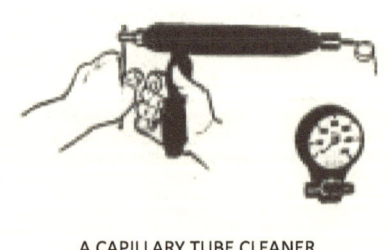

A CAPILLARY TUBE CLEANER
WITH ITS PRESSURE GAUGE

There are some capillary tube replacements on the market called patented tubes. Some of them are available with different sized strainers while some are fitted with a calibrated wire inside to control the flow of refrigerant (see fig. 45c).

Figure 45c

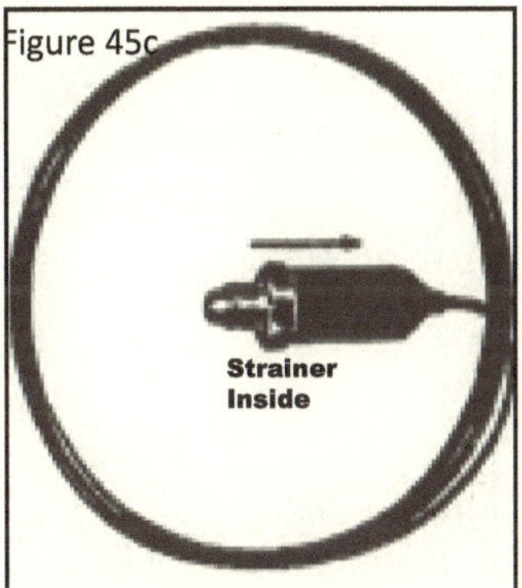

Strainer Inside

Patented tubes can be used for capillary tube replacement.

If a new capillary tube is needed, it must always be replaced with one having the same inside diameter and the same length; otherwise, evaporator temperature will be affected.

For better understanding, you should know that there are three types of compressors made today: *high-temperature types*, which produce temperatures down to about 0°F; *medium-temperature types*, which produce temperatures of about 0°F to -10°F; and *low-temperature* models which produce temperatures below -10°F.

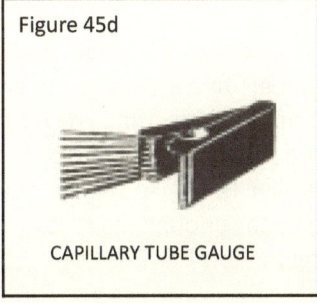

Figure 45d

CAPILLARY TUBE GAUGE

A capillary tube sizing gauge can be purchased from a major local refrigeration supply house. This is a tool similar in appearance to a spark plug gapping tool used by auto mechanics. It consists of a number of different sized wires to measure the inside diameter of capillary tubes (see fig. 45d).

See the chart on the next page for the required length of tubing based on its inside diameter, the horsepower, and temperature rating of the compressor.

When handling capillary tubes, it is important to remember that

1. capillary tubes are connected to the sealed system mostly by silver brazing (a flared connection is seldom used) and,
2. since capillary tubes are too small in diameter to be cut with a tubing cutter, the usual practice is to score them with the edge of a file. It is then bent carefully until it breaks.

Figure 45e

**CAPILLARY TUBE LENGTH (IN FEET)**

| hp | COMPRESSOR TEMPERATURE | ID: | 0.031 | 0.036 | 0.040 | 0.042 | 0.049 | 0.055 | 0.065 |
|---|---|---|---|---|---|---|---|---|---|
| 1/3 | Medium | | | | | | | | |
| | Low | | 1.75 | 3.5 | 5.6 | 7 | 17 | 2.5 | |
| 1/4 | High | | | | | | | 5 | |
| | Medium | | 1.1 | 2.2 | 3.5 | 4.5 | 9 | 15 | |
| | Low | | | | | | | | 7.5 |
| 1/5 | High | | | | | | | 10 | |
| | Medium | | 2.2 | 4.4 | 7 | 9 | 18 | 31 | |
| | Low | | 5.2 | 10.5 | 17 | 21 | 42 | 73 | |
| 1/8 | High | | 1.1 | 2.2 | 3.5 | 4.5 | 9 | 15 | |
| | Medium | | 4 | 8 | 13 | 16 | 32 | 56 | |
| | Low | | 9 | 18 | 29 | 36 | 72 | 126 | |

To silver-braze a small capillary tube to a large tubing, place the small tube at least two inches inside and against one wall of the larger tubing. Using a pair of pliers, crimp the opposite wall of the larger tubing until it fits snugly around the capillary tube. Then clean and silver-braze the joint as instructed earlier. (Because small tubing absorbs heat very rapidly, be careful no solder gets far enough inside to block the opening of the small tube and cause a restriction.)

**COPPER OR BRASS ACCESS VALVES**

These are the strongest and most dependable because they are brazed on the tube instead of clamped on with screws.

1. Fit the correct size valve body (number 1) on the tube and crimp the saddle tabs around the tube with pliers. Then braze in place. Apply heat from below being careful not to overheat the body. Always sand the surfaces to be brazed and use flux.
2. When cooled, insert the piercing shaft (number 2) into the body being careful not to damage the O-ring.
3. Insert the valve core (number 3) into the shaft and screw into place with core driver (number 4). Do not overtighten.
4. Screw on access valve cap (number 5). Tighten securely with one wrench while holding the body nut with another. Do not overtighten.

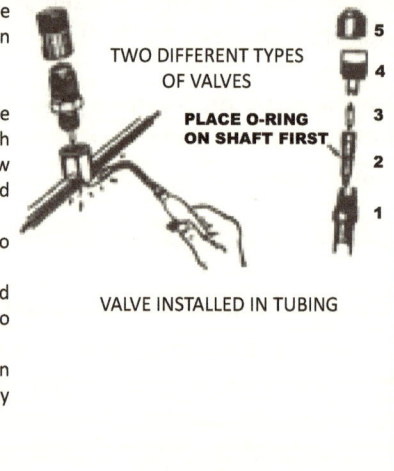

TWO DIFFERENT TYPES OF VALVES

PLACE O-RING ON SHAFT FIRST

VALVE INSTALLED IN TUBING

Figure 45f

Courtesy of Wagner Products Corp.
Miami, Florida

## PINCH-OFF TOOL

In refrigeration repairs, a pinch-off tool is used to seal off copper tubing up to three-eighth inch in diameter.

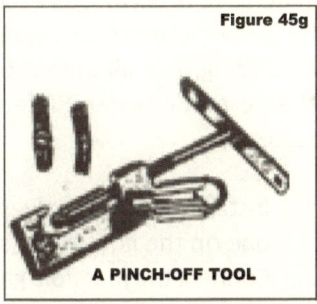

Figure 45g

A PINCH-OFF TOOL

Figure 45g shows a typical pinch-off tool. To use it, put the tubing through the opening to the point where it is to be sealed. As the shaft is turned by the T-handle, the tubing is compressed between a ball bearing at the end of the shaft and the die at the base of the tool. A permanently pinched line is made by turning the handle slowly and not overtightening.

This too becomes necessary when there is a need to test the compressor efficiency or in an emergency, such as a severe leak, when a section of the sealed system must be isolated for repair. For example, in commercial units where there are multiple evaporators, the one with a leak can be isolated while allowing the rest of the system to operate during the repair work. Thus, the contents of the unit can stay cold and be saved.

## ADDING OIL TO THE COMPRESSOR

Systems with hermetic compressors seldom requires charging with oil. When a leak occurs, oil escapes with the refrigerant and must be replaced for proper lubrication. Loud compressor noises can sometimes be remedied by adding a small amount of oil (usually no more than one-half cup) to the hermetic system. Use oil compatible with the system and of proper viscosity.

1. Install two access valves on the refrigerator, one on the suction line, and one on the liquid line (see figs. 45f, 45k, 21, 79).
2. Connect the manifold gauges to the valves. Compound gauge to the valve on the suction line, and high pressure gauge to the liquid line. By opening the valves on the manifold, freon can be released through the middle hose on the manifold. By law, the middle hose must be connected to a recovery machine (p. 139) to recover the freon into an empty cylinder to be used again. Details about this can be obtained from any refrigeration supply house where these tools are sold.
3. As shown in fig. 84, connect a vacuum pump to the unit, and pull a vacuum to 29" of mercury, close the valves on the manifold, and turn off the vacuum pump.
4. As shown in fig.45i, submerge the manifold middle hose in the compressor oil.
5. Open the manifold low pressure valve and allow oil to be drawn into the compressor.

CAUTION: Shut off the low-side manifold valve while the end of the hos is still submerged in oil to prevent air from entering the system.

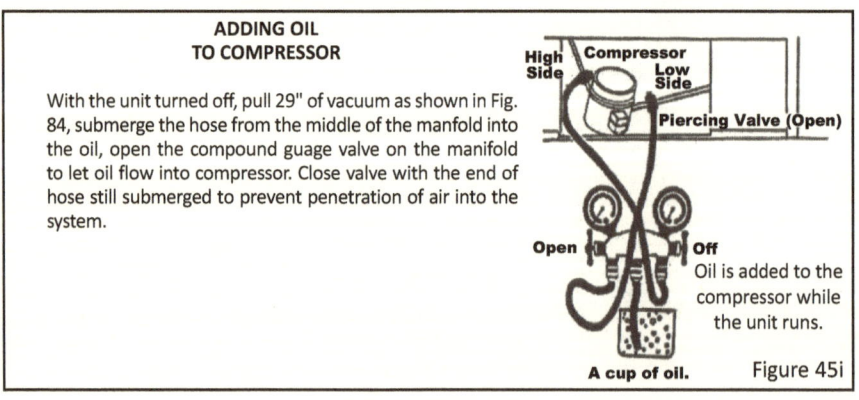

Figure 45i

Access valves come in a variety of sizes and styles.

# AIR CONDITIONING AND REFRIGERATION REPAIR MADE EASY

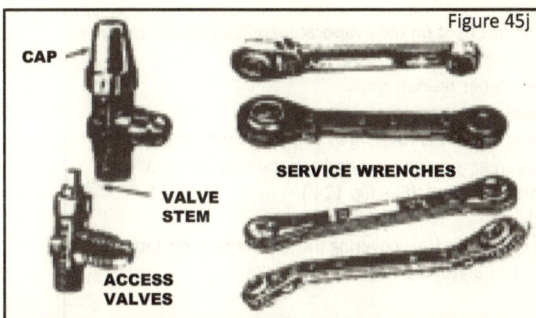

Figure 45j

In residential units, piercing valves must be installed on the discharge and suction lines prior to charging the unit with oil. In commercial units, the service valves on the compressor discharge side (or on the receiver) and on the suction side may be opened and closed with a service wrench. (More about service valves on page 155.)

Courtesy of Henry Valve Co.

## ACCESS VALVES

Valves in figure 45k can be installed on the compressors having high- and low-side access tubes or on the tubing. Remove the valve cap and stem from the valve, clean the joints with acetone as well as sandpaper, braze the correct-size valve on the tube, and allow it to cool before replacing the valve stem and cap.

Numbers 1, 2, and 3 are access valves that are silver-brazed to copper tubing. Number 1 can be used on various diameters of tubing. Number 4 shows the removal of the valve core from the body. Numbers 5, 6, and 7 are T-fittings installed on high- or low-side refrigerant lines. Numbers 5 and 7 are silver-brazed, and number 6 is connected by flared connections. Number 8, shown with its cap, has an extended tube. It can be installed on the compressor access tube as numbers 1, 2, and 3, or on larger sized tubing where in-line fittings may not be available. Drill a hole in the wall of the tubing, insert the valve tube in the hole, and silver-braze the joint. Take great care to prevent foreign particles from entering the system.

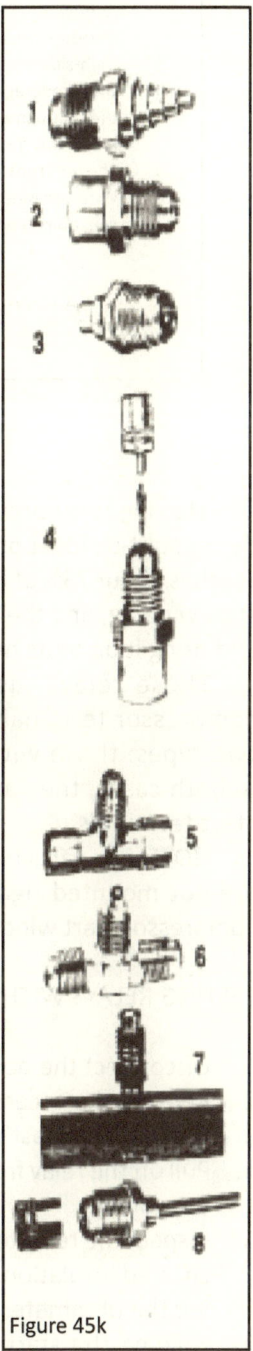

Figure 45k

Courtesy of Robinair Division, SPX Corporation

**INDICATIONS OF A LOW-CAPACITY COMPRESSOR**

> 1. A heavy accumulation of soft frost on the evaporator coil that can be easily removed.
> 2. Low-side pressure reads higher than normal.
> 3. High-side pressure reads lower than normal.
> 4. Low amperage reading when unit is running as compared with the FLA on the unit nameplate. Get an amperage reading from the wire connected to either the run or common terminal. (See fig. 124.)
> 5. The unit runs constantly.
> 6. When a residential unit is running, frost covering the accumulator disappears very quickly when the accumulator is held in the hand.
>
> A LOW-CAPACITY COMPRESSOR MUST BE REPLACED

## TESTING STARTING RELAYS

The starting relay provides power to the start winding of the compressor during start-up for approximately three seconds. At this time, the compressor reaches about 75% of its running speed, the relay shuts off the power to the start winding, and the compressor run winding continues to run the motor and bring it up to its normal speed. (See fig. 46.)

These relays, used in residential units, are mounted on the compressor terminals under the terminal cover. Basically, they are of two types: those with two electrical terminals and those with three. In both cases, they are connected directly to the compressor run and start terminals.

NOTE: Pay particular attention to the word TOP on start relays. If relays are not mounted right side up, they will never open the circuit to the compressor start winding, causing the winding to burn.

TESTING RELAYS WITH TWO TERMINALS: (See figs. 46 and 13)

1. Disconnect the power supply.
2. Remove the access cover in the back of the unit.
3. Remove the plastic compressor terminal cover.
4. Pull off the relay from the compressor and disconnect its terminals from the wiring.
5. Inspect the relay terminals for burn discoloration and the lead wires for charred insulation. If so, replace the relay. If not:
6. Set the ohmmeter to the RX1 scale and zero the meter.
7. Holding the starting relay in one hand with the TOP up, place the ohmmeter probes in the relay terminals. The meter should register a

continuity reading. Otherwise, the starting relay is defective and should be replaced.
8. Turn the relay upside down and place the ohmmeter probes in the terminals. The meter should register an open-circuit reading. Otherwise, the relay should be replaced.

## TESTING RELAYS WITH THREE TERMINALS

1. Disconnect the power supply, remove the compartment cover of the unit, and remove the plastic cover.
2. Remove the relay from the compressor terminals and inspect the relay terminals for burn discoloration and the wires for charred insulation. If so, replace the relay.
3. Holding the starting relay with the *top* up, touch one probe to the terminal that connects to the compressor run terminal and the other one to the terminal that connects to the compressor start terminal. The ohmmeter should show an open circuit reading on the RX1 scale; otherwise, the relay is defective.
4. Touch one probe to the relay terminal on the outside and the other probe to the relay terminal that connects to the compressor start terminal. The ohmmeter should register an open circuit. Otherwise, the relay is bad. (See fig. 46a)
5. Now turn the starting relay upside down and touch one probe to the relay terminal that is connected to the compressor run terminal and the other one to the relay terminal that connects to the compressor start terminal. The meter should register a zero reading on the ohmmeter. Otherwise, replace the relay.

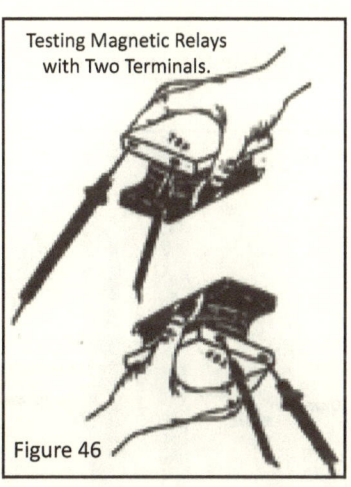

Testing Magnetic Relays with Two Terminals.

Figure 46

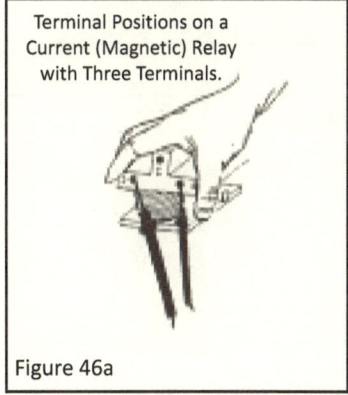

Terminal Positions on a Current (Magnetic) Relay with Three Terminals.

Figure 46a

## STARTING RELAYS

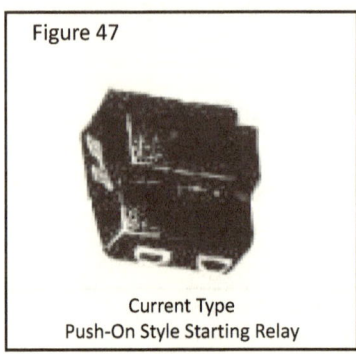

Figure 47

Current Type
Push-On Style Starting Relay

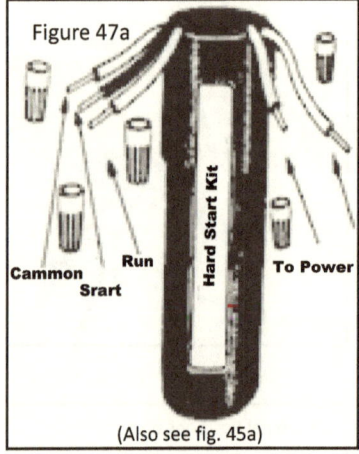

Figure 47a

Cammon, Run, Srart, Hard Start Kit, To Power

(Also see fig. 45a)

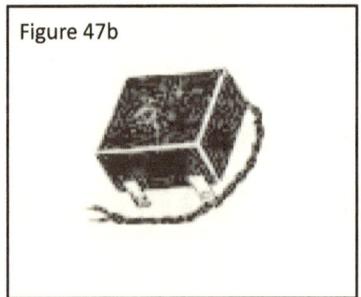

Figure 47b

Courtesy of Johnson Controls

Starting relays in commercial units (unlike residential types) are not always fastened directly to the compressor terminals. They are usually installed in a box next to the capacitors and are wired to the compressor terminals. (See fig. 151) The relays shown in figure 47 are connected directly to the compressor terminals.

There are four types of relays:

1. Current (magnetic) relay
2. Potential (voltage) relay
3. Thermal relay
4. Solid-state relay

1. *Current (magnetic) relays.* The compressor draws up to 600% more current when it is in its starting stage than what it draws during its normal speed. This causes a high amperage flow into the motor (through the starting relay) when the compressor is energized. The high amperage flowing through the relay winding creates a magnetic field, causing the contacts within the relay to close and supply power to the compressor start winding. As the compressor motor picks up speed, it draws less current. By the time it reaches two-thirds of its normal run speed, the reduction in current flow causes the magnetic field in the relay winding to become weak and then gravity separates the contacts and disconnects power to the compressor start winding. At this point, the compressor run winding takes over and brings the motor to its full running speed. Current-type relays are widely used in residential units.

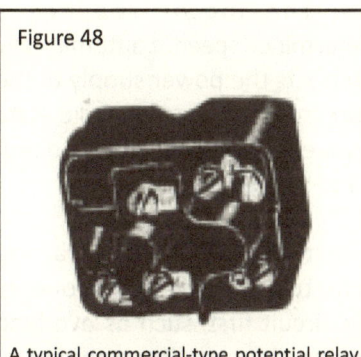

Figure 48

A typical commercial-type potential relay, which is not mounted on the compressor terminals. The contacts in a potential relay are normally closed.

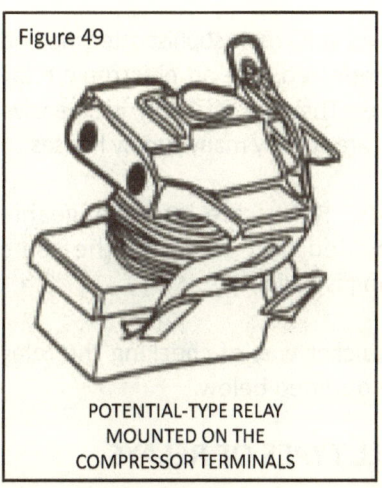

Figure 49

POTENTIAL-TYPE RELAY MOUNTED ON THE COMPRESSOR TERMINALS

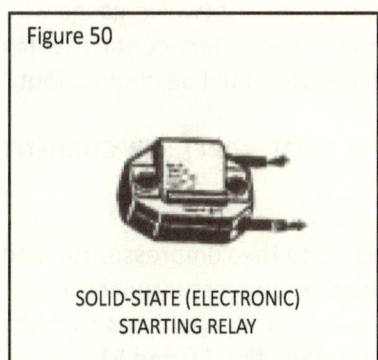

Figure 50

SOLID-STATE (ELECTRONIC) STARTING RELAY

Courtesy of White-Rodgers, Division of Emerson Electric Company

2. *Potential relays* are often called voltage relays. These relays are used with larger capacitor-start compressor motors which require more torque to start. They are widely used in commercial refrigeration and appear similar to the current (magnetic) relays but work by voltage fluctuation. As the speed of the motor increases, voltage is increased. This causes the winding in the relay to create more magnetism, forcing its contacts to open, disconnecting the power to the compressor start winding. By the time the motor reaches two-thirds of its normal speed, the contacts within the relay are opened. (The contacts are normally closed.) (See figs. 48 and 49.)

3. *Thermal relays.* A bimetal-metal strip is installed within the relay. A resistance wire is mounted close to the bimetal strip. As current flows to the compressor start winding through the resistance wire, the resistance wire heats the bimetal strip causing it to flex and open the contacts within the relay, interrupting the flow of current to the compressor start winding. By the time the contacts open, the motor reaches its operating speed. (Its contacts are normally closed.)

4. *Solid-state electronic relays.* These relays are not sensitive to motor size; therefore, they can be used for a variety of motors from 1/12 to 1/3 hp.

Diodes and triacts are used in their construction.

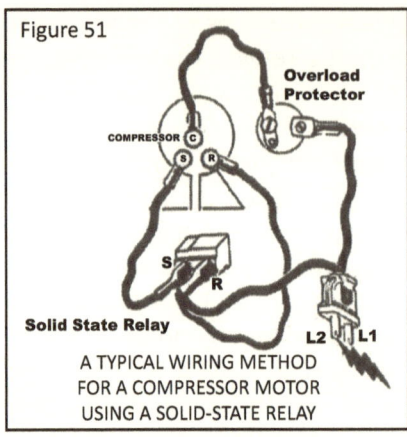

A TYPICAL WIRING METHOD FOR A COMPRESSOR MOTOR USING A SOLID-STATE RELAY

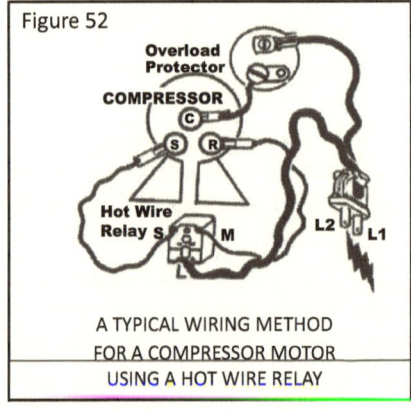

A TYPICAL WIRING METHOD FOR A COMPRESSOR MOTOR USING A HOT WIRE RELAY

When the motor reaches a predetermined speed, it automatically disconnects the power supply to the compressor start winding. Solid-state relays are replaced and not repaired. (See fig. 50.)

The best way to test the starting relay is to test other components in the circuit first, such as overload protector, the capacitor, thermostat (or pressure control if used in the system). If everything else checks out good, then replace the relay.

However, if a more sophisticated method is preferred, use an electronic relay tester. They are relatively inexpensive and are sold by many supply houses.

NOTE: Relays are interchangeable provided that one with the same rating is used.

A quicker way of checking the relay that experienced technicians use today is outlined below.

## METHODS FOR TESTING ALL TYPES OF RELAYS

First, prepare a special test cord. This should be at least twelve-gauge wire with insulated alligator clips at each end and a momentary-contact switch (normally open) in the center. Total overall length should be about a foot.

I. **TESTING A SOLID-STATE, THERMAL (HOT WIRE), OR CURRENT (MAGNETIC TYPE) RELAY**

These relays are directly or indirectly attached to the compressor run and start terminals. The relay terminal attached to the compressor run terminal is marked R or sometimes M (for motor). The relay terminal, which is attached to the compressor start terminal, is marked S. (See figs. 50 and 51.)

This test is conducted by bypassing the relays.

Testing Procedure

1. Disconnect power to the unit.
2. Disconnect the relay from the compressor by pulling it off the compressor terminals or by pulling off the wires, connecting it to the compressor run and start terminals. (In most commercial units, the relay is not directly attached to the compressor terminals, instead it is installed in a control box next to the capacitor near the compressor and wired to the compressor terminals.)
3. Disconnect the only power line to the relay and connect it directly to the compressor run terminal.
4. Using a length of wire with alligator clips at each end and a momentary-contact switch in the middle, connect the compressor R and S terminals. (The wire should be twelve gauge or heavier.)
5. Reconnect power to the unit and depress the switch for no longer than three seconds and release it. If the compressor starts and remains running, the relay must be replaced.

(See figs. 50 and 51. Also see "Testing the Compressor with a Test Cord").

## II.   TESTING A POTENTIAL-TYPE RELAY

Potential relays are wired to all three compressor terminals and not just to the compressor run and start terminals.

When a compressor motor fails to operate and the potential relay is suspected to be the cause, the best way to test it is to use a compressor test cord. (See fig. 35,36) Start the compressor by bypassing the relay. If the compressor starts running, the starting relay must be replaced. (Potential relay testers may be obtained from major refrigeration supply dealers if that method is preferred. They all come with complete instructions.)

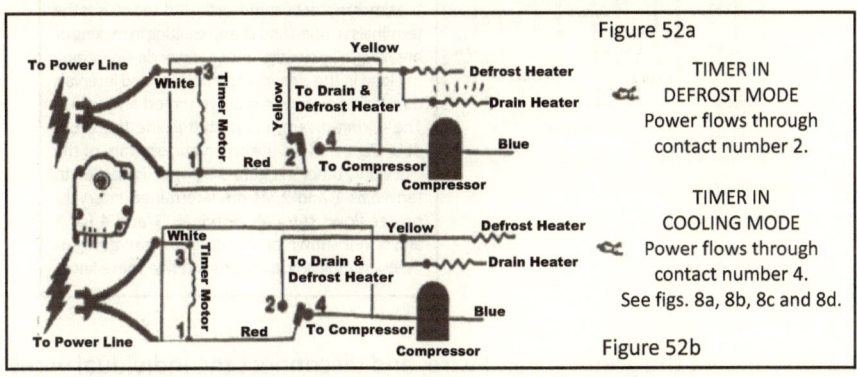

Figure 52a

Figure 52b

## TESTING TIMERS IN FROST-FREE RESIDENTIAL AND COMMERCIAL REFRIGERATING UNITS

In residential units, timers are usually located behind the toe plate (front grille at the bottom of the unit) near the front or back of the unit, or sometimes next to the cold control (thermostat) inside the refrigerator compartment behind the control panel, and sometimes in the back of the unit. Pay particular attention to figures 11a-11e and 54c-54l. Different manufacturers use different timers with different terminals to energize the compressor and the defrost timer. This test assumes that terminal 4 is the defrost and terminal 2 is for the compressor. Before running a test, check the terminals on any particular unit being tested.

1. Disconnect the power supply.
2. Disconnect the timer.
3. Disconnect the block from the defrost timer by pulling it apart.

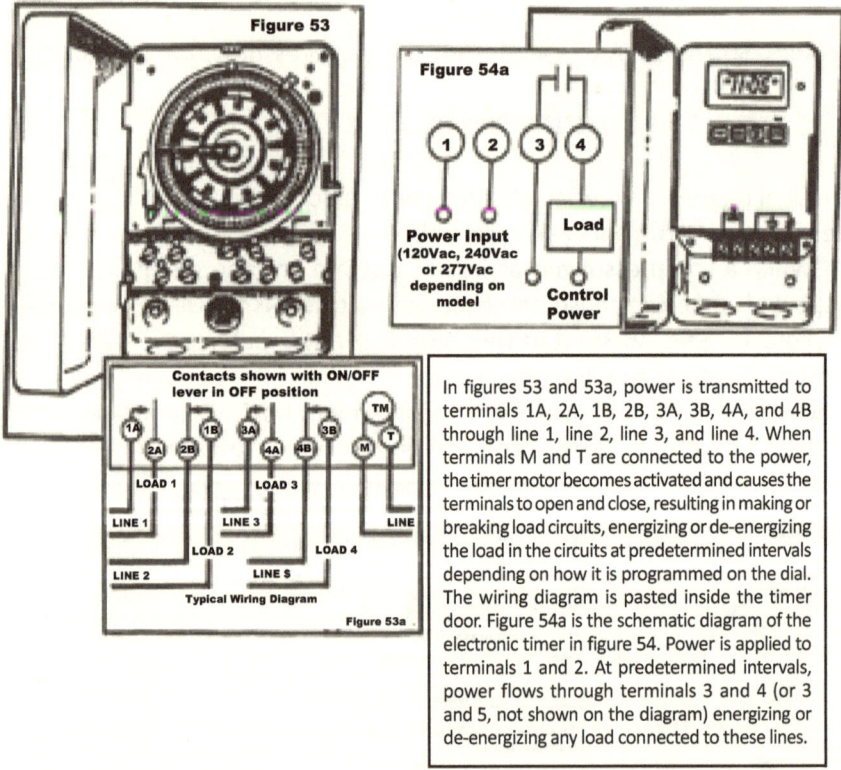

In figures 53 and 53a, power is transmitted to terminals 1A, 2A, 1B, 2B, 3A, 3B, 4A, and 4B through line 1, line 2, line 3, and line 4. When terminals M and T are connected to the power, the timer motor becomes activated and causes the terminals to open and close, resulting in making or breaking load circuits, energizing or de-energizing the load in the circuits at predetermined intervals depending on how it is programmed on the dial. The wiring diagram is pasted inside the timer door. Figure 54a is the schematic diagram of the electronic timer in figure 54. Power is applied to terminals 1 and 2. At predetermined intervals, power flows through terminals 3 and 4 (or 3 and 5, not shown on the diagram) energizing or de-energizing any load connected to these lines.

4. Disconnect the green ground wire, and disconnect the individual wires from the timer if there is no connecting block. Be sure to mark every

wire so it may be reconnected correctly. (Method: make a little "flag" for each wire with masking tape and a fine line marker.)
5. Set the ohmmeter to the RX1 scale and zero it.
6. Connect the ohmmeter probes to the timer terminals 1 and 4; then using a screwdriver, coin, or putty knife blade, turn the cam on the timer clockwise very slowly. The meter should register a zero reading until a click is heard. At that precise moment, stop; the meter should register an open-circuit reading. If not, the timer should be replaced.
7. Place the two ohmmeter probes on terminals 1 and 2 and continue turning the cam. A zero reading should be registered on the meter until a click is heard. Stop. The meter should register an open-circuit reading. If not, the timer should be replaced.
8. Set the ohmmeter on the RX1000 scale and zero it.
9. Connect the ohmmeter probes to the timer terminals 1 and 3. The meter should register a continuity reading. Otherwise, the timer motor is bad and the timer must be replaced.

This test can determine if the timer has any electrical defects. Timers can also develop mechanical problems. Occasionally, a defrost timer motor freezes in one particular cycle, and it no longer advances because of mechanical problems (or a short or a disconnection in the timer motor).

To test the timer for mechanical problems, turn the cam very slowly clockwise. If you feel any "snag" anywhere while turning it, the timer should be replaced.

While the timer is disconnected from its electrical wires, visually check all of the terminals for burns. If there are any brown spots or burns on any of the terminals, replace the timer. When connecting the timer terminals to the system, make sure every connection is secure. Tighten loose connections and replace broken or frayed wires.

---

**CHECKING THE TIMER MOTOR BY DIRECT CONNECTION**

This test can determine if the timer motor has any electrical defects such as a short or a disconnection in the timer motor causing it not to advance. (Occasionally, a timer motor freezes in one particular cycle and no longer advances because of a mechanical defect).

1. Disconnect the power supply.
2. Turn the timer shaft to a point just before a click is heard and leave it there.
3. Using a test cord with insulated alligator clips, connect timer terminals 1 and 3 directly to the power supply (where the unit is normally connected).
4. If the timer motor starts turning, it is good. Otherwise, disconnect it from the power and replace the timer. (In some timers, you cannot see the motor rotation through its housing. In which case, wait for about fifteen minutes; if you hear the click, the timer is good. Otherwise, replace it).

## HOW A COMMERCIAL TIMER WORKS

As you can see in figure 54b, only one of the power lines is connected directly to the loads (compressor, fans, and defrost heater). The other power line runs through the timer terminals before it reaches the loads to energize them. The timer motor, operating on a twenty-four-volt power supply, opens and closes contacts at regular predetermined intervals, causing the loads to become energized or de-energized (taking the unit to the defrost or run cycle). The frequencies and lengths of these intervals are regulated on the timer dial. There are two sets of contacts in the timer: (1) The normally closed contacts (terminals 1 and 2) and (2) the normally open contacts (terminals 3 and 4). When the normally open contacts are closed, the defrost heater becomes energized (only for a short time). The normally closed contacts energize the compressor(s) and the condenser fan(s). Terminal 1 connects power to the compressor(s) and the condenser fan(s). Terminal 4 connects power to the defrost heater. Twenty-four-volt power connected to terminals 5 and 6 energizes the timer motor. When normally closed contacts open, the normally open contacts close simultaneously and vice versa.

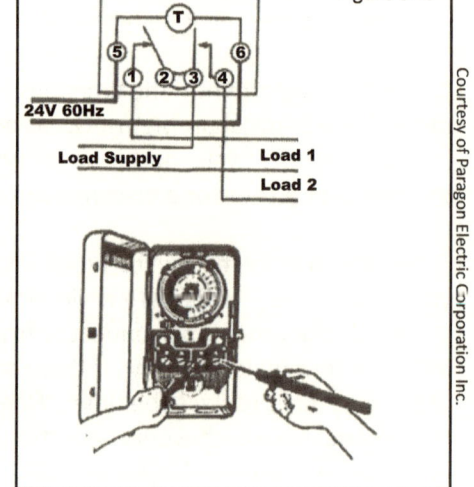

Figure 54b

Courtesy of Paragon Electric Corporation Inc.

## TESTING A COMMERCIAL TIMER

1. Disconnect the power and wires connected to the timer terminals.
2. Check the diagram inside the timer door for normally open (NO) and normally closed (NC) contacts such as the ones in figures 53a, 54a, and 54b.
3. Remove the wires connected to the terminals.
4. Set ohmmeter scale on X1 and zero it. (No adjustment is necessary if you are using a digital meter.)
5. Turn the dial clockwise until it reaches the zone where the distance between the clicks is longer. (The unit is now in the cooling cycle.) Touch the probes to the normally closed contacts (numbers 1 and 2). The meter should register continuity. Otherwise, replace the timer.

6. Turn the timer dial clockwise to the area in which the distance between the clicks (start/stop time tabs) is shorter. This is the defrost cycle.
7. Touch the ohmmeter probes to contacts 3 and 4; the meter should register a continuity reading. If not, replace the timer.
8. Touch the two terminals connected to the timer motor (numbers 5 and 6). The meter should register an ohm reading. If not, replace the timer.
9. Turn the dial slowly clockwise. If you feel any snag, replace the timer.
10. If you notice any charring or burn marks on the terminals, replace the timer. Also, if the timer fails to change cycles, check the terminals connected to the timer motor (numbers 5 and 6) with a test light to make sure proper voltages reaches the timer. In cases where the timer motor operates on low voltage (40, 24, or 12 VAC), if the test light doesn't glow, check the transformer.

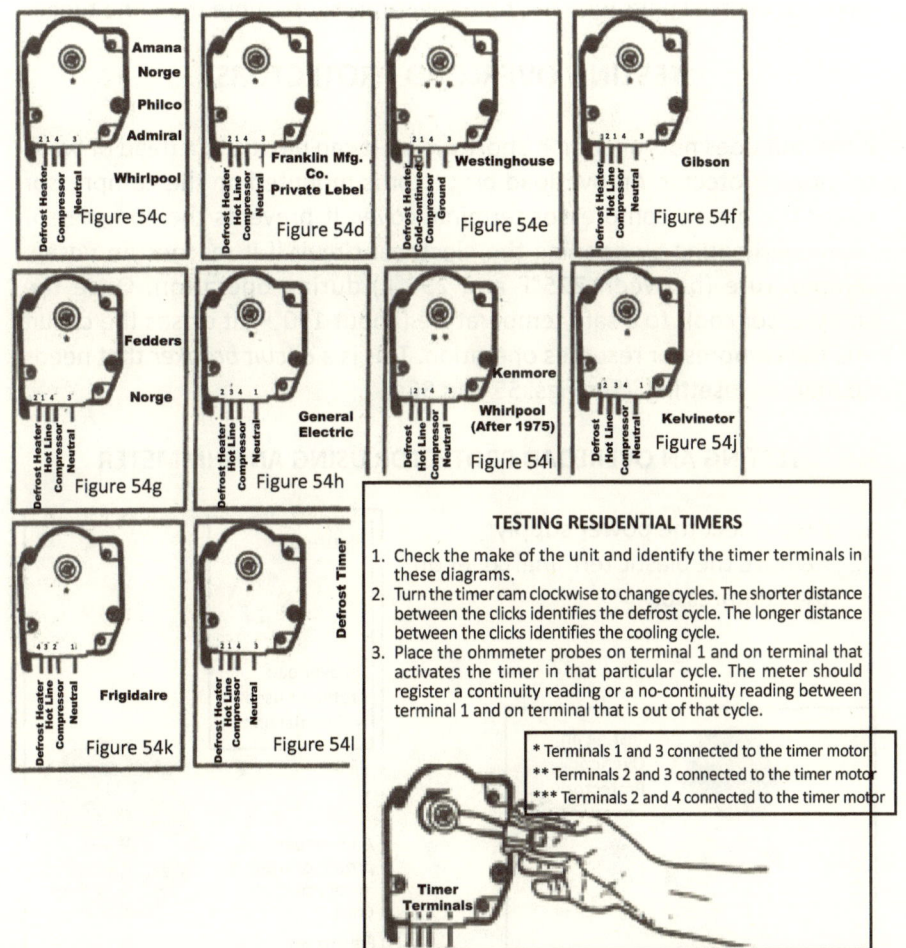

A good way to find out if the timer motor fails to change cycles is by visual observation. Watch the timer dial while it is connected to the power for a short time. If the timer dial turns, there is nothing wrong with the timer motor.

Some commercial-type timers are equipped with a sensing bulb, which is attached to the evaporator coil. When enough ice accumulates on the evaporator coil, the sensing bulb transmits the changes in temperature to the timer bellows and causes the timer to take the unit into the defrost cycle by stopping the compressor motor, the evaporator fan(s), and energizing the evaporator heater or the hot gas solenoid. Then, when sensing a rise in evaporator temperature, power is restored and the unit restarts.

The defrost period on these timers can be regulated by turning an adjusting knob inside the unit cabinet to lengthen or shorten the defrost period. Some large commercial units with multiple evaporators use more than one timer.

## TESTING OVERLOAD PROTECTORS

If the unit does not run, or if it short-cycles, it can be due to a dead or weak overload protector. An overload protector is mounted on the compressor under the plastic compressor terminal cover. It prevents the compressor from overheating by opening the electrical circuit if it reaches an unsafe temperature (between 225°F and 250°F) during operation. Once the compressor cools to a safe temperature (about 150°F), it closes the circuit and the compressor resumes operation. This is a *circuit breaker* that needs no manual resetting. (See figs. 55 and 56.)

### I. TESTING AN OVERLOAD PROTECTOR USING AN OHMMETER

1. Disconnect the power supply.
2. Remove the plastic terminal cover on the compressor.

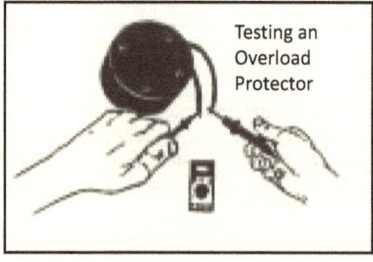

Testing an Overload Protector

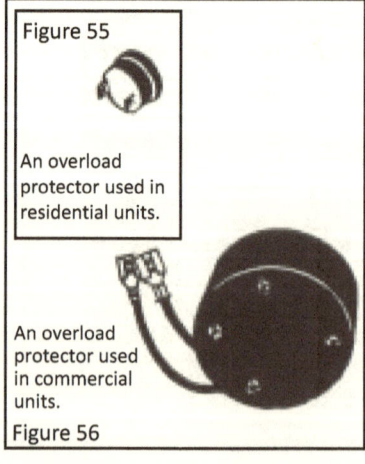

Figure 55

An overload protector used in residential units.

An overload protector used in commercial units.

Figure 56

3. Remove the two wires connected to the terminals of the overload protector.
4. Set the ohmmeter to the RX1 scale and zero the meter.
5. Touch one probe to each of the wire terminals on the overload protector to get a reading of zero on the meter. If not, replace the defective overload protector.

Also, visually check for cracks on the overload protector. If there are any, the protector is bad and it must be replaced. Sometimes, in spite of the fact that they test OK with the ohmmeter, when they are cool, they may have become weak and open the circuit once they are warm.

## II. TESTING AN OVERLOAD PROTECTOR USING THE BYPASS METHOD

(A test wire with two alligator clips will be needed.)

1. Disconnect the power supply.
2. Clip the two test wire alligator clips to the two terminals on the overload protector. Be careful not to allow the clips to touch any other object. Be sure the insulating covers on the clips completely enclose them. (See fig. 56a).
3. Reconnect the power for a short time. Be careful not to touch any part of the unit as there is always the danger of a shock hazard.
4. If the compressor motor starts running without drawing an unusual amount of amperage (this can be checked by a clamp-on ammeter clamped around the wire supplying power to the overload protector), the overload protector is defective and will need replacing.
5. Disconnect the power supply, disconnect the test wire and replace the overload protector.

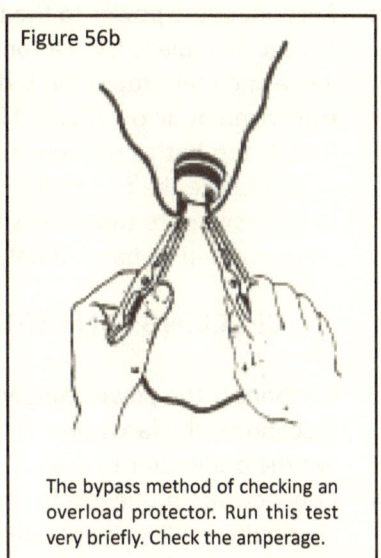

Figure 56b

The bypass method of checking an overload protector. Run this test very briefly. Check the amperage.

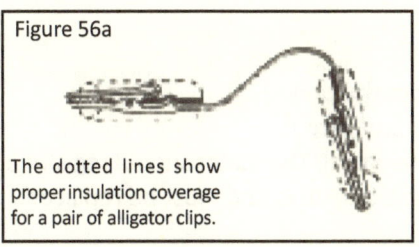

Figure 56a

The dotted lines show proper insulation coverage for a pair of alligator clips.

## TESTING THE CONDENSER FAN MOTOR

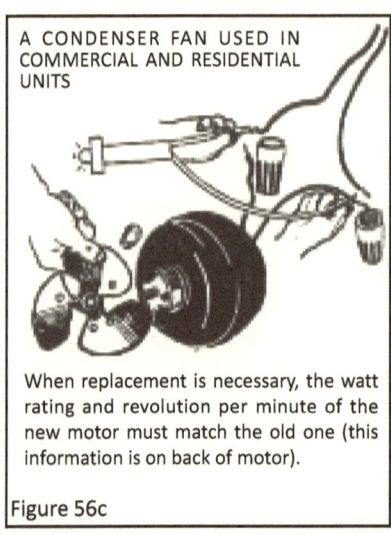

A CONDENSER FAN USED IN COMMERCIAL AND RESIDENTIAL UNITS

When replacement is necessary, the watt rating and revolution per minute of the new motor must match the old one (this information is on back of motor).

Figure 56c

Condenser fan motors are used to circulate air through the condenser to cool it. They are located next to the condenser (in fan-forced condenser types) behind the compressor-compartment cover. (See fig. 56c and 57).

1. Remove the rear, lower access cover with the unit running to determine if the fan is working.
2. If not, disconnect the power supply.
3. Remove the two condenser fan motor wires.
4. Before testing the condenser fan motor, make sure that the failure is not due to lack of power reaching the fan motor.

Checking Procedure

a. Connect a test lamp equipped with alligator clips (see page 21) to the two wires removed from the fan motor. Be sure the clips are well insulated and are touching no other part of the unit.
b. Reconnect the power to the unit. If the lamp glows, the fan motor will have to be replaced. Occasionally, even a bad motor starts and runs for a while and then stops. This is due to an internal electrical short, which is evidenced by an overheated fan motor (after a few minutes of running, it feels too hot to the touch) or due to worn shaft bearings. In either case, replace the fan motor. Sometimes the fan motor shaft jams and becomes hard to turn. The shaft should turn freely without squeaking when rotated by hand; if not, replace the fan motor.

## CHECKING FAN MOTORS WITH AN OHMMETER

1. Disconnect the power supply.
2. Disconnect the fan motor wires from the unit.
3. Set the ohmmeter to the RX1 scale and zero it.
4. Connect each probe to a fan motor terminal. The meter should register a continuity reading. If not, the motor is defective and must be replaced.

In the case of capacitor-run or capacitor-run-capacitor-start fan motors in commercial units, the problem can also be due to a defective capacitor in which case the capacitor will have to be checked. (See the pages about Capacitors.)

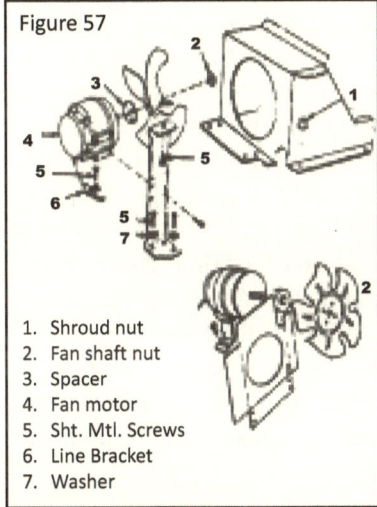

Figure 57

1. Shroud nut
2. Fan shaft nut
3. Spacer
4. Fan motor
5. Sht. Mtl. Screws
6. Line Bracket
7. Washer

Courtesy of Marvel Industries
Division of Northland Corporation

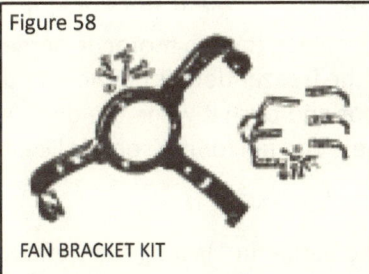

Figure 58

FAN BRACKET KIT

This condenser fan bracket is a popular style for residential units. The ring is mounted on the back of the fan motor, and the legs are fastened to the fan shroud.

a. Sometimes the blade must be removed before removing the motor. Hold the blade with one hand and, with a nut driver or a pair of pliers, remove the stamped nut and gently remove the blade. Note the pitch of the blade. If it is remounted backward on the new motor, it will move the air in the wrong direction.
b. Remove the motor and its bracket by removing the screws connecting the bracket to the refrigeration unit.
c. Remove the bracket from the defective motor (see fig. 58).
d. Mount the bracket to the new motor in the same location it was on the old motor.
e. Mount the fan blade to the new motor ensuring that its pitch is in the same direction as on the old motor.
f. Install the assembly, reconnect the wires and, replace the cover panel.

When replacing condenser or evaporator fan motors in commercial refrigeration units, always pay close attention to the *watt rating* of the motor to be replaced. The new fan must have at least the same wattage and revolution per minute. A fan motor with lower revolution per minute turns slower, causing a considerable loss in the efficiency of the evaporator or the condenser. Consequently reducing the efficiency of the unit.

Fan motors do not usually come with blades. Buying them separately will facilitate getting an exact match for the replacement. An instruction sheet is included with the new motor. Before removing the old fan, note the direction of the motor rotation and the angle of the blade. Incorrect assembly changes the direction of the airflow, causing the unit to malfunction.

## TESTING EVAPORATOR FAN MOTORS

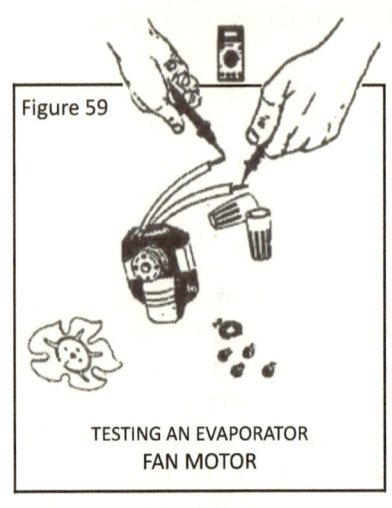

Figure 59

TESTING AN EVAPORATOR FAN MOTOR

The function of the evaporator fan is to force cold air from the freezer compartment into the refrigerator compartment and to circulate air over the cold evaporator coil. It is located in the freezer compartment behind the evaporator cover and fan support plate.

If the unit fails to cool and you notice an accumulation of ice on the evaporator panel, and you are suspicious of a fan motor that does not operate, or operates at a lower speed, in general, everything said about checking a condenser fan motor holds true for an evaporator fan motor. (See fig. 59)

## HOW TO TEST AN EVAPORATOR FAN MOTOR

1. Open the freezer door and listen for the fan motor. If the fan is not heard, or if there is any unusual noise, replace the fan motor. In some units, the fan motor will not operate if the freezer door is opened; find the push button around the freezer door and push it while the door is open and listen to the fan. If you can hear the fan, or if it sounds like it is operating at a reduced speed.
2. Disconnect the power.
3. Remove the evaporator cover (explained earlier in "Testing the Defrost Limit Switch") and remove the fan support plate. (A long running time with an inoperative fan can be evidenced by an accumulation of ice on the evaporator plate).
4. If the fan does not run, check to see if there is power at the motor terminals with a test light. (As explained in "Testing the Condenser Fan Motor," step 4). Take care not to touch any live wires while the unit is plugged in.

If there is power at the fan motor and it is not running, replace the fan motor.

Sometimes, defective fan motors run for a while, but due to an internal short, after a brief period of running, heat up and stop operating. Sometimes the fan runs, but it appears not to be running at full speed. This ohmmeter test will determine whether or not it is defective. If the ohmmeter registers a reading ±5% beyond the factory ohm rating, or if no-continuity reading is registered, replace the fan motor. Refrigeration units manufactured by General Electric, in particular, and many other brands come with schematic wiring diagrams (covered in detail in the "Basic Electricity" chapter) that indicate the ohm value of each electrical component.

When replacing fan motors, check for the watt rating and the revolution per minute of the replacement part. The new fan motor should be rated as close to the original part as possible. This is particularly important in commercial refrigeration considering the fan speed variation in those units.

Often, replacement commercial fan motors come with fan blade shafts that are too long. Measure the length needed and cut off the excess with a hacksaw. During installation, pay close attention to the direction of its rotation. It is stamped CW for clockwise or CCW for counterclockwise on the back of the motor or on the nameplate. The new fan motor must rotate in the same direction as the original.

This is made simple in commercial fan motors (see fig. 61). Usually, two extra wires come out of the motor. By connecting these two wires together, the shaft rotation can be reversed. Instructions to this effect are supplied with new motors.

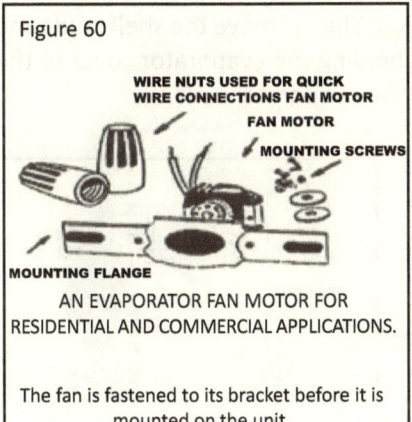

Figure 60

WIRE NUTS USED FOR QUICK WIRE CONNECTIONS FAN MOTOR
FAN MOTOR
MOUNTING SCREWS
MOUNTING FLANGE

AN EVAPORATOR FAN MOTOR FOR RESIDENTIAL AND COMMERCIAL APPLICATIONS.

The fan is fastened to its bracket before it is mounted on the unit.

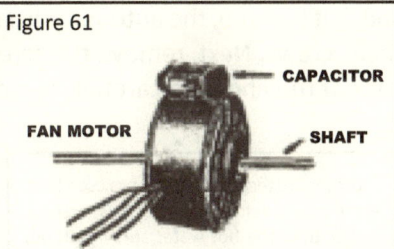

Figure 61

CAPACITOR
FAN MOTOR
SHAFT

Many PSC fan motors (see fig. 137) come with a long double shaft. The shaft is cut with a hacksaw to the length needed. Blades may be mounted on one or both ends. These fan motors can have several uses such as blowers, ventilators, condenser, or evaporator fans.

## TESTING THE DEFROST LIMIT SWITCH
### (Defrost Bimetal or Defrost Thermostat)

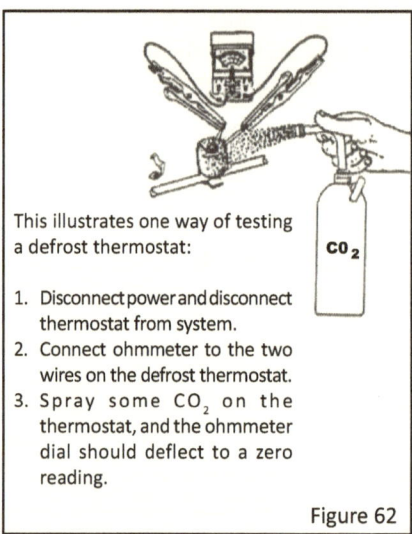

This illustrates one way of testing a defrost thermostat:

1. Disconnect power and disconnect thermostat from system.
2. Connect ohmmeter to the two wires on the defrost thermostat.
3. Spray some $CO_2$ on the thermostat, and the ohmmeter dial should deflect to a zero reading.

Figure 62

A defrost limit switch is a bimetallic device that opens and closes the circuit to the evaporator defrost heater during the defrost cycles in frost-free units. In this cycle, the timer disconnects the power supply to the compressor and at the same time connects power to the evaporator defrost heater to melt the ice built up on the evaporator surface (see fig. 62). The current to the evaporator heater passes through this bimetal device, which controls the on/off function of this circuit. Since this device is mounted on the evaporator coil, once sufficient frost builds up to cause it to contract, it closes the circuit to the heater. As the timer takes the unit into the defrost cycle, the heater melts the ice, the coil becomes warmer, causing the device to expand and open the circuit to de-energize the defrost heater.

Because the bimetal switch is clipped to the evaporator in the freezer compartment, the evaporator cover must be removed to gain access to it. To do this, remove all food from the freezer compartment, all the shelves, and, if it has one, the automatic ice maker. Then remove the shelf studs and their screws. Next, remove the screws holding the evaporator cover to the back of the liner and carefully remove the evaporator cover.

Most service technicians prefer to use a heat gun for rapid defrosting of freezers and coils rather than the hot water spray method, which generates an excessive amount of drain water. These guns operate similar to a hair drier except for their temperatures that are considerably higher (between 350°F and 750°F with high-velocity air). They are also used to apply shrink tubing for insulation, drying out moisture in hard-to-reach places, loosening rusted nuts and bolts, speed drying paint, and bending tubing.

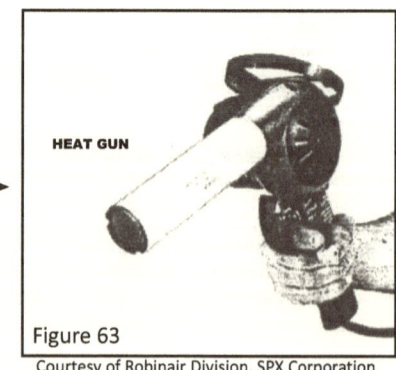

HEAT GUN

Figure 63

Courtesy of Robinair Division, SPX Corporation

## HOW TO TEST THE DEFROST THERMOSTAT (TERMINATION SWITCH) WITH AN OHMMETER

The termination thermostat is clipped to the evaporator. When the evaporator temperature rises to about 50°F, the contacts within the thermostat open, causing the defrost heater to become de-energized. When the evaporator temperature drops to about 20°F, the contacts within the thermostat close and the defrost heater becomes energized when the unit is in the defrost mode.

Buy a small container of $CO_2$ (a fire extinguisher) before starting this test.

1. Disconnect power to the unit.
2. Remove the evaporator panel.
3. Disconnect the two wires from the termination thermostat and remove it from the unit.
4. Set the ohmmeter on the RX1 scale and zero it. Then connect the two meter probes to the two thermostat wires with alligator clips.
5. Spray the thermostat with $CO_2$ for about five seconds. Be sure the $CO_2$ does not touch the ohmmeter. The ohmmeter should register a zero reading. Otherwise, replace the termination thermostat.
6. Hold the termination switch in your hand for about one and a half minutes or long enough to warm it. With the alligator clips holding the probes to the thermostat wires, the meter should register an open-circuit reading. Otherwise, replace the thermostat (see fig. 62).

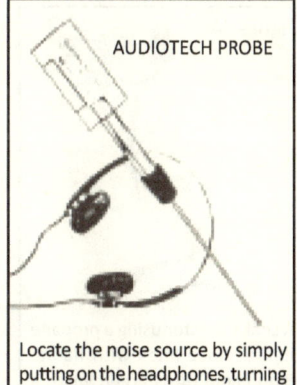

AUDIOTECH PROBE

Locate the noise source by simply putting on the headphones, turning on the amplifier, and touching the end of the stainless steel probe to the suspected problem area.

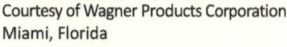

Courtesy of Wagner Products Corporation
Miami, Florida

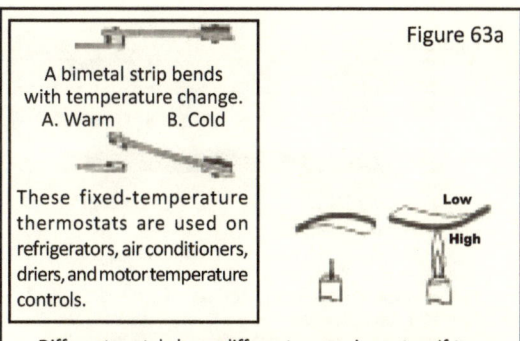

Figure 63a

A bimetal strip bends with temperature change.
A. Warm   B. Cold

These fixed-temperature thermostats are used on refrigerators, air conditioners, driers, and motor temperature controls.

Different metals have different expansion rates. If two metals of different alloys are bonded together and heated, they tend to expand toward the metal with the lower expansion rate. Conversely, as they cool, they tend to contract toward the metal with the higher expansion rate. By employing this principle, bimetal switches can open and close circuits through heat sensitivity.

## A QUICK TERMINATION THERMOSTAT CHECK
## (USING THE BYPASS METHOD)

1—Disconnect the power to the unit.
2—Disconnect the termination thermostat from the unit.
3—Take the two disconnected wires leading into the termination thermostat and connect them together, making sure this new bypass connection is well insulated.
4—Turn on the unit and turn the timer screw to put the unit into the defrost cycle. If the defrost heaters begin heating in this bypassed circuit, the termination thermostat is bad and it must be replaced.

## COMPRESSOR NAME PLATE

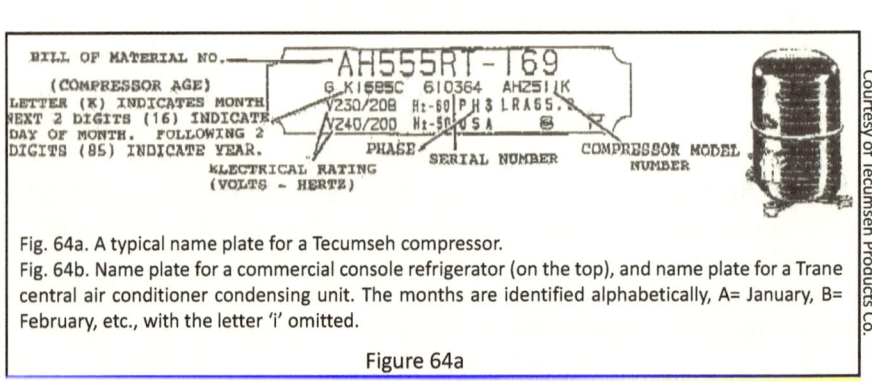

Fig. 64a. A typical name plate for a Tecumseh compressor.
Fig. 64b. Name plate for a commercial console refrigerator (on the top), and name plate for a Trane central air conditioner condensing unit. The months are identified alphabetically, A= January, B= February, etc., with the letter 'i' omitted.

Figure 64a

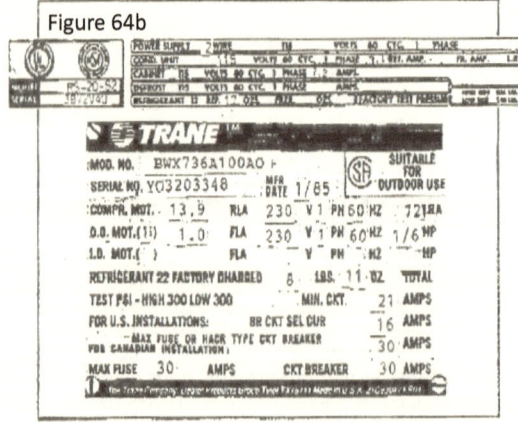

Figure 64b

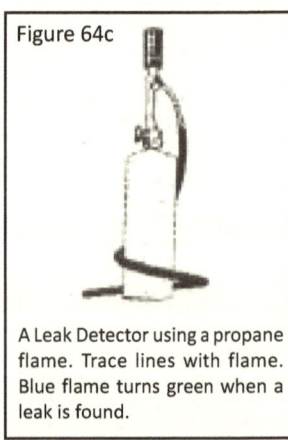

Figure 64c

A Leak Detector using a propane flame. Trace lines with flame. Blue flame turns green when a leak is found.

Courtesy of Robinair Div., SPX Corp.

# CAPACITORS

Capacitors (See Fig. 65) increase the power factor and efficiency of electric motors. Voltage is accumulated on one pole and when charged to capacity, it is released to the other. This is done in rapid succession. Start capacitors provide extra voltage to enable the motor to start easier while drawing less amperage. Run capacitors provide more energy to the motor run winding to help it run easier. Capacitor capacitance is measured in Micro-Farads.

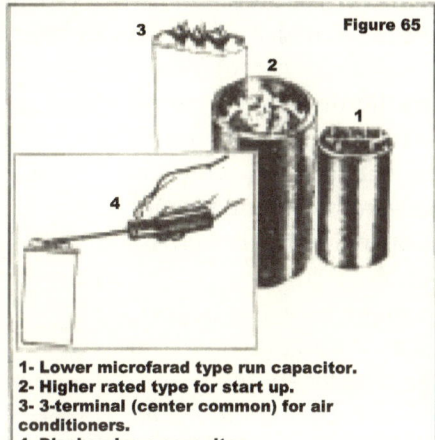

Figure 65

1- Lower microfarad type run capacitor.
2- Higher rated type for start up.
3- 3-terminal (center common) for air conditioners.
4- Discharging a capacitor.

## TO PREVENT THE POSSIBILITY OF A SEVERE SHOCK, DISCHARGE THE CAPACITOR BEFORE TOUCHING IT!

Discharge capacitors by shorting the terminals with an insulated screwdriver described in Testing Capacitors. Very often a compressor may not start or short-cycle due to a bad capacitor. Some capacitors are used to help motors start. They are called start capacitors. They are wired in series with the compressor start winding and the start relay. (See Figs. 43 & 44). The run capacitor is in the starting circuit too. It helps the motor run smoother and more efficiently (See Fig. 45). To accurately check a capacitor, use a capacitor tester like the one shown in Figure 66. Follow the instructions furnished with the instrument. Use the same micro-farad rated capacitor to replace one with another. Never replace a capacitor with a lower rated one. In the event an exact duplicate cannot be obtained, one with up to a 10% higher rating is permissible.

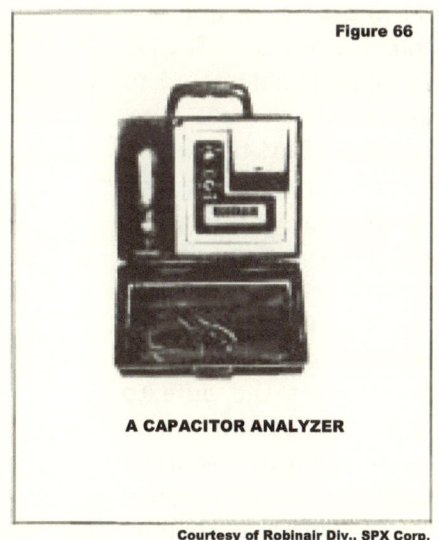

Figure 66

A CAPACITOR ANALYZER

Courtesy of Robinair Div., SPX Corp.

As an example, suppose a 100 mfd capacitor cannot be found to replace a faulty one. Any capacitor rated 101 to 110 mfd may work, but never use one rated under 100 mfd. All capacitors are labeled with their voltage and mfd rating.

The liquid used in some capacitors is bipheny dielectric fluid, which is dangerous. Do not attempt to open the shell or burn it.

In many air conditioners, two capacitors are incorporated into one. These capacitors are seen with three terminals. One marked HERM (hermetic compressor motor), another terminal marked C (common terminal), and one marked FAN. The common terminal is connected to both the fan motor wire and the compressor motor. The fan terminal is connected to the second fan motor wire, and the HERM terminal is connected to the second compressor wire (see fig. 65). There is a red dot (or sometimes other symbols to indicate the grounded terminal) on the capacitor. It indicates the pole to which the hot wire is to be connected. In the event of a short, the current is drained off harmlessly protecting personnel and equipment.

>  **A RULE OF THUMB FOR SELECTING CAPACITORS**
>
> When replacing a capacitor, the voltage rating must be at least as high as the original, and the microfarad (mfd) rating must be as near as possible to the original.
>
> Sometimes it is impossible to identify a defective capacitor. The chart in figure 66a can help you select a capacitor for any unit up to 3 hp. Simply determine the horsepower rating of the motor (compressor) and check the columns next to it for the mfd @ AC volts for the start or run capacitor you wish to replace.
>
> Remember: the motor-operating voltage has little or no bearing to the voltage rating of the capacitor, i.e., a motor that operates on 110 VAC may use a 330 VAC capacitor.

| HP Motor Rating | Capacitor Start MFD @ VAC | Capacitor Run MFD @ VAC |
|---|---|---|
| 1/8 | 72-88@110<br>75-90@110 | 5@370<br>6@370 |
| 1/6 | 86-100@110 | 7.5@370<br>7.5@440 |
| 1/4 | 108-130@110<br>124-149@110 | 10@370<br>10@440 |
| 1/3 | 161-193@110 | 15@370<br>15@440 |
| 1/2 | 200-240@110<br>216-259@110 | 15@370<br>15@440 |
| ¾ | 324-388@110<br>340-408@110 | 20@440<br>20@370 |
| 1 | 378-440@110<br>400-480@110 | 20@370<br>20@440 |
| 1½ | 540-648@110<br>75-90@250<br>81-97@250<br>108-130@250<br>121-145@250 | 25@370<br>25@440<br>25@370<br>25@440 |
| 2 | 127-152@330<br>135-162@330 | 20@370<br>to<br>35@370 |
| 3 | 130-162@ | 40@440 |

Figure 66a  Courtesy of White-Rodgers, Division of Emerson Electric Company

Make it a habit to make a pencil sketch or tag the wires with a piece of tape before removing a capacitor. By doing this, there will be no doubt about the proper connections when installing a new capacitor.

# TESTING CAPACITORS

(The best way to test a capacitor is to replace it with a new one of the same rating).

I. **TESTING CAPACITORS USING AN OHMMETER:**

   a. Disconnect the power to the unit.
   b. Use an insulated screwdriver to shunt between the capacitor terminals to discharge it. Then remove the capacitor from the circuit.
   c. Set the ohmmeter on its highest scale.
   d. Touch the two capacitor terminals with the two leads from the meter. If the capacitor is in good condition, the meter should register zero ohms and then move slowly back toward infinity.
   e. Switch the leads and repeat step d to get the same results. Otherwise, replace the capacitor.

The needle should deflect to infinity and stay there from the very beginning if the capacitor is open. The needle should register 0 Ω and stay there if the capacitor is shorted.

II. **TESTING CAPACITORS USING A DIGITAL CAPACITOR TESTER (They are inexpensive and very accurate.)**

   a. Disconnect power from the unit.
   b. Discharge the capacitor as in Ib above.
   c. Turn on the digital tester and connect its leads to the capacitor terminals. This will show the exact capacitance of the capacitor in microfarads. Compare the reading shown on the meter with the capacitor rating printed on its side.
   d. If it is below what it should be, the capacitor is weak and it must be replaced.
   e. If the meter registers no reading, it will mean there is a disconnection in the capacitor, and it must be replaced.

III. **TESTING CAPACITORS USING A TEST LIGHT (See fig. 66)**

   a. Disconnect power to the unit.
   b. Discharge the capacitor as in Ib above.
   c. Check the mfd rating on the capacitor and set the test light selector switch for that range.
   d. Connect the test leads to the capacitor terminals.
   e. Turn on the tester. The indicator light should glow dimly and go out.

f. If the light stays on, it will mean that the capacitor is shorted, and it must be replaced.
g. If the light never comes on, there is a disconnection in the capacitor and it must be replaced.

## TESTING THERMOSTATS

In residential units, thermostats are located inside the refrigerator either in the side behind the control panel or at the back. Generally, the thermostat is behind the cold control knob. (See Fig. 67)

## TESTING THERMOSTATS USING AN OHMMETER

1. Disconnect the power source.
2. Remove the control knob(s) by pulling it (them) off.
3. Remove the control panel.
4. Using a screwdriver (or nutdriver) remove the screws holding the thermostat. Do not damage the thermostat bulb and very gently pull the bulb from where it is seated.
5. Disconnect every wire from the thermostat.
6. Set the ohm meter on the RX1 scale and turn the thermostat to its warmest setting.
7. While holding the thermostat bulb in your hand for about 2 minutes, touch the two ohm meter probes to the two thermostat terminals. You should get a 0 reading on the meter. If not, the thermostat is bad and should be replaced.
8. Turn the thermostat to the off position and touch the probes again to the two terminals. You should get an open circuit reading. If not, the thermostat must be replaced.
9. Turn the thermostat to its mid position and put the thermostat bulb in a mixture of crushed ice and a little water for about 3 minutes while touching the ohm meter probes to the two thermostat terminals. The indicator needle should move from 0 to an open circuit reading. If not, the thermostat must be replaced.

---

**REPLACING THERMOSTATS**

Thermostats used in residential units are installed in the fresh food or freezer compartment while the sensing bulb is attached to the evaporator coil. There are occasions when the sensing bulb is routed through a maze of holes in the cabinet making it appear impossible to replace once the original one is removed. There is a way to make this job much easier. When all the screws securing the defective thermostat, sensing bulb and/or sensing bulb line are removed, tie a piece of strong cord (a few feet longer than the sensing bulb line) securely to the sensing bulb. Pull out the old thermostat and its bulb from the other end and the cord will follow. When the old line is removed, remove the cord from the old bulb and tie it to the new one. With the new sensing bulb line carefully straightened, pull the new bulb through with the cord and replace the screws securing the thermostat and its bulb.

## TESTING THERMOSTATS USING THE BYPASS METHOD

In cases where the unit will not run at all and a defective thermostat is suspected, simply by-pass the thermostat to see if the compressor will resume running.

This method has a potential shock hazard, so proceed cautiously and observe safety rules.

1. Disconnect the unit from its power source.
2. By-pass the two thermostat terminals with a well insulated piece of wire with insulated "alligator" clips at each end. (The wire should be at least the same gauge as the original wire on the thermostat.) Make sure the clips touch no other part of the unit!
3. Connect the power supply. If the unit starts running, the thermostat is defective.
4. Disconnect the unit again to shut it off. Remove the by-pass wire and replace the defective thermostat.

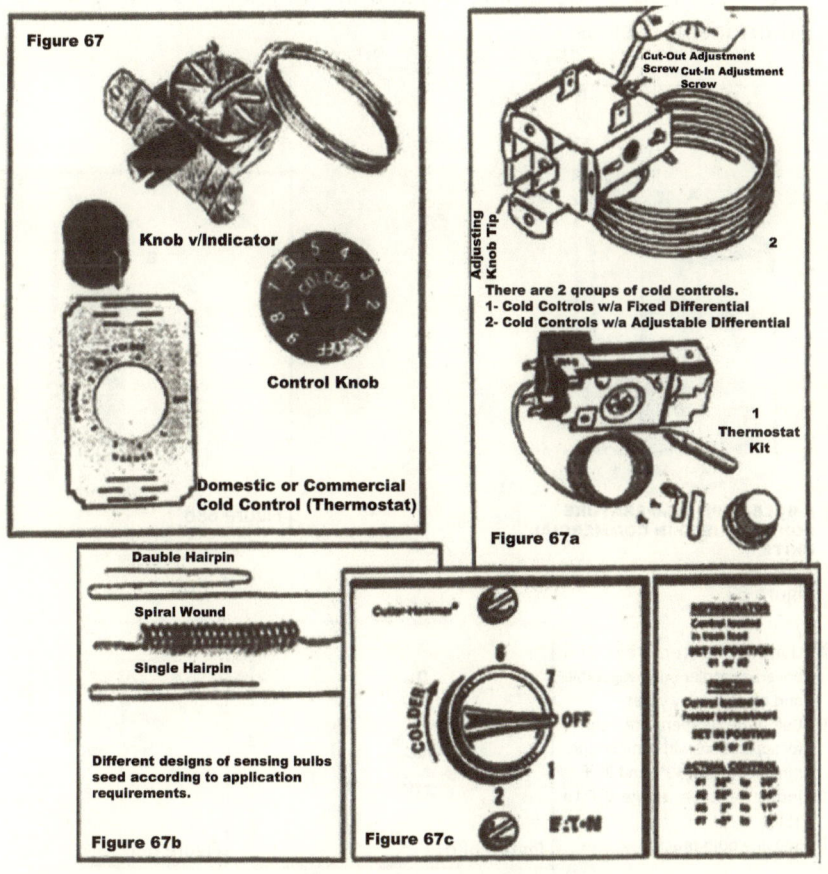

In commercial refrigeration, there are primarily two types of thermostats in use:

1. The *bulb type* (see fig. 68). In this type, a bulb attached to the thermostat is fastened to the evaporator coil. The bulb is filled with refrigerant. As the bulb senses temperature changes, the refrigerant in it expands or contracts and transmits these changes through the line to the thermostat, which in turn causes the bellows to flex and open or close the contacts in the mechanism. The thermostat may be located anywhere as long as the bulb touches the evaporator coil and the line is not exposed to extremes of temperature to prevent transmission of erroneous readings to the thermostat. Be sure that the sensing bulb bracket is positioned in a way that the bulb makes full contact with the evaporator coil, and the nuts and bolts are securely fastened.

When coiling or uncoiling the line on bulb-type thermostats, care must be exercised to prevent kinking or breaking it to preclude refrigerant from escaping or becoming trapped in the line. This will render the thermostat useless.

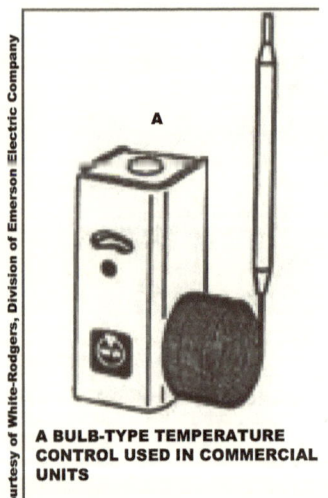

A BULB-TYPE TEMPERATURE CONTROL USED IN COMMERCIAL UNITS

Figure 68

Figure 68b

A. Used on water, beer and beverage coolers, and vegetable and meat display cases
B. Bulb-type temperature control for applications with temperature range between 40°F and 90°F
C. Industrial type: range 0°F to 150°F
D. Sensing bulb bracket

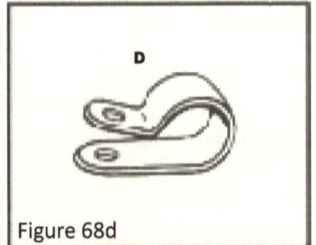

Figure 68d

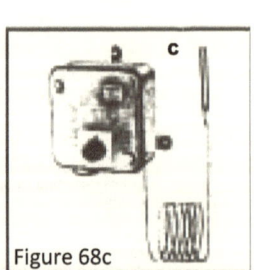

Figure 68c

2. The *air-coil type* (see fig. 69). This type of thermostat is installed somewhere within the refrigerated area, preferably not near a door that is opened frequently, or in the path of the air currents in the unit, to prevent false reactions.

The refrigerant in the coil expands and contracts with temperature changes and transmits signals to the thermostat to control the operation of the unit. When an air-coil thermostat is purchased, determine its cut-in and cut-out range from the instructions (or the dealer) to see if it is suitable for the particular application intended.

As it will be shown later, there is a relation between saturated refrigerant, vapor pressure in the sealed system, and the temperature produced. The operation of the temperature-regulating-pressure controls is based on this principle. Most commercial units use this type of temperature control. Some units employ both thermostat(s) and pressure control(s). Manually operated thermostats are still widely used in commercial kitchen refrigeration where those owners prefer to adjust temperatures periodically to meet their needs. Pressure controls require adjustment by qualified technicians and cannot be manually adjusted by the customer.

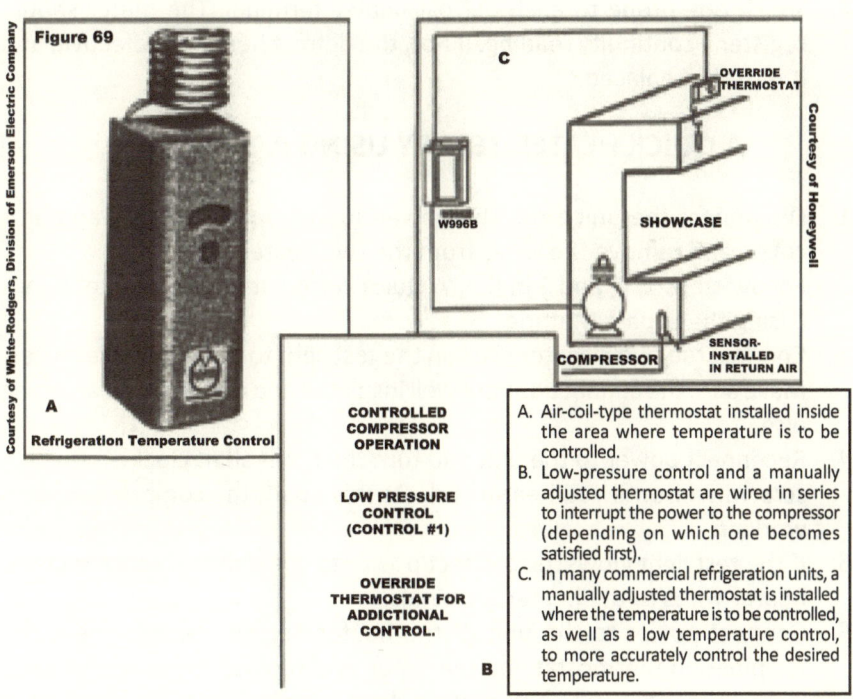

Figure 69

A. Air-coil-type thermostat installed inside the area where temperature is to be controlled.
B. Low-pressure control and a manually adjusted thermostat are wired in series to interrupt the power to the compressor (depending on which one becomes satisfied first).
C. In many commercial refrigeration units, a manually adjusted thermostat is installed where the temperature is to be controlled, as well as a low temperature control, to more accurately control the desired temperature.

## TESTING THE DEFROST HEATER

One possible cause of frost buildup on the evaporator plates or fins can be a burnt defrost heater. Defrost heaters are pressed into the fins adjacent (parallel) to the coil. It is best to test them with an ohmmeter when the ohm rating of the heater is known. This ohm rating is then compared with the reading registered on the meter. If the difference is more than ±10%, the heater must be replaced. If the ohm rating is not available, a continuity test can do the job in the majority of cases.

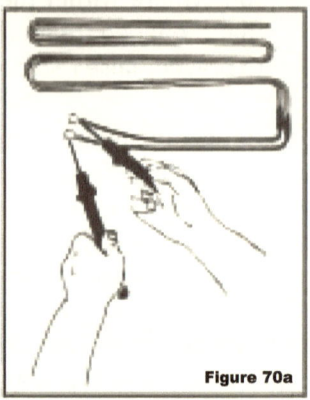

Figure 70a

## TESTING THE DEFROST HEATER WITH AN OHMMETER

1. Disconnect the unit from the power source and remove the evaporator cover.
2. Disconnect the two defrost heater terminals from their electrical circuit.
3. Set the ohmmeter on its lowest scale (RX1) and zero it.
4. Touch one probe to each defrost heater terminal. The meter should register a continuity reading. If not, the defrost heater is defective, and it must be replaced.

## A QUICK HEATER TEST BY USING A TEST LIGHT

1. Disconnect the unit from the power source, remove the evaporator cover, and remove the wires from the two heater terminals.
2. Follow steps 1, 2, and 3 in the "A Quick Termination Thermostat Check (Using the Bypass Method)."
3. Connect the two alligator clips on the test light to the two heater wires. Make sure the connections are well insulated and do not touch anything else.
4. Reconnect power to the unit and turn the timer shaft clockwise with a screwdriver until you hear a click. At this point, the compressor stops working.
5. If the test light glows, disconnect power to the unit and reconnect the heater wires to the heater terminals.
6. Connect power to the unit. If the heater begins heating, check the termination thermostat and the timer, and replace the defective part. If the heater doesn't heat, replace the heater.

## TESTING A DEFROST HEATER WITH A CLAMP-ON AMMETER

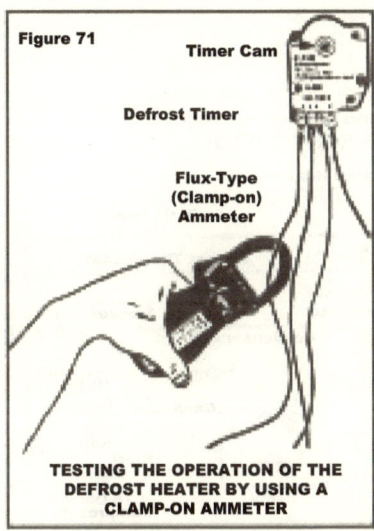

Figure 71

Timer Cam

Defrost Timer

Flux-Type (Clamp-on) Ammeter

TESTING THE OPERATION OF THE DEFROST HEATER BY USING A CLAMP-ON AMMETER

This simple method of checking the defrost system will benefit you the most after studying the chapter on "Basic Electricity," as this test requires some understanding of schematic wiring diagrams and the heating-element ohm variation in different units. You will need a clamp-on ammeter to run this test.

### HOW TO WORK WITH A CLAMP-ON AMMETER

By pushing a small lever, the tongs of the ammeter are opened to slip around the wire to be tested. If the reading registered on the ammeter dial is too low to be precisely read, wrap the wire around the jaw once—this will double the reading. Divide the reading by two to make it accurate.

## TESTING PROCEDURE

1. Find the schematic wiring diagram of the unit.
2. Pinpoint the defrost heater and make a note of its ohm rating shown on the diagram.
3. Determine the supply voltage for the unit.
4. Divide the supply voltage by the heater ohm rating reflected on the diagram. This determines the approximate reading you should expect to get when using the meter.
5. Trace the wire connecting the defrost heater to the defrost timer (in fig. 72, it is the wire connected to timer terminal number 4).
6. Place the jaws of the clamp-on ammeter around the wire connected to defrost timer terminal number 4. Turn the timer shaft clockwise and stop when you hear a click. You should expect to get a reading of about 5 A (4.8 A as calculated in step number 4). If the meter registers a significantly lower reading (such as 1/2 or 1½ A), you will know that the accumulation of ice on the evaporator plate must be due to a defective defrost system (the defrost heater, the defrost timer, or the defrost thermostat). Check them one by one as instructed in the chapter on "Testing Residential and Commercial Refrigeration Units" to determine the root of the problem.

CAUTION: The foregoing method has a potential *shock hazard*. Proceed cautiously. Be sure not to touch any wire without its insulation. If the insulation is cracked or frayed, or if for any reason it does not seem right, unplug the unit, put the meter in place, and then reconnect the unit and wait for about three minutes until the compressor starts, then proceed.

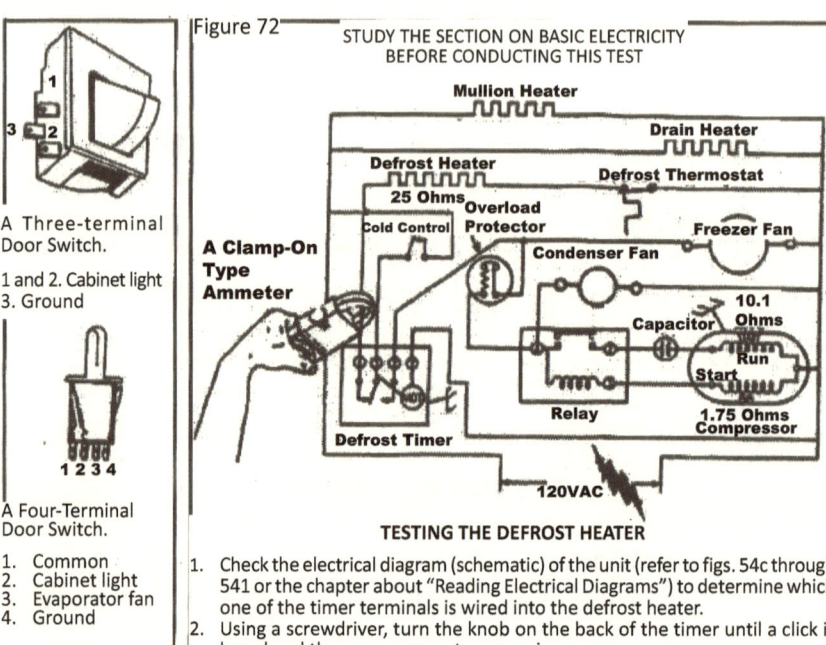

Figure 72

A Three-terminal Door Switch.

1 and 2. Cabinet light
3. Ground

A Four-Terminal Door Switch.

1. Common
2. Cabinet light
3. Evaporator fan
4. Ground

TYPICAL DOOR SWITCHES

**TESTING THE DEFROST HEATER**

1. Check the electrical diagram (schematic) of the unit (refer to figs. 54c through 54l or the chapter about "Reading Electrical Diagrams") to determine which one of the timer terminals is wired into the defrost heater.
2. Using a screwdriver, turn the knob on the back of the timer until a click is heard and the compressor stops running.
3. Place the clamp of the clamp-on ammeter around the proper wire and check the reading. The meter should register about 5½ A. If the ammeter registers a considerably lower reading (such as 1 or 2 A), the defrost system must be inoperative. Check the defrost bimetal and the defrost heater and replace the defective part.

Sometimes, due to a defective light switch, the light in the fresh-food compartment will not turn off when the door is closed. The heat created by the bulb inside the unit cabinet will not let the cabinet temperature drop low enough to satisfy the thermostat. When this happens, the compressor never shuts off. Check the door switch by opening the door and pushing the light switch. If the light stays on, replace the light switch.

## CHECKING A DOOR GASKET

A leaky door gasket allows outside air to penetrate the cold compartment and cause frost to collect on the evaporator coil and longer running time. In the fresh-food compartment, moisture appears on the walls and floor.

1, 2, and 4. Compression-type gaskets
3. Magnetic-type gasket
Gaskets for residential units are purchased through suppliers of those units. Commercial gaskets can be purchased from commercial refrigeration hardware supply houses. They can be ready-made or purchased by the yard. In some commercial units, gaskets are secured by cement.

## DOOR GASKET

Gaskets are installed to prevent air from getting into the cabinet. Magnetic and compression gaskets are the two types used in refrigerators and freezers today. They are secured to the door by screws (or sometimes by cone or dart clips in chest-type freezers) that pass through the gaskets, the door liner, then threaded into the door panel. In some models, retainer strips are used. In which case the screws pass through the gasket, the retainer strip, and then screwed into the main door panel. Chest-type freezers are generally equipped with compression gaskets having three or four sides with a magnetic strip along the front to give the lid a leakproof seal. If these gaskets are to be

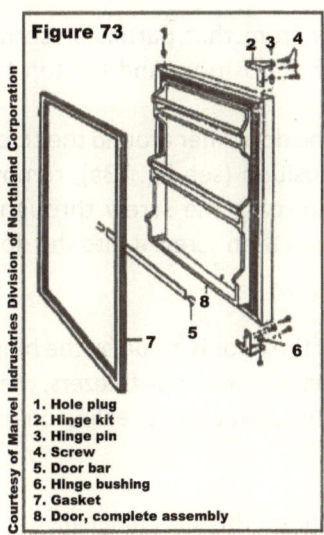

**Figure 73**

1. Hole plug
2. Hinge kit
3. Hinge pin
4. Screw
5. Door bar
6. Hinge bushing
7. Gasket
8. Door, complete assembly

Courtesy of Marvel Indrustries Division of Northland Corporation

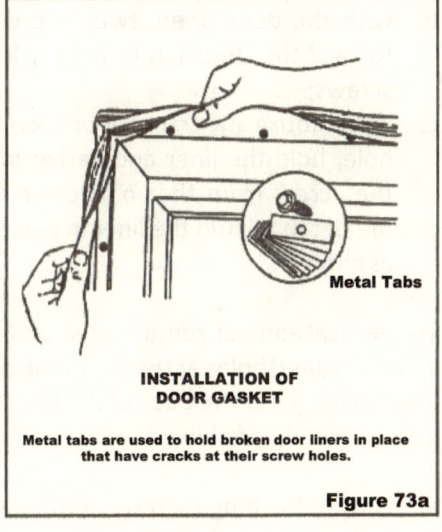

**INSTALLATION OF DOOR GASKET**

Metal tabs are used to hold broken door liners in place that have cracks at their screw holes.

**Figure 73a**

removed, use a screwdriver to pry out the clips. Accumulation of dirt and grease on a gasket makes it sticky, and as the door (or lid) is opened and closed, the gasket tears or pulls loose, or sometimes the door liner cracks. Door liners can also crack because of excessive pressures caused by incorrectly positioned food products that prevent the door from closing properly.

Clean sticky door gaskets with a mixture or vinegar and water or glass cleaner. Rinse with clear water and dry with a towel. Vinegar-water mixture is also a good food compartment cleaner and deodorizer.

The upright unit should be tilted a little to the rear to allow the door(s) to swing closed from a 45° angle. A poorly fitted gasket in most freezers and refrigerators can be adjusted by the hinges and the door latch. Warped doors can be straightened by twisting or adjusting the gasket screws.

## DOOR SEAL INSPECTION AND ADJUSTMENTS

1. Visually inspect the condition of the door seal. Pick a spot and place a piece of paper between the gasket and door jamb. With the door closed, pull it around the perimeter of the door. If the gasket is fitted properly, the paper will offer some resistance as it is being pulled. An air leak can occur where the paper offers no resistance. If a corner is not sealing,

    a. loosen half the gasket screws starting from that point to the two nearest corners;
    b. with the door open, twist the door from that particular corner toward the door jamb, hold it in that position, and tighten the screws;
    c. if you notice any cracks or breaks in the door liner around the screw hole, hold the liner and gasket in position (see fig. 73a), remove the screw from that particular hole, pass the screw through a metal tab to hold the liner in place, and then screw it into the door assembly.

2. The vertical and horizontal adjustment of the door is made by the hinge with elongated holes at the top of the door. (In chest-type freezers, there are two hinges at the back of the lid.) If the gasket is not sealing on one side, proceed as follows:

    a. Loosen the hinge screws at the top (or on the back hinges).
    b. Force the door into its proper position. To do this, you will have to force the door against the jamb in a way that all four corners of the gasket are flush and sealing. Hold the door in that position while tightening the screws on the hinge. (You may have to repeat step 2b for proper door adjustment.) If you still observe a poor seal on one side of the door gasket, proceed to the next step.
    c. Remove the top hinge and the door and place a metal washer on the bottom hinge. Reinstall the door and follow step 2b to get a

good gasket seal. (A washer is placed between the bottom of the door and the bottom hinge to compensate for any slight bend of the hinge caused by the weight of the door).

When ordering a new gasket to replace an old one, it is *set* in an angular pattern from being in the box. If it is mounted in this condition, the gasket will come out *wavy* instead of being straight. To make it soft and pliable, simply put it in a clothes drier or warm water for four or five minutes prior to installation. It will then become easier to handle and provide a neat installation.

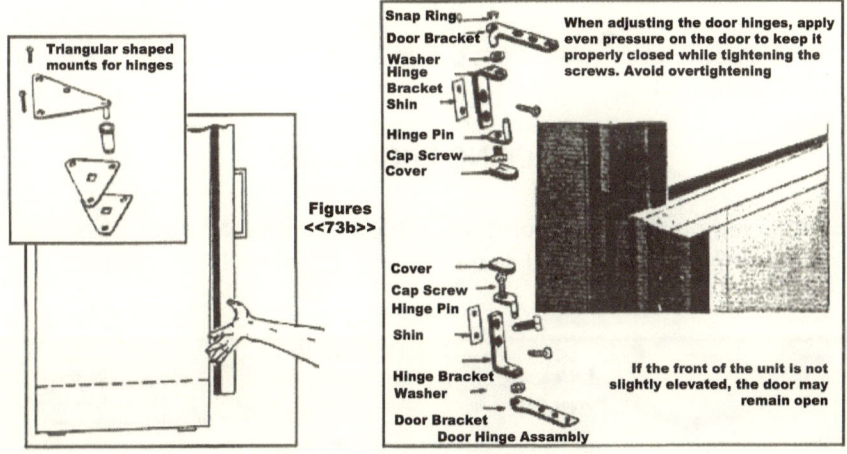

Figures <<73b>>

Start with the corners first when installing a gasket. Place the next screws in the center between the corners, and the next screws between the centers of the last ones, and so on until all of the screws are replaced. New gaskets may come with shims to improve gasket fit should the door be only slightly warped.

Cracked or broken compartment liners can be neatly repaired with a Fiberglas repair kit bought from a local auto parts store or a boat dealer. When the repair dries, it can then be sanded and painted to match the rest of the interior.

For exterior repair of chipped or scratched paint, touch-up kits are sold by the dealers for any particular brand. The year of manufacture and the color of the unit are needed. After applying the touch-up paint, it may take a day or two for the color to dry to the original shade.

You may place a 150-watt flood light inside the cabinet aimed at one area of the door gasket. Close the door and inspect that area. A poorly fitted gasket is indicated by light penetration between the door jamb and gasket. Inspect the entire gasket by directing the light at consecutive areas of the gasket.

## TESTING THE POWER CORD

Disconnect the power cord from the wall outlet and inspect it. It should be solid, supple, and free of cracks, splits, drying, or exposed inner wires or insulation. Especially in the case of older units, even if the cord is only dry and stiff, replacing it is strongly recommended. In addition to making the circuit safer, it is inexpensive and will save the customer problems in the future.

If there is power at the wall outlet but none is getting to the unit, the power cord must be bad.

## CHECKING PROCEDURE

1. After removing the cord from the wall socket, remove the lower access cover in the rear of the unit and disconnect the cord from the junction box.

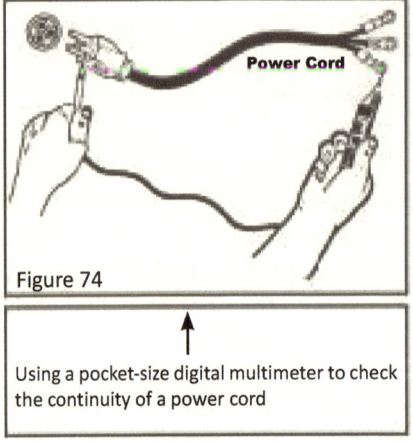

Figure 74

Using a pocket-size digital multimeter to check the continuity of a power cord

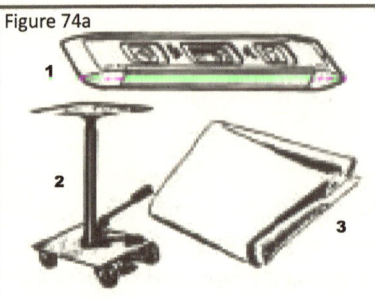

Figure 74a

1. An odinary carpenter's level can be used to adjust the leveling legs
2. Hydraulic tables help when moving wal-mounted units
3. Floor covers protect floors and help with cleanup

Courtesy of Robinair Division, SPX Corporation

2. Set the ohmmeter on its lowest scale and zero it.
3. Hold the ohmmeter probes on each end of each wire in the cord. The meter should register a zero reading in each case. If not, it will mean that there is a break in one of the inner wires and the cord will have to be replaced.
4. If you are using a digital multimeter, simply set the meter on its ohm setting to read the amount of resistance. Or set it on Continuity, touch

the two sides of each individual line as indicated in figure 74. If there is no break in the line being tested, you will hear a light buzzing sound.

## TIPS ON INSTALLING REFRIGERATORS AND FREEZERS

1. When transporting a refrigerator, never lay it on its side or back because the compressor oil leaves the compressor. Then when in place and turned on, the compressor operates without lubrication, and it ends up having to be replaced. If for any reason it must be laid down for five minutes or longer, be sure it is in an upright position for a minimum of twelve hours before being turned on. This will allow sufficient time for the oil to return to the compressor crankcase.

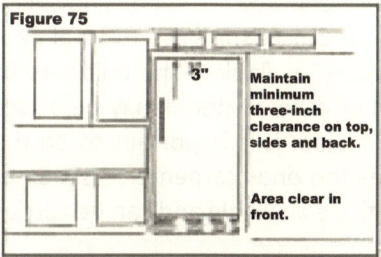

MAINTAIN ADEQUATE CLEARANCE AROUND UNIT FOR AIR CIRCULATION

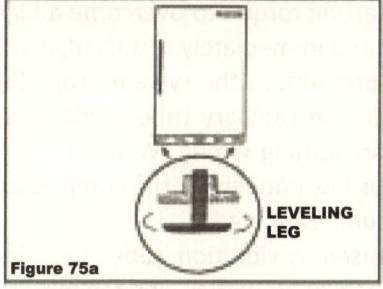

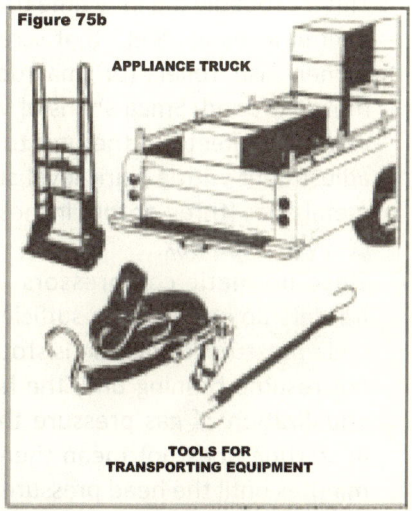

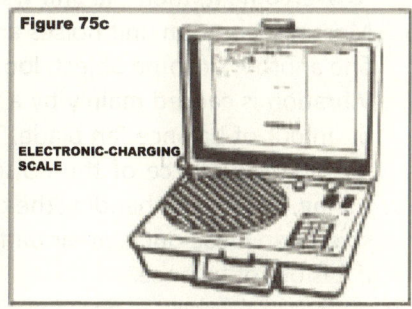

Fig. 75. A minimum clearance around a unit
Fig. 75a. Leveling legs under the unit
Fig. 75b. A two-wheel appliance truck, sixteen-inch nylon webbing with ratchet, and a length of bungee cord used for cargo tie-down.

The programmable electronic charging scale in figure 75c can be used for weighing R-12, R-22, R-500, and R-502. It is designed for refrigerant tanks up to 50 lbs and has a standard scale up to 80 lbs. When programmed for a particular charge, a solenoid stops the charge, audible and visual signals show that the procedure is complete.

Courtesy of Robinair Division, SPX Corporation

2. Electrical power supply should provide enough voltage to start the compressor motor. For residential units, 115 volts ± 10% is the

allowable fluctuation limit to start the motor. In other words, the voltage at the wall receptacle must be at least 108 volts. Otherwise, it will not be able to start the compressor motor. A voltmeter is needed to check the voltage. (Refer to "Testing the Supply Voltage" and see fig. 34.)

3. Sufficient ventilation is important where the refrigeration unit is installed. In the case of units that have a fan-cooled condenser, be sure nothing is blocking air circulation at the bottom and in the front of the unit. For units with their condenser in the back, a wall clearance of at least two and a half inches is required for proper air circulation. Make certain that adequate clearance is provided on top and bottom as well as all sides (see fig. 75).

4. Refrigeration units should be carefully leveled at the time of installation. This is probably the only time the level is checked. Most units now have built-in leveling "feet" that screw in or out of the base to level each corner. Also, rollers (or small dollies) are available if the unit is to be moved around. Small shims of wood or other material may be placed under the "feet" of the unit to be leveled. It is important to do this adjustment with a spirit level such as the ones carpenters use. Small metal ones (three or four inches long) are available and can be carried easily in a toolbox.

5. Since hermetic compressors used in residential refrigerators and freezers do not produce sufficient starting torque to overcome a high head pressure, if the unit is stopped and immediately restarted, it will not resume running until the head pressure in the system drops (by equalization of gas pressure through the capillary tube). This delay in starting does not mean there is something wrong. Wait for a few minutes until the head pressure drops low enough for the compressor to overcome it, then the unit will resume running.

6. Most refrigeration unit noises are caused by vibration, tubes touching one another or some object, loose ducts or air baffles, and fan motors. Vibration is caused mainly by a unit that is not level, uneven flooring or an out-of-balance fan blade. These noises can be eliminated by (a) locating the source of the noise, (b) placing rubber blocks between tubing, or carefully bending the tubing out of the way, (c) tightening the screws on loose ducts or air baffles and/or, (d) replacing the fan blade if it is out of balance.

## TEST YOUR KNOWLEDGE ON TESTING COMPONENTS

1. What type of compressors are used in wall air-conditioning units? (p.62)
2. On what scale is an ohmmeter set when testing a starting relay? (p.70)
3. How is a three-terminal starting relay used on residential units tested? (p.71)
4. On what scale is an ohmmeter set when testing defrost heaters? (p.96)
5. How many types of commercial thermostats are there? (pp. 94,95))
6. Describe the operation of termination thermostats. (p.87)
7. Describe the bypass method of checking an overload protector. (p.81)
8. Name four different types of starting relays. (p.72)
9. What is the horsepower range of compressors that can use solid-state relays. (p.73)
10. Under what conditions are start relays interchangeable? (p.74)
11. Where would a defrost timer be located in a household refrigerator or freezer? (p.76)
12. How is a timer for a residential unit tested? (pp.76,77,79)
13. What requires extra attention when replacing fan motors? (p. 85)
14. How many ways are there to test a capacitor? (p. 91)
15. When is a hard start kit required to be installed? (p. 63)
16. What is a unit of capacitance? (p. 89)
17. How can a cracked refrigerator door liner be repaired? (p. 101)
18. How do you check for leaks in a door gasket? (p. 100)
19. Where are commercial air-coil thermostats installed? (p.89)
20. Where is a thermostat sensing bulb attached? (p. 94)
21. What precaution should you take with the sensing bulb line? (p.94)
22. What is the maximum permissible voltage drop for a unit that operates on 115 VAC? (p.52)
23. What instrument is used to test a power cord? (p.102)
24. Why do fan motors for commercial units come with longer shafts? (p.85)
25. What is the function of timer terminal 1 on a Philco residential refrigerator? (p.79)

# PROCEDURES FOR THE EVACUATION AND CHARGING OF COMMERCIAL AND RESIDENTIAL REFRIGERATION UNITS

This section covers the most common techniques of opening, evacuating, and charging a sealed system in addition to repairing leaks in commercial and residential refrigeration units.

## PREPARING THE SYSTEM FOR CHARGING

There are many repairs that require the opening of the sealed system in a refrigeration unit, such as replacing a compressor or a filter-drier, repairing a puncture in the tubing, replacing a capillary tube, etc. When the repair is done, take these three steps prior to charging and operating the unit:

1. Remove the old filter-drier and install a new one.
2. Evacuate the air from the system by using a vacuum pump or by the purging method where the compressor is used as the pump.
3. Make certain there is no refrigerant leak in the system by running a leak detector over the lines and components (and also by checking for a pressure rise in the system after evacuation).

---

### REPLACEMENT OF A FILTER-DRIER

1. Remove the old filter-drier. (If it is silver-brazed, use a tubing cutter and cut about one-fourth inch from the drier. If it is a flared connection, unscrew the nuts.) Do not use the same flare again as it can become weak from tightening and loosening the nut. Cut the old flared end off and make a new one.
2. Install the new drier paying heed to the flow-direction arrow on the body of the new drier. When using a flared connection, wrap the male threads of the connector (on each end of the drier) with Teflon tape to get a good, leakproof seal.

Residential units do not use flared connections because of their smaller size of tubing on which brazing works easily. Some of the driers used in residential units also come with access valves (see figs. 76e and 76f). Commercial filter-driers are sized according to the tonnage of the unit. Before buying one, take a look at the nameplate on the unit to see what type of refrigerant is used. Filter-driers are made for different types of refrigerant. Whether a flared or a brazed connection is used, determine what size tubing it has to be connected to and buy a filter-drier for that particular tubing size (see figs. 76b and 76c). Some filter-driers are installed on liquid lines and some on suction lines. Suction-line filters are always installed after a compressor burnout as close the compressor as possible. (See fig. 76a for the brazing technique on a commercial type drier.) When installing a filter-drier on a residential unit, be sure to insert the capillary tube about one and a half to two inches into the filter-drier entry tube. The capillary tube should not touch the element inside the drier, but should be in far enough to preclude any possibility of melted solder running in over the end of the capillary tube, plugging it. Before brazing is begun, the filter-drier entry tube has to be crimped along one side to form a snug fit around the capillary tube.

---

### EVACUATING A SYSTEM BY USING A VACUUM PUMP:

1. Shut off the unit.
2. Connect the vacuum pump to the center connection of the gauge manifold, and the gauge manifold to the high side and low side of the compressor through piercing (or any access) valves (see fig. 84).

3. Run the vacuum pump until you get a vacuum reading of about thirty inches (thirty inches below the zero) on the compound gauge. It is a good practice to heat the system by using an electric heater or heat lamp(s) and blower fans as the system is being evacuated.
4. Shut off all valves and the vacuum pump and leave everything as it is for ten minutes.
5. If at the end of this period the gauges show a pressure rise, either there is moisture in the system, or the compressor has an internal leak, or there is a leak in the system (or in the valves on the manifold gauge) that must be repaired before proceeding. (See "Repairing a Leak.")
6. Close the valves on the suction and discharge lines and disconnect the vacuum pump from the system and connect the middle hose on the manifold gauge to a refrigerant cylinder. Connect the blue hose on the compound gauge to the suction line (see fig. 82), start the compressor, and charge the unit until the high-pressure gauge reads about 75 psi.
7. Run a leak detector over the joints and the valves. If no leak is found, complete the charging procedure. Otherwise, seal the leak and do steps 1 through 4 again.

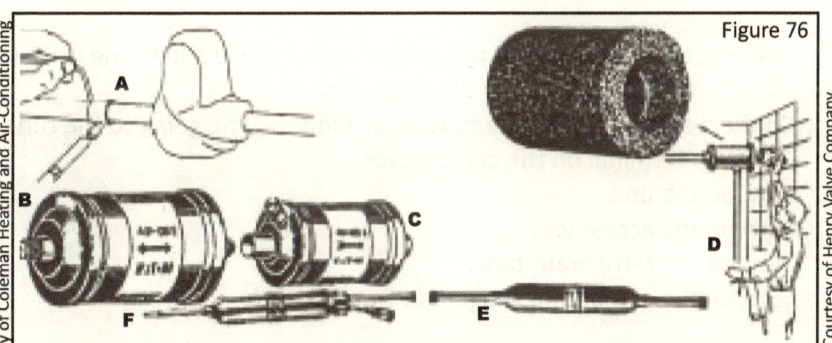

Figure 76

A. Before silver-brazing the filter-drier into the sealed system, remove the valve core(s) of the access valve(s) and wrap the drier in a wet cloth to prevent any heat damage. Do the brazing as quickly as possible to avoid excessive heat buildup.
B. A bidirectional filter-drier used in heat pumps.
C. A suction-line filter-drier. Its purpose is to collect all foreign matter to keep it from entering the compressor that may cause internal damage. It should always be installed after a compressor burnout.
D. Larger commercial systems use bolted-type suction-line filter-driers to allow a quick replacement of the filter element. They are also equipped with an access valve through which pressure is checked (with the compressor running) and then compared with the compressor suction pressure. If the gauge registers a pressure drop of more than 2 psi, the filter-drier should be replaced.
E. A liquid-line filter-drier used in residential and small commercial units.
F. A liquid-line filter-drier equipped with an access valve used in residential and small commercial units.

E and F are installed in systems using a capillary tube.

Courtesy of Eaton Corporation Controls Division

## THE FOUR METHODS OF CHARGING A REFRIGERATION UNIT

* Charging by ammeter reading.
* Charging by the amount of refrigerant shown on the nameplate (critical charge).
* Charging by the evaporator frost pattern.
* Charging by the low side gauge pressure.

### CHARGING THE SYSTEM BY AMMETER READING

When measuring the charge by an ammeter, bear in mind that with the unit running, the more refrigerant is charged into the unit, the higher the ammeter will read and vice-versa. If the unit is operating properly, any reading other than the FLA amperage (which is printed on the nameplate of the unit) will either be an undercharge or an overcharge.

In any of the above methods, a piercing valve must be installed on the discharge line and the suction line of the compressor. Commercial units already have service valves installed. (Instructions on installation of piercing valves are provided with the valves.)

a. Connect a refrigerant tank to the compound gauge and to the valve (as shown in fig. 78) and leave all valves closed.
b. Connect a clamp-on type ammeter around the line going to the run or common terminal on the compressor.
c. Turn on the unit.
d. Turn on the access valve.
e. Turn on the refrigerant tank.
f. Turn on the compound gauge and watch the ammeter.

As the system is being charged, the amperage will increase. When it reaches the FLA amperage indicated on the nameplate of the unit, the system is charged with the correct amount of refrigerant. Immediately close the compound gauge valve, the piercing valve, and the refrigerant tank. Replace the cap on the piercing valve. Use a little Teflon tape on the threads to prevent a refrigerant leak.

### HOW TO PUT REFRIGERANT INTO A CHARGING CYLINDER

A charging cylinder is a tool that provides the means to place a specific amount of refrigerant into a unit (see p. 112). It is a calibrated cylinder which can measure, in ounces, the exact amount as specified on the nameplate of the unit or in a reference book mentioned on page 112. Before starting to put refrigerant into a charging cylinder, make sure all the valves are closed on the refrigerant cylinder, the charging cylinder, and those on the gauges (see fig. 83).

1. Connect the yellow (center) hose of the manifold gauge to the charging cylinder.
2. Connect the suction (blue) hose to the refrigerant cylinder valve.
3. Place refrigerant canister upside down and turn on its valve. Then turn on the bottom valve of the charging cylinder.
4. Turn on the valve on the compound gauge while watching the scale on the charging cylinder. Keep a hand on that valve. You will see the level of liquid refrigerant rise in the cylinder.
5. Turn off the valve when the cylinder is filled to the desired level (which is the amount specified on the nameplate of the unit being charged). Be certain to use the scale on the charging cylinder that corresponds with the ambient temperature.
6. Turn off the rest of the valves and disconnect the refrigerant tank and the charging cylinder.

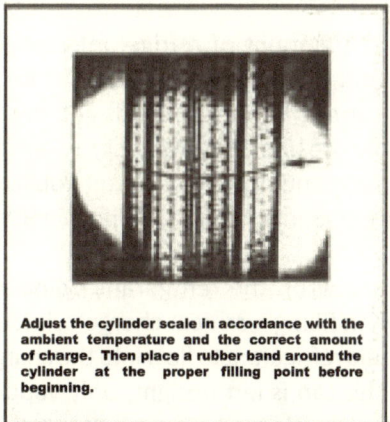

Adjust the cylinder scale in accordance with the ambient temperature and the correct amount of charge. Then place a rubber band around the cylinder at the proper filling point before beginning.

HINT: By placing a rubber band around the cylinder at the level of refrigerant you desire before filling it, it will be much easier to spot the place to stop. (See fig. 77).

The valve on top of the charging cylinder is used as a pressure-relief valve. As the cylinder fills, pressure builds inside until it becomes equal to the pressure inside the refrigerant cylinder. At this point, the cylinder stops filling. Relieve some of the pressure by opening the top (relief) valve very little, then close it again. The cylinder will resume filling. (When buying a critical-charge cylinder, be sure to get one with a pressure gauge on top.) When the cylinder is adequately charged, plug in its electrical cord to a 115 V outlet. The cord is connected to the cylinder-heating element. As the element heats, pressure is built up in the cylinder. This pressure is used to force the refrigerant into the unit being charged. NEVER FILL THE CHARGING CYLINDER ALL THE WAY TO THE TOP!

Generally, building up a pressure between 150 and 200 psi is enough for most units. DO NOT ALLOW THE PRESSURE IN THE CHARGING CYLINDER TO EXCEED 275 PSI!

WARNING: When the cylinder is plugged in, do not leave it unattended. It will take a few minutes to build up pressure. Too much pressure will cause the cylinder to explode.

If the amount of refrigerant charge cannot be found on the nameplate, or if for any reason it cannot be determined, get the information from the book *Tech Master for Refrigerators and Freezers* published by Master Publications, Euclid Street, Santa Monica, California 90404.

The reason the refrigerant cylinder is inverted when filling a critical-charge cylinder is to allow liquid refrigerant to flow into the charging cylinder. If the can is left upright, only vapor refrigerant can be released. Some other refrigerants are quite the opposite; they flow out of the cylinder in the form of liquid when the can is in upright position. Make it a habit to read the instructions printed on the cylinder first.

This electronic portable charging station has a built-in vacuum pump and gauge manifold. It greatly simplifies charging any unit as the exact amount of refrigerant charge can be electronically programmed. It operates on 115 VAC and is capable of pumping R-12, 22, 500, or 502 at a rate of 4 ft$^3$/min from either a 30 or 50 lb refrigerant tank. Manually operated stations have a charging cylinder such as pictured below.

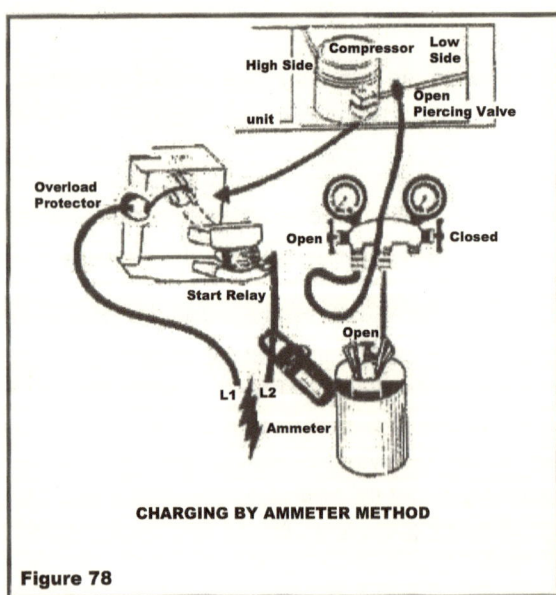

**CHARGING BY AMMETER METHOD**

**Figure 78**

A unit can also be charged by building up pressure in the charging cylinder to force liquid refrigerant into the system through the high-side access valve while the unit is shut down.

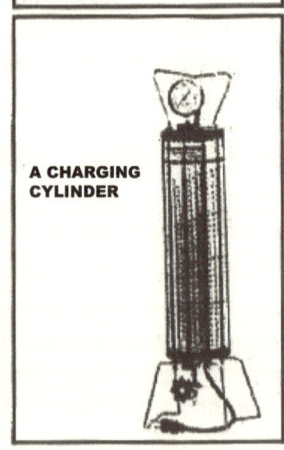

A CHARGING CYLINDER

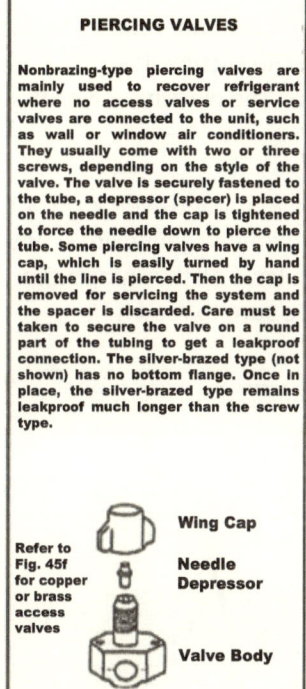

### PIERCING VALVES

Nonbrazing-type piercing valves are mainly used to recover refrigerant where no access valves or service valves are connected to the unit, such as wall or window air conditioners. They usually come with two or three screws, depending on the style of the valve. The valve is securely fastened to the tube, a depressor (specer) is placed on the needle and the cap is tightened to force the needle down to pierce the tube. Some piercing valves have a wing cap, which is easily turned by hand until the line is pierced. Then the cap is removed for servicing the system and the spacer is discarded. Care must be taken to secure the valve on a round part of the tubing to get a leakproof connection. The silver-brazed type (not shown) has no bottom flange. Once in place, the silver-brazed type remains leakproof much longer than the screw type.

Refer to Fig. 45f for copper or brass access valves

Wing Cap
Needle Depressor
Valve Body

Figure 79

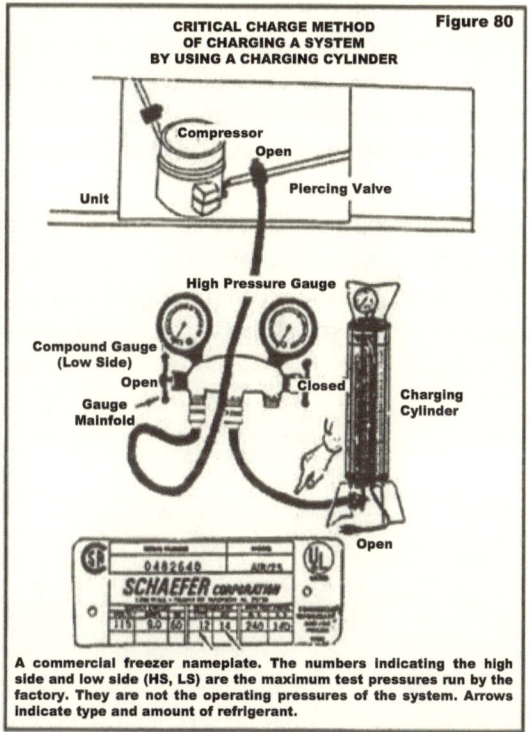

A commercial freezer nameplate. The numbers indicating the high side and low side (HS, LS) are the maximum test pressures run by the factory. They are not the operating pressures of the system. Arrows indicate type and amount of refrigerant.

## CHARGING A UNIT BY THE AMOUNT OF CHARGE SHOWN ON THE NAMEPLATE (CRITICAL CHARGE)

a. Check the nameplate on the unit to determine the exact amount of charge. This figure is reflected in ounces. (Some larger commercial units indicate this figure in pounds [fig. 64b]. In this case, the refrigerant tank must be weighed. Subtract the amount of charge and then leave the tank on the scale. Charge the unit until the correct amount of refrigerant leaves the tank.) A more precise method is to use an electronic charging scale (See fig. 75c).
b. Fill the charging cylinder with the exact amount of refrigerant as described earlier and in figure 80.
c. If there is no service valve on the suction line, connect a piercing valve and leave it closed.
d. Connect the charging cylinder to the gauge manifold and to the piercing valve as shown in figure 80 and leave it closed.
e. Turn on the unit.
f. Turn on the piercing valve, the compound gauge, and the valve on the charging cylinder until the Freon in the charging cylinder gets to zero on the scale of the charging cylinder. Then,

g.  1. Shut off the charging cylinder valve.
    2. Shut off the piercing valve and the valve on the compound gauge.
    3. Remove the compound gauge hose from the suction line.
    4. Remove all of the hoses and put the cap on the piercing valve after wrapping its threads with Teflon tape.

When installing a piercing valve, tighten the screws gently and evenly. Do not overtighten. Also be careful when removing the manifold gauge lines. Any movement of the valve on the tubing may cause a leak. All this information is given in the instructions with piercing valves. Be sure to read it.

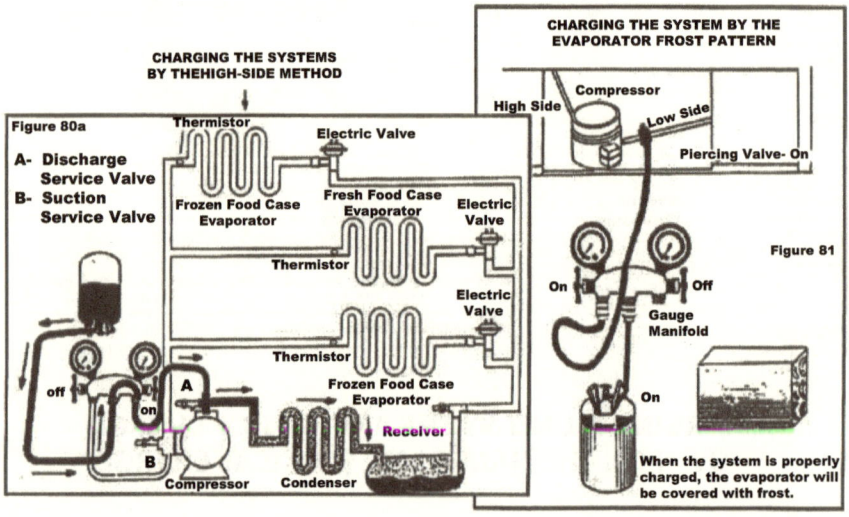

Commercial units can be charged by the high-side method. Connect the manifold gauge to a refrigerant container (never use a disposable container with this method, as it may explode). Then connect the gauge to the suction and discharge service valves as shown in figure 80a.

   a. Using a service wrench, turn the stem on the discharge service valve all the way clockwise with the refrigerant container upright. (See figs. 101 and 102.)
   b. Run the compressor no longer than a few seconds, just enough to build a pressure in the refrigerant container, about 35 to 40 psi above the pressure in the condenser.
   c. Turn off the compressor and turn the refrigerant container upside down to allow liquid Freon flow from the cylinder. Read the printed instructions on the refrigerant cylinder for the type of Freon you are using first.
   d. Turn the stem on the discharge service valve counterclockwise until you hear a gurgling sound (which means liquid refrigerant is flowing into the receiver). Read page 117 before using this method.

## CHARGING THE UNIT BY THE EVAPORATOR FROST-PATTERN METHOD

This method of charging is used where the evaporator can be readily seen, such as in manual-defrost units, cycle-defrost units, and most commercial units where evaporator temperature is designed to be below 32°F. In residential frost-free units where the evaporator is covered, removal of the cover would be too time-consuming to use this method.

a. If there is no service valve on the suction line, install one and leave it closed.
b. Remove the evaporator cover. Using this charging method becomes necessary when working on a unit where the charging information is not available, or where an ammeter or a charging cylinder (or a charging scale) is not available.
c. Connect a refrigerant tank to the compound gauge on the manifold and also to the piercing valve as indicated in figure 81.
d. Turn on the unit.
e. Turn on the piercing valve, the gauge manifold, and the refrigerant tank as in figure 81.
f. Keep a hand on the valve of the refrigerant tank while watching the frost pattern appear on the evaporator. Let the refrigerant flow into the system while the unit runs.
g. The frost pattern will appear and advance on the evaporator coil. Once the frost pattern covers all of the evaporator coil, close the valve on the refrigerant tank, the gauge manifold, and the piercing valve.
h. Remove the hoses from the unit. Put some Teflon tape on the threads of the piercing valve and replace the cap.

## CHARGING THE UNIT BY THE LOW-SIDE GAUGE PRESSURE METHOD

This method is widely used in commercial refrigeration since the exact back-pressure requirement is known. (This is explained in "Saturated Vapor Refrigerant Pressure and Temperature Relationship.") The correct low-side pressures of all residential units (made in the USA) can be obtained from the *Tech Master* publication mentioned on page 112. All that is needed is the name and model number of the unit.

a. Install a piercing valve on the compressor suction side and leave it closed.

b. Connect the piercing valve to the compound gauge manifold and the middle line in the manifold to the refrigerant tank as shown in figure 82. Leave all valves closed.
c. Turn on the unit.
d. Turn on the piercing valve, the compound gauge valve, and the refrigerant tank.

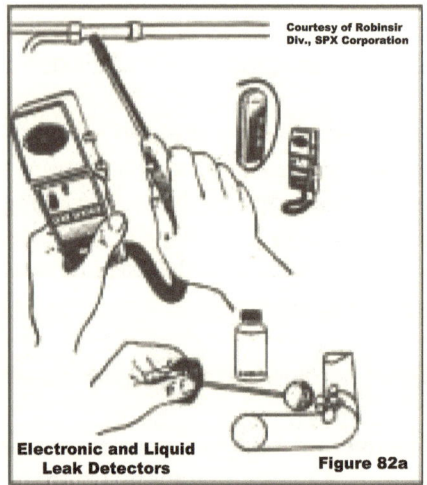

Electronic and Liquid Leak Detectors

Figure 82a

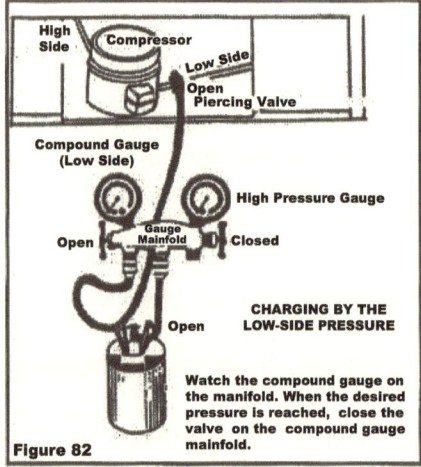

Figure 82

e. Keeping a hand on the compound gauge valve, watch the compound gauge, and every few seconds, turn off the valve and check the gauge for the desired low-side pressure.
f. Upon reaching the correct low-side pressure, turn off the manifold valve, the piercing valve, and the refrigerant tank valve.
g. Disconnect the lines, put a little Teflon tape on the threads of the piercing valve, and replace its cap.

Make it a habit of running an electronic leak detector over all the tubing and valves in the system after recharging. Make certain there is no refrigerant leakage. A mixture of one-half cup of liquid soap to one gallon of water can also be used as a leak detector when a leak is suspected in a particular location. Simply apply the mixture to the tubing or the connections and watch for bubbling.

# AIR CONDITIONING AND REFRIGERATION REPAIR MADE EASY 117

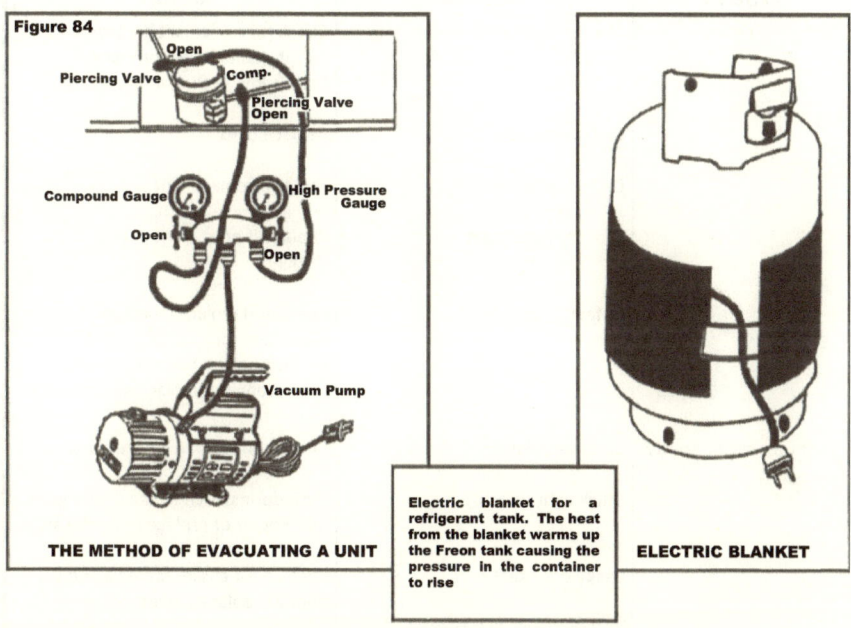

THE METHOD OF EVACUATING A UNIT

Electric blanket for a refrigerant tank. The heat from the blanket warms up the Freon tank causing the pressure in the container to rise

ELECTRIC BLANKET

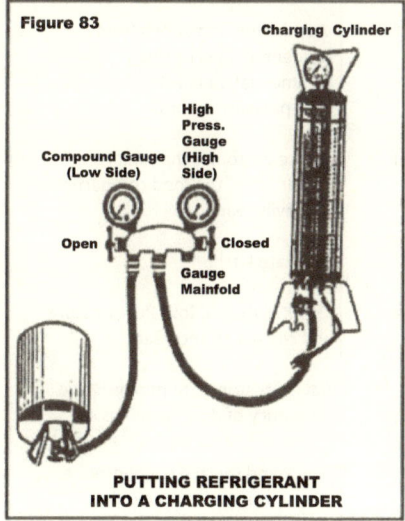

PUTTING REFRIGERANT INTO A CHARGING CYLINDER

To increase the pressure and cause the liquid (or gas) to be expelled faster, hold the container under your coat next to your body or put it on a condenser to warm it. In severely cold weather, use a refrigerant-container heating blanket such as the one pictured above. If an electric blanket is not available, it is also possible to increase the pressure in the refrigerant container by placing it in a tub of warm water at the time of charging.

Before charging a unit, purge all charging hoses to rid them of moisture and air. Commercial systems may also be charged by the high-side method as shown in figure 80a. The system must be turned off when it is being charged through the high side. This method is used only if the entire amount of refrigerant has been removed from the system. The inverted refrigerant container is placed on a scale, and the unit is charged until the proper amount of liquid refrigerant is transferred into the system.

## TROUBLESHOOTING RESIDENTIAL REFRIGERATING UNITS

| PROBLEM | POSSIBLE CAUSE | REMEDY |
|---|---|---|
| Unit won't start. | Low voltage or no power. | Check fuse or circuit breaker. Check outlet. (Call power company if there's low or fluctuating voltage.) |
| | Bad power cord. | Replace cord. |
| | Bad starting relay. | Test and replace if necessary. |
| | Broken thermostat. | Replace if no current passes through. |
| | Defective timer. | Check and replace if necessary. Also check for loose or broken connections on thermostat and timer; repair or replace as necessary. |
| Unit runs continuously. | Bad thermostat. | Check and replace if necessary. |
| | Stuck light switch. | Push door switch to see if light goes out. Repair or replace as necessary. |
| | Leaky door seal. | Check door gasket for leaks. Adjust door or replace gasket. |
| | No air circulation around unit. | Move unit or objects around it to create proper air circulation. |
| | Linted condenser coil. | Vacuum lint and debris from residential units or clean commercial units with alkali and high-pressure air or $CO_2$. |
| | Too much warm food loaded at one time. | Advice customer that unit runs longer when the food compartment is heavily loaded. |
| | Kitchen too warm. | Ventilate kitchen. |
| | Improper refrigerant charge. | Check high- and low-side pressure and correct as necessary. |
| Freezer and fresh-food compartments too warm. | Door opened frequently. | Advise customer to minimize the frequency of door openings. |
| | Defective fan motor. | Check and replace condenser or evaporator fans as necessary. |
| | Defective or incorrectly set thermostat. | Check thermostat; replace if necessary. |

| PROBLEM | POSSIBLE CAUSE | REMEDY |
| --- | --- | --- |
| Freezer and fresh-food compartments too warm. (cont.) | Linted or poor air circulation around condenser. | Clean condenser and check condenser fan and replace if necessary. |
| | Poor air circulation around unit. | Provide sufficient clearances for proper air circulation. |
| | Accumulation of ice on evaporator. | A bad thermostat, timer, or defrost heater could cause the problem. Check each and replace as necessary. |
| | Refrigerant shortage. | Check low-side pressure. If too low, check for leak. Repair, evacuate system and recharge it. |
| | Leaky door seal. | Adjust door or replace seal as necessary. |
| | Door opened excessively. | Advise customer to minimize frequency of door openings. |
| Freezer compartment too cold. | Thermostat set too cold. | Adjust thermostat. |
| | Defective thermostat. | Check and replace if necessary. |
| | Thermostat sensor not clamped tightly to evaporator. | Tighten clamp. |
| Frequent accumulation of ice on evaporator. | Defrost drain plugged. | Inspect and clean drain line. |
| | Leaky door gasket. | Check and replace if brittle, cracked or worn. Check and adjust door if needed. |
| | Defective defrost heater. | Check and replace if necessary. |
| | Defective defrost bimetal. | Check and replace if necessary. |
| Refrigerator compartment too warm. | Defective evaporator fan. | Check fan and wiring. Repair or replace as necessary. |
| | Compartment light stays on. | Repair or replace light switch. |
| | Poor air circulation due to overloading of shelves. | Train customer. |
| | Hot food in refrigerator. | Food must cool to room temperature first. |
| | Thermostat set too warm. | Adjust thermostat. |
| | Leaky door seal. | Replace door gasket if necessary. |

| PROBLEM | POSSIBLE CAUSE | REMEDY |
| --- | --- | --- |
| Refrigerator compartment too warm. (cont.) | No or poor airflow from freezer to refrigerator. | Damper should open. If not, replace it. (Frost-free units cool by this method.) |
| Unit runs all the time and is too cold. | Defective thermostat. | Test thermostat, replace if necessary. |
| | Thermostat bulb not in contact with evaporator. | Fasten bulb to evaporator. |
| Unit runs all the time and is too warm. | Accumulation of ice on the evaporator surface. | Check for leaky door gasket, replace if necessary. Check for defective defrost heater on the evaporator coils. Check for bad defrost timer. Replace these as necessary. If it's a normal defrost, unit needs defrosting. |
| | Low amount of refrigerant. | Check for leak; repair and recharge. |
| Unit cycles on overload. | Defective overload protector. | Check and replace if necessary. |
| | Defective starting relay. | Check and replace if necessary. |
| | Defective compressor. | Check and replace if necessary. |
| | Low line voltage. | Check outlet for proper voltage (107-126). Unit may require separate or new circuit. New line must be installed by licensed electrician. |
| | Defective condenser fan. | Check and replace if necessary. |
| | Extension cord undersized and too long. | Replace cord with one of proper size. |
| Freezer alternately cools and warms. | Restriction in capillary tube due to moisture in the system. | Discharge and evacuate the system. Install new filter-drier and recharge the system. |
| Fresh-food compartment gets too cold. | Cold-control knob set too high. | Turn knob to warmer setting. |
| | Defective airflow heater. | Check and replace if necessary. |
| | Broken or defective airflow control. | Check and replace if necessary. |
| | Object stuck in airflow control causing it to remain open. | Remove object obstructing control. |
| Reduction of freezing ability of the unit | Obstruction in capillary tube by accumulation of wax. | Check capillary tube and clear obstruction or replace tube. |
| Accumulation of ice on finned evaporator. | Defective timer. | Check and replace if necessary. |
| | Defective defrost heater. | Check for continuity; replace if necessary. |
| | Defective thermostat. | Check and replace if necessary. |

| PROBLEM | POSSIBLE CAUSE | REMEDY |
| --- | --- | --- |
| Unit makes noise when operating. | Flooring loose or unstable. | Brace flooring. |
| | Loose compressor mounting nuts and bolts. | Tighten or replace mounting nuts and bolts. |
| | Unit not level. | Level the unit by adjusting legs. |
| | Tubing vibrating or touching cabinet. | Move tubing, wedge, or tape to stop noise. |
| | Drip tray vibrating. | Place tray on pad or tape in place. |
| | Fan blades striking object. | Adjust fan mounting or move object. |
| Freezer door freezes shut. | Defective gasket or seal. | Check gasket for hardening, cracks or dirt. |
| | Defective case heater or mullion heater. | Check both heaters and replace if necessary. |
| Unit runs continuously but does not cool. | Accumulation of ice in the cabinet insulation. | Unplug unit for forty-eight hours and allow ice to melt. Dry insulation, seal cabinet leaks, and reassemble. |
| Compressor no longer runs. | Broken starting winding. | Replace compressor. |
| | Broken run winding. | Replace compressor. |
| | Broken compressor valve. | Replace compressor. |
| | Overheated compressor. | Replace compressor. |
| Accumulation of water at bottom of unit liner. | Leaky door gasket. | Inspect and replace if necessary. |
| | Clogged drain hose. | Clean inside of hose so water will drain. |
| | Evaporator drain hole plugged. | Clean drain passage. |
| Nothing works except the cabinet light. | Timer stuck on defrost cycle. | Turn timer knob. If unit starts running, replace the timer. |
| Ice built up on bottom of freezer (freezer liner). | Defrost drain plugged. | Clean defrost drain. |
| | Cabinet not level, causing poor drainage. | Adjust legs to level unit. |
| Sweating inside the cabinet. | Blocked air ducts. | Remove object(s) obstructing cold airflow from freezer compartment to fresh-food compartment. |
| | Kitchen humidity too high. | Sweating will stop when humidity lowers. |
| | Defective door gasket. | Check door gasket; adjust or replace as necessary. |

## NAMEPLATE LOCATION
## (UNIT IDENTIFICATION PLATE)

The nameplate can be a piece of metal riveted to the body of a refrigeration unit, a thin metal label with a sticky back or a piece of paper glued to the unit. In smaller refrigerators and freezers, it is sometimes inside at the bottom of the cabinet, under or next to the crisper drawers, or seldom in the back of the unit. In side-by-side units, it is normally hidden behind the kick panel (grille or toe plate) and can be easily found when the panel is removed. In commercial units, it is most commonly riveted to the side wall inside the cabinet or somewhere close to the compressor. In window air conditioners, it can be found under the front panel. In central air conditioners, it is attached to the condensing unit (or to the compressor compartment panel in the case of console units).

Information found on the nameplate includes the following:

a. Model number
b. Serial number
c. Type of refrigerant used and amount of charge (in console units)
d. Amperage (free-load amperage)
e. LRA (locked-rotor amperage; indicates the amperage the unit draws at the instant of starting)
f. Electrical rating of the unit (maximum and minimum voltage and hertz requirement)
g. Phase

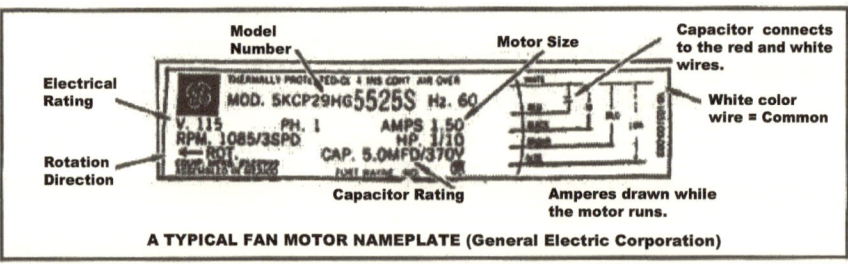

A TYPICAL FAN MOTOR NAMEPLATE (General Electric Corporation)

As explained earlier, the model number is needed to replace a part. The serial number usually indicates the age of the unit. Next to the type of refrigerant (R-12, R-22, etc.), the charge requirement may be shown in pounds and/or ounces.

The FLA and LRA amperage on the nameplate indicate the amount of current drawn by a good compressor under normal conditions. These are the figures by which comparisons are made in diagnosing the situations mentioned earlier. (Also see figs. 64a, 64b, and 80.)

## TRACING REFRIGERANT LEAKS IN THE SEALED SYSTEM

When lower-than-normal pressures are observed on both the high and low sides and the system does not cool properly, look for a refrigerant leak. This is done with a leak detector (see fig. 82a). Pass the probe of the detector over the suspected joints in the system. As the leak is approached, the detector gives off a piercing sound. Some leak detectors also come with a light that flashes. Some are sensitive enough to detect a leak of as little as 0.05 ounce per year.

Before getting into the repair procedures, note that there are systems of either copper or aluminum. Should you receive a call from a home owner stating that he/she has poked a hole in the evaporator in an attempt to more quickly defrost it, it may be concluded that it is a manual- or cycle-defrost refrigerator and that the evaporator is probably made of aluminum. Advise the owner to immediately unplug the unit to prevent any further damage to the compressor due to moisture penetrating the system. If the unit continues to run, the compressor will be ruined in no time.

Any repair of an aluminum evaporator is made with an epoxy instead of brazing. The high temperatures produced by brazing cannot be used on aluminum that has a low melting point.

Aluminum repair kits are available at regular supply houses. They contain a tube of resin, a tube of hardener, a vial of cleaner (acetone), emery cloth, mixing spatulas (popsicle sticks), and a piece of aluminum foil patching. These kits contain easy-to-follow instructions; but most important of all is the absolute thorough cleaning of the surface prior to applying the epoxy. It cannot be stressed too much how important it is to use the acetone to thoroughly clean the area to be repaired. Any trace of dirt, oil, or residue does not allow the repair substance to adhere.

Any time the sealed system is opened, the filter-drier *must* be replaced with a new one.

When a large leak is found in copper tubing, it is sometimes easier to cut the tubing and reconnect it by flaring or silver brazing. Prior to silver brazing, use a short length of copper tubing with an inside diameter equal to the outside diameter of the tube being repaired to join the two ends. This makes an excellent silver-brazed joint. Square off the ends to be joined and clean them thoroughly. Clean the inside of the piece of tubing being used as a connector, flux the joint areas, center the short piece of tubing over the joint, and braze it.

During the preparation for charging a unit, a few drops of a liquid chemical leak detector can be placed in the inlet hose before connecting it to the compound gauge manifold. (Follow the manufacturer's instructions on the package.) When the system is charged and resumes running, in case of a leak, the substance appears as a red ring around the hole as it escapes with the refrigerant, making the leak easy to find.

## Repairing a Leak in Copper Tubing

1. Connect piercing valves (if there are no access valves in the system) and discharge the system leaving the valve open.
2. Clean the puncture with sandpaper and wipe it with a rag soaked in acetone to remove all contaminants from the area to be silver-brazed.

   Be thorough in this step; otherwise, the solder will not adhere properly.

Epox-A-Leak is formulated to repair leaks on aluminum evaporators and condensors. It is ideal for joining all metal-to-metal parts. Epox-A-Leak will have shear strength in excess of 2000 psi within thirty minutes. Refrigeration systems can be recharged within one hour after repair.

3. Seal the hole by silver brazing.
4. When the leak is sealed, replace the filter-drier (see fig. 76) and connect a manifold gauge to the valves and recharge the unit with the proper refrigerant. (About 3 oz for residential units, and for commercial units, just enough to raise the pressure in the tubing above atmospheric pressure.) Then close the valves; start the unit and check the system using a leak detector to ensure there are no more leaks.

NOTE: It is best to check for refrigerant leaks in the low side when the unit is not running as the low-side pressure rises during the off cycles.

5. Follow the procedures given for evacuating and recharging the system.
6. Sandpaper all the new silver-brazed joint to ensure no flux residue is left in the tubing. Flux contains acid and corrodes copper if left on the tubing. Then vacuum the system to force out the moisture that penetrates the

system during the repair and recharge the unit with the proper amount of refrigerant.

## AIR LEAKS IN THE CABINET WALLS AND DOORS

Residential units with freezers cause trouble when there is a puncture in the inside or outside wall of the unit through which air penetrates the walls.

This becomes apparent when the owner notices a swelling in the side or back of the unit. What happens is that air, with moisture in it, penetrates inside through the puncture and condenses on the inside surface of the cold walls of the unit and develops a gradual ice buildup.

To remedy this problem, unplug the unit and place a very low output heater inside the cabinet with the door closed. This causes the ice to melt and gradually evaporate after about forty-eight hours. A heat gun (Fig. 63) or a regular hair drier may also be used to more quickly serve the same purpose. The puncture may then be sealed by brazing or by sealing it with the same epoxy used for repairing evaporators. Fiberglas kits (available from the supplier) can also be used. Make sure the unit is moved to a place away from the extreme cold and humid ambient air during this procedure.

## WATER LEAKS INSIDE THE CABINET

The moisture drain tube to the water evaporator pan may become plugged with debris and gummy substances. When that happens, water created in the defrost cycle cannot drain out. It overflows and leaks down, covering the freezer floor, and then during the refrigeration cycle, it turns to ice. To remedy the problem, unplug the unit and let it defrost. Clear the drain line of debris by shooting a stream of air into it or simply by blowing into it.

A similar thing happens if the unit is not seated level and does not allow the drain water to reach the opening of the drain tube. In that case the water accumulates to the point that it overflows and runs down the insides of the walls, misses the drain pan, and forms a puddle under the unit. (See fig. 3)

Most models come with a heating element around the top of the drain tube. If that element stops working, ice will form in the opening and cause one or more of the symptoms mentioned above. Test the heating element with an ohmmeter as described in "Testing the Defrost Heater."

## WATER LEAK UNDER THE UNIT

The formation of a puddle of water under the refrigeration unit is a very common problem. In figure 85, note that the suction line going into the compressor runs next to the capillary tube coming from the filter-drier and condenser. When the warm capillary tube is placed next to the cold suction line, its heat is transferred to the suction line. This causes the capillary tube to become cooler, and the suction-line temperature to rise at the same time.

The purpose of this heat exchanger is to help the capillary tube start cooling down on its way to the evaporator, and to help the suction line to warm up on its way to the compressor. Usually, in residential and light commercial units, these two lines are brazed together at the factory or placed together and insulated on the assembly line to form a permanent heat exchanger.

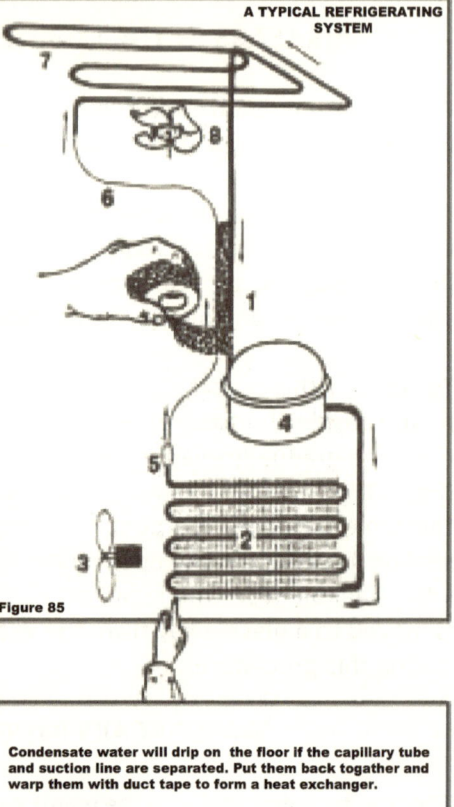

A TYPICAL REFRIGERATING SYSTEM

Figure 85

Condensate water will drip on the floor if the capillary tube and suction line are separated. Put them back together and warp them with duct tape to form a heat exchanger.

1. Heat Exchanger
2. Condenser
3. Condenser Fan
4. Compressor
5. Filter-drier
6. Capillary tube
7. Evaporator
8. Evaporator Fan

The suction line coming from the evaporator is cold; when it is not insulated, it causes condensation when coming in contact with surrounding warmer air. This moisture drips on the floor. If water is found on the floor, inspect the heat exchanger. More than likely the insulation is damaged or unwrapped. Put these two lines together and bind them with an insulation such as duct tape, and the condensation will stop.

In commercial refrigeration and air-conditioning units, the suction line is covered with a heavy sleeve for insulation to prevent condensation.

# AIR CONDITIONING AND REFRIGERATION REPAIR MADE EASY

## REFRIGERATION SYSTEM DIAGNOSTIC CHART
### (For Residential Refrigerators and Freezers)

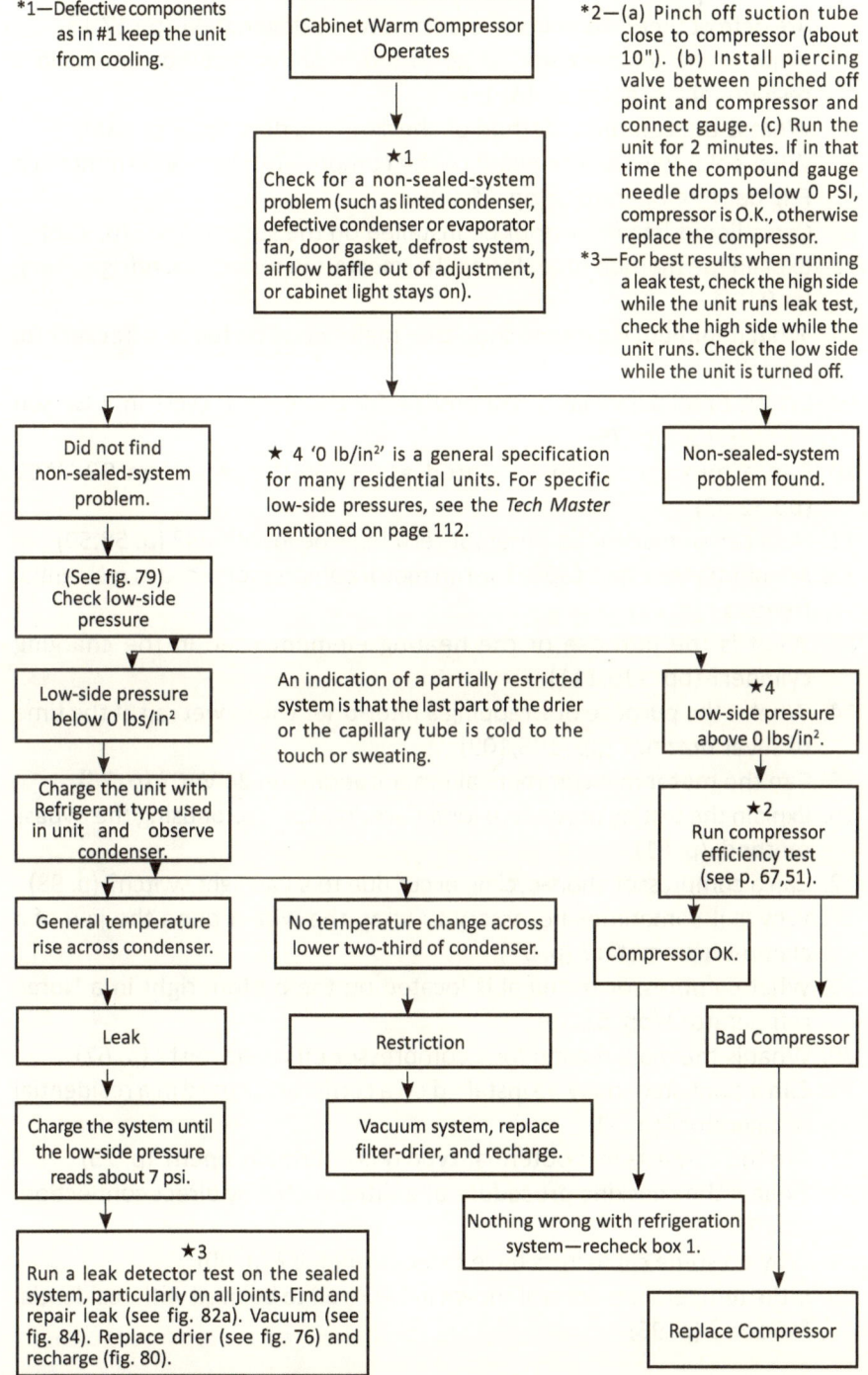

## A TEST OF KNOWLEDGE OF REFRIGERATION IN GENERAL

1. What purpose does a heat exchanger serve? (p. 126)
2. Are defrost timers used in cycle-defrost type refrigerators? (pp.17,18)
3. What is the temperature range in the fresh-food compartment in a residential refrigerator? (p. 16)
4. What is the common method of checking the door seal? (p. 100)
5. What color wire is connected to the compressor starting terminal in a Frigidaire refrigerator? (pp. 59,60)
6. Can a liquid-line filter-drier be installed on the suction line? (p. 108)
7. Explain the method of replacing thermostats in residential refrigerators. (p. 92)
8. What minimum clearance should be maintained on top of a freezer? (p. 104)
9. Which timer terminal is responsible for the defrost cycle in a Gibson refrigerator? (p. 79)
10. Can a motor compressor rated at 115 VAC operate on 99 VAC? (pp.52,53)
11. How can unmarked compressor terminals be identified? (p. 58,59)
12. Are permanent-split-capacitor-run motor compressors used in residential freezers? (p. 62)
13. What is the purpose of the heating element used in the charging cylinder? (pp. 110,111)
14. What is the purpose of wrapping a filter-drier with a wet rag at the time of silver brazing? (pp. 108,109)
15. Can the motor in a commercial timer operate on 24 VAC? (p. 78)
16. Explain the testing procedure for an overload protector using the bypass method. (p. 81)
17. Can a compressor short-cycling occur due to a bad light switch? (p. 98)
18. Why is it sometimes necessary to wrap the wire around the jaw of a clamp-on ammeter? (p. 97)
19. What compressor terminal is located on the bottom right in a Norge refrigerator? (pp. 58,59)
20. What is the main reason for a compressor efficiency test? (p. 67)
21. Can a solid-state relay be installed on a compressor used in a residential refrigerator? (p. 73)
22. Are the contacts in a potential-type relay normally open? (p. 73)
23. Explain the checking procedure of a timer motor by direct connection. (p. 77)
24. Why do some capacitors have three terminals? (p. 90)
25. Is the temperature control shown in figure 69 to be used with residential freezers? (p. 95)

26. What are the major causes of water leaks inside residential refrigerator cabinets? (p. 125)
27. Is it OK to maintain a five inches clearance on the sides of a refrigerator? (p. 103)
28. What type of compressor motors are used for light commercial applications (such as a salad bar)? (pp. 61,62)
29. What other terms are used for "defrost thermostat"? (p. 28)
30. What are the first things to be checked when a unit fails to cool? (p. 127)

## A TEST OF KNOWLEDGE ON EVACUATING AND CHARGING REFRIGERATION SYSTEMS

1. Name the four methods of charging a unit. (p. 110)
2. When should a suction-line filter-drier be installed? (pp. 108,109)
3. How is a unit tested for a leak before charging? (pp. 108,109)
4. How can it be made easier to spot the exact point to shut off the charging cylinder valve? (p. 111)
5. When would the frost-pattern method of charging a unit be used? (p. 115)
6. What are the possible causes for a refrigeration unit to run all the time without cooling? (pp. 119,120)
7. Can a defective condenser fan motor cause a compressor to short-cycle? (pp. 119,120)
8. What are the symptoms of a leaky door seal? (p. 118)
9. Why is it a good habit to check the sealed system with a leak detector after a repair? (p. 116)
10. How is the correct amount of refrigerant metered into large commercial units? (p. 113)
11. When is liquid leak detector usually used? (p. 116)
12. Why is it necessary to use a charging cylinder? (p. 111)
13. Why is a refrigerant tank turned upside down when filling a charging cylinder? (p. 112)
14. What three steps should always be taken before charging a system (when a sealed system is opened for any reason)? (p. 108)
15. What method is generally used to charge a commercial refrigeration unit? (p. 115)

# ADDITIONAL CONTROLS FOR COMMERCIAL UNITS AND COMPRESSOR SERVICE VALVES

This section covers the controls used in commercial and residential units. It includes the temperature-pressure relationship, common types of refrigerant, temperature and pressure controls, their operation and placement in sealed systems.

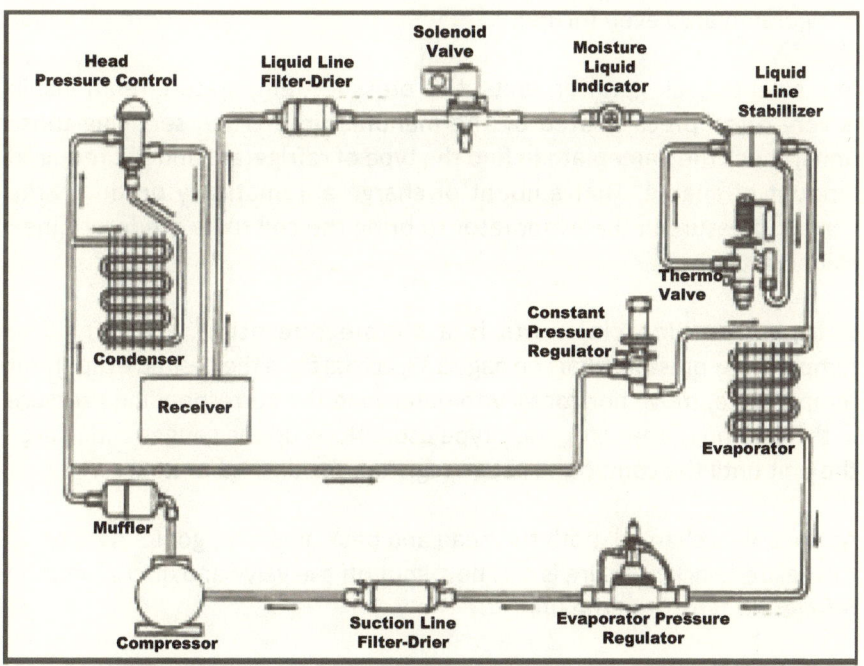

Courtesy of ALCO Controls

## SATURATED VAPOR REFRIGERANT PRESSURE AND TEMPERATURE RELATIONSHIP

There are several types of refrigerant with different qualities and boiling points. These different boiling points affect pressures in the evaporator and the condenser coils.

By increasing or reducing the pressure in the evaporator, an increase or decrease in temperature can be achieved. By looking at the chart on page 124, you'll see how each type of refrigerant produces different temperatures under various pressures.

It is possible to achieve a desired evaporator temperature by manipulating the low-side pressure. The lower the pressure under a saturated condition (holding as much vapor as it can), the lower the temperature of the evaporator coil. Under a lowered pressure, liquid refrigerant in the evaporator coil vaporizes more rapidly and absorbs more heat from the walls of the evaporator coil, reducing the temperature of the freezer or the refrigerated area even further.

In residential refrigeration units, this pressure-temperature relationship is very often precalculated by the manufacturer. When servicing these units, check the nameplate to find the type of refrigerant and the required amount of charge. That amount of charge automatically produces the correct pressure in the evaporator to bring the coil to its predetermined temperature.

Charging commercial units is a simple operation. By using the temperature-pressure chart on page 133, across from the desired evaporator temperature, move horizontally to determine the corresponding pressure in the column of the refrigerant type used. Hook up the gauges and charge the unit until the compound gauge registers the desired pressures.

As the unit is charged, both the head and back pressures go up. As soon as the desired back pressure is reached, shut off the valve and disconnect the refrigerant tank from the unit.

# Air Conditioning and Refrigeration Repair Made Easy

## Saturated Vapor Refrigerant Pressure and Related Temperature Chart

| Temperature °F | Pressure (psi) Refrigerant Type | | | | |
|---|---|---|---|---|---|
| | R-12 | R-22 | R-500 | R-502 | R-717 |
| -60 | 19.0 | 12.0 | — | — | 7.0 | 18.6 |

| Temperature °F | R-12 | R-22 | R-500 | R-502 | R-717 |
|---|---|---|---|---|---|
| -60 | 19.0 | 12.0 | — | — | 7.0 | 18.6 |
| -55 | 17.3 | 9.2 | — | 3.6 | 16.6 |
| -50 | 15.4 | 6.2 | — | 0.0 | 14.3 |
| -45 | 13.3 | 2.7 | — | 2.1 | 11.7 |
| -40 | 11.0 | 0.5 | 7.9 | 4.3 | 8.7 |
| -35 | 8.4 | 2.6 | 4.8 | 6.7 | 5.4 |
| -30 | 5.5 | 4.9 | 1.4 | 9.4 | 1.6 |
| -25 | 2.3 | 7.4 | 1.1 | 12.3 | 1.3 |
| -20 | 0.6 | 10.1 | 3.1 | 15.5 | 3.6 |
| -18 | 1.3 | 11.3 | 4.0 | 16.9 | 4.6 |
| -16 | 2.0 | 12.5 | 4.9 | 18.3 | 5.6 |
| -14 | 2.8 | 13.8 | 5.8 | 19.7 | 6.7 |
| -12 | 3.6 | 15.1 | 6.8 | 21.2 | 7.9 |
| -10 | 4.5 | 16.5 | 7.8 | 22.8 | 9.0 |
| -8 | 5.4 | 17.9 | 8.9 | 24.4 | 10.3 |
| -6 | 6.3 | 19.3 | 9.8 | 26.0 | 11.6 |
| -4 | 7.2 | 20.8 | 11.0 | 27.7 | 12.9 |
| -2 | 8.2 | 22.4 | 12.1 | 29.4 | 14.3 |
| 0 | 9.2 | 24.0 | 13.3 | 31.2 | 15.7 |
| 1 | 9.7 | 24.8 | 13.9 | 32.2 | 16.5 |
| 2 | 10.2 | 25.6 | 14.5 | 33.1 | 17.2 |
| 3 | 10.7 | 26.4 | 15.1 | 34.1 | 18.0 |
| 4 | 11.2 | 27.3 | 15.7 | 35.0 | 18.8 |
| 5 | 11.8 | 28.2 | 16.4 | 36.0 | 19.6 |
| 6 | 12.3 | 29.1 | 17.0 | 37.0 | 20.4 |
| 7 | 12.9 | 30.0 | 17.7 | 38.0 | 21.2 |
| 8 | 13.5 | 30.9 | 18.4 | 39.0 | 22.1 |
| 9 | 14.0 | 31.8 | 19.0 | 40.0 | 22.9 |
| 10 | 14.6 | 32.8 | 19.8 | 41.1 | 23.8 |
| 11 | 15.2 | 33.7 | 20.5 | 42.2 | 24.7 |
| 12 | 15.8 | 34.7 | 21.2 | 43.2 | 25.6 |
| 13 | 16.4 | 35.7 | 21.9 | 44.3 | 26.5 |
| 14 | 17.1 | 36.7 | 22.6 | 45.4 | 27.5 |
| 15 | 17.7 | 37.7 | 23.4 | 46.6 | 28.4 |
| 16 | 18.4 | 38.7 | 24.2 | 47.7 | 29.4 |
| 17 | 19.0 | 39.8 | 24.9 | 48.9 | 30.4 |
| 18 | 19.7 | 40.8 | 25.7 | 50.1 | 31.4 |
| 19 | 20.4 | 41.9 | 26.5 | 51.2 | 32.5 |
| 20 | 21.0 | 43.0 | 27.3 | 52.4 | 33.5 |
| 21 | 21.7 | 44.1 | 28.2 | 53.7 | 34.6 |
| 22 | 22.4 | 45.3 | 29.0 | 54.9 | 35.7 |
| 23 | 23.2 | 46.4 | 29.8 | 56.2 | 36.8 |
| 24 | 23.9 | 47.6 | 30.7 | 57.4 | 37.9 |
| 25 | 24.6 | 48.8 | 31.6 | 58.7 | 39.0 |
| 26 | 25.4 | 49.9 | 32.4 | 60.0 | 40.2 |
| 27 | 26.1 | 51.2 | 33.3 | 61.4 | 41.4 |
| 28 | 26.9 | 52.4 | 34.3 | 62.7 | 42.6 |
| 29 | 27.7 | 53.6 | 35.2 | 64.1 | 43.8 |
| 30 | 28.4 | 54.9 | 36.1 | 65.4 | 45.0 |
| 31 | 29.2 | 56.2 | 37.0 | 66.8 | 46.3 |
| 32 | 30.1 | 57.5 | 38.0 | 68.2 | 47.6 |
| 33 | 30.9 | 58.8 | 39.0 | 69.7 | 48.9 |
| 34 | 31.7 | 60.1 | 40.0 | 71.1 | 50.2 |
| 35 | 32.6 | 61.5 | 41.0 | 72.6 | 51.6 |
| 36 | 33.4 | 62.8 | 42.0 | 74.1 | 52.9 |
| 37 | 34.3 | 64.2 | 43.1 | 75.6 | 54.3 |
| 38 | 35.2 | 65.6 | 44.1 | 77.1 | 55.7 |
| 39 | 36.1 | 67.1 | 45.2 | 78.6 | 57.2 |
| 40 | 37.0 | 68.5 | 46.2 | 80.2 | 58.6 |
| 41 | 37.9 | 70.0 | 47.2 | 81.8 | 60.1 |
| 42 | 38.8 | 71.4 | 48.4 | 83.4 | 61.6 |
| 43 | 39.8 | 73.0 | 49.6 | 85.0 | 63.1 |
| 44 | 40.7 | 74.5 | 50.7 | 86.6 | 64.7 |
| 45 | 41.7 | 76.0 | 51.8 | 88.3 | 66.3 |
| 46 | 42.6 | 77.6 | 53.0 | 90.0 | 67.9 |
| 47 | 43.6 | 79.2 | 54.2 | 91.7 | 69.5 |
| 48 | 44.6 | 80.8 | 55.4 | 93.4 | 71.1 |
| 49 | 45.7 | 82.4 | 56.6 | 95.2 | 72.8 |
| 50 | 46.7 | 84.0 | 57.8 | 96.9 | 74.5 |
| 55 | 52.0 | 92.6 | 64.1 | 106.0 | 83.4 |
| 60 | 57.7 | 101.6 | 71.0 | 115.6 | 92.9 |
| 65 | 63.8 | 111.2 | 78.1 | 125.8 | 103.1 |
| 70 | 70.2 | 121.4 | 85.8 | 136.6 | 114.1 |
| 75 | 77.0 | 132.2 | 93.9 | 148.0 | 125.8 |
| 80 | 84.0 | 143.6 | 102.5 | 159.9 | 138.3 |
| 85 | 91.8 | 155.7 | 111.5 | 172.5 | 151.7 |
| 90 | 99.8 | 168.4 | 121.2 | 185.8 | 165.9 |
| 95 | 108.2 | 181.8 | 131.3 | 199.8 | 181.1 |
| 100 | 117.2 | 195.9 | 141.9 | 214.4 | 197.2 |
| 105 | 126.6 | 210.8 | 153.1 | 229.8 | 214.2 |
| 110 | 136.4 | 226.4 | 164.9 | 245.8 | 232.3 |
| 115 | 146.8 | 242.7 | 177.4 | 262.7 | 251.5 |
| 120 | 157.6 | 259.9 | 190.3 | 280.3 | 271.7 |
| 125 | 169.1 | 277.9 | 204.0 | 298.7 | 293.1 |
| 130 | 181.0 | 296.8 | 218.2 | 318.0 | |
| 135 | 193.5 | 316.6 | 233.2 | 338.1 | |
| 140 | 206.6 | 337.2 | 248.8 | 359.1 | |
| 145 | 220.3 | 358.9 | 265.2 | 381.1 | |
| 150 | 234.6 | 381.5 | 282.3 | 403.9 | |
| 155 | 249.5 | 405.1 | 300.2 | 427.8 | |

To convert degree centigrade to degree Fahrenheit, use this formula:

$$(°C \times 9/5) + 32°F = °F$$

To convert degree Fahrenheit to degree centigrade, use this formula:

$$(°F - 32) \times 5/9 = °C$$

EXAMPLE: To determine the corresponding low-side pressure to produce the desired evaporator temperature of 33°F in a walk-in cooler using R-12 refrigerant, refer to the temperature-pressure chart.

ANSWER: Across the line on which 33°F is shown, in the R-12 column find 30.9 as the pressure needed to produce 33°F. Charge the system until the compound gauge (low-pressure side) reaches 30.9 lbs/in$^2$; turn off the valve and stop charging.

The cabinet temperatures in different types of refrigeration units are fairly standard. Figure 86 shows recommended cabinet temperatures for different units.

As a rule of thumb, evaporator temperature should be set 20°F lower than the desired cabinet temperature.

EXAMPLE: To determine the required low-side pressure in a unit (using R-12) to maintain a general cabinet temperature of 45°F, subtract 20°F from the cabinet temperature to get 25°F (45-20=25).

ANSWER: Referring to the chart, across the line from 25°F in the R-12 column, 24.6 psi is the required low-side pressure to maintain the general cabinet temperature at 45°F.

The evaporator size should match the rest of the system. If an oversized evaporator is used, it will starve; and if an undersized evaporator is selected, it will become flooded. In both cases, the temperatures produced in the evaporators will never drop to a desired point.

## RECOMMENDED REFRIGERATOR TEMPERATURES

Figure 86

| Cabinet | Temperature (°F) |
|---|---|
| Back bar or beverage cooler | 37-40 |
| Beverage precooler | 35-40 |
| Candy case (display or storage) | 59-66 |
| Dairy cases (single or double) | 35-40 |
| Delicatessen case | 35-40 |
| Dough refrigerator | 35-39 |
| Floral display case | 40-50 |
| Floral storage case | 36-45 |
| Frozen-food cabinet (closed) | -10/-5 |
| Frozen-food cabinet (open) | -8/-4 |
| Grocery refrigerator | 35-39 |
| Pastry display | 45-50 |
| Reach-in refrigerator | 35-39 |
| Restaurant storage cooler | 34-39 |
| Top display case (closed) | 35-42 |
| Vegetable display (open or closed type) | 36-41 |

## DETERMINING HIGH-SIDE PRESSURES

The high-side pressure is related to the ambient air temperature. When servicing a unit with the condenser inside an air-conditioned building, expect the head pressure to be considerably lower than a unit with its condenser on the roof in the hot summer sun. The warmer the surrounding air, the higher the head pressure, and the more difficult it becomes for the unit to cool.

As a general rule, to determine the expected high-side pressure in a unit with an air-cooled condenser, add 30°F to the current ambient (surrounding the condenser) temperature; check the chart on page 133 for the pressure reading at that added temperature for the type of refrigerant used in the unit. This figure indicates the expected head pressure in the unit. For water-cooled condensers, the head pressures should be calculated by adding 20°F to the exhaust-water temperature. (See the section on Water-Cooled Condensers.)

EXAMPLE: Suppose there is a commercial central air-conditioning unit using R-22 operating in an ambient temperature of 90°F. Assume also that this is a new unit with no malfunctioning parts. What head pressure should you expect to read on the high-pressure gauge? For the answer, add 30°F to the ambient temperature (90 + 30 = 120°F), find 120°F on the chart, and read across to the R-22 column. Hence, 259.9 lb/in$^2$ will be the expected high-side pressure.

If the system is overcharged, if the condenser is dirty and/or linted, if there is a slow or inoperative condenser fan, or if there is a restriction in the system, the head pressure increases dramatically, causing a higher than normal head pressure. A temperature-pressure chart should always be carried with you on service calls. The amount and type of refrigerant used in the unit can be determined by the unit nameplate.

PROBLEM: Determine the head and the back (low side) pressures of an ice cream freezer with a cabinet temperature of -20°F using refrigerant 502. The unit operates in a room with an ambient temperature of 80°F.

SOLUTION: The evaporator temperature must be adjusted to 20°F below the desired temperature produced. Therefore, 20°F - 20°F = -40°F. Referring to the temperature-pressure chart, at -40°F, the corresponding pressure for units using refrigerant 502 is 4.3 psi, which should be the low-side pressure. To determine the high-side pressure, add 30°F to 80°F, which will be equal to 110°F. The corresponding pressure on the chart in the R-502 column at 110°F is 245.8 psi = the proper head pressure for this unit in this environment.

EXAMPLE 1: What evaporator temperature is required to maintain the cabinet temperature of a floral display case at 40°F?

ANSWER: 40°F - 20°F = 20°F.

EXAMPLE 2: Suppose service is needed on a delicatessen case with a cabinet temperature of 35°F using refrigerant R-12. The ambient temperature is 75°F. Determine the high- and low-side pressures.

ANSWER: The evaporator temperature should be regulated at 15°F. (35°F-20°F=15°F.) According to the chart, 17.7 psi back pressure is required for R-12 to create 15°F.

To determine the high-side pressure: 75°F + 30°F = 105°F. On the chart, the corresponding pressure for 105°F in the refrigerant R-12 column is 126.6 psi.

EXAMPLE 3: Determine the head and back pressures of a frozen-yogurt machine, which uses R-502 with evaporator temperature of -10°F. The condenser is cooled by water, which is in tubes that run through the condenser coil. (Many commercial units have water-cooled condensers for improved heat exchange. See figure 87 and the pages concerning water-cooled condensers for more detail).

ANSWER: In the case of water-cooled condensers, measure the water temperature as it leaves the condenser through the return pipe by placing a thermometer on that pipe. Consider that the ambient temperature and add 20°F to that reading. Assume that this exhaust water temperature is 60°F. To determine the expected head pressure, 60°F + 20°F = 80°F. On the chart, 80°F creates 159.9 lbs/in$^2$ of head pressure in the system when 502 refrigerant is used.

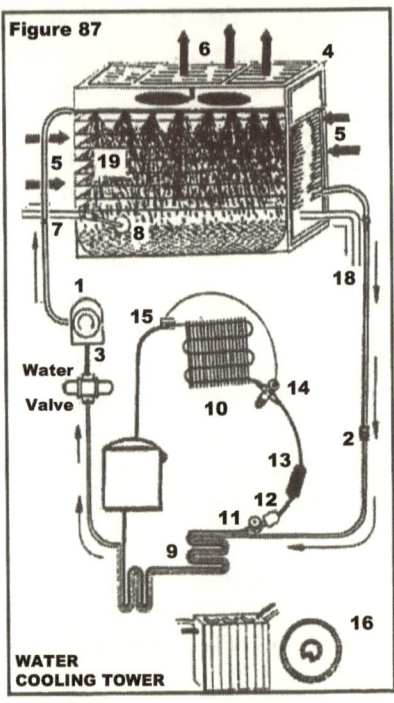

Figure 87

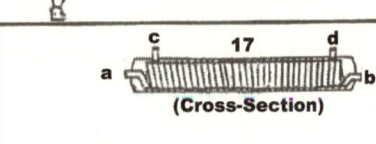

(Cross-Section)

1. Water out
2. Water in
3. Water pump
4. Cooling tower with water sprayer
5. Cooler air in
6. Warmer air out
7. Make up water supply
8. Reservoir float
9. Condenser
10. Evaporator
11. Sight glass
12. Filter-drier
13. Expansion valve
15. Expansion valve-sensing bulb
16. Tube-within-a-tube
17. Shell and coil condenser
   a. Water out
   b. Water in
   c. Refrigerant in
   d. Refrigerant out
18. Water drain line
19. Levels of thin, corrosion-resistant material

To determine the low-side pressure: the chart shows that under R-502, 22.8 psi produces -10°F.

## REFRIGERANT

When dealing with refrigerant, always bear in mind that, practically, every refrigerant has a different boiling point that averages about 59°F below zero! Should any refrigerant in its liquid state come in contact with skin, it will cause severe pain and damage to that area.

Since refrigerant is heavier than air, when released, it displaces air at ground (floor) level. When handling refrigerant or discharging a unit, be sure to provide adequate ventilation. By failing to do this, the refrigerant accumulates and starts to fill the working space from the floor up. Then, during brazing, the flame breaks down the refrigerant it is exposed to and generates toxic fumes, creating a breathing problem with a sensation of burning.

Should these symptoms ever be experienced, leave the work area immediately, get some fresh air, and wait until the toxic fumes have dispersed.

There are several types of refrigerant used in residential and commercial refrigeration for different applications. They are odorless, colorless, and tasteless.

## SOME OF THE MOST COMMON TYPES OF REFRIGERANT

*R-12* is very popular. It is used mostly in residential refrigerators and freezers, in commercial display cases and in walk-in coolers. It is colorless, more or less odorless, and boils at -21.7°F. This substance is nonflammable, noncorrosive, and nonirritating. It is sold in various sizes of cylinders and in 12 oz cans. The containers are color-coded white.

*R-22* is used mainly in refrigeration installations that operate with low-evaporating temperatures and also in air-conditioning units. With a boiling point of -41°F, it produces low temperatures without having to generate below atmospheric pressures in the sealed system. It comes in cylinders or cans color-coded green.

*R-11.* This refrigerant is widely used in very large units with large centrifugal compressors; it is also used as a flushing agent. With a boiling point of -50.1°F at sea level, it is used where temperatures of -60°F to 0°F are required. It is mainly used in frozen-food display cases, frozen food processing plants, ice cream freezers, and ice machines that work with reciprocating compressors. It is considered a low-pressure type of refrigerant. R-11 cylinders are orange colored.

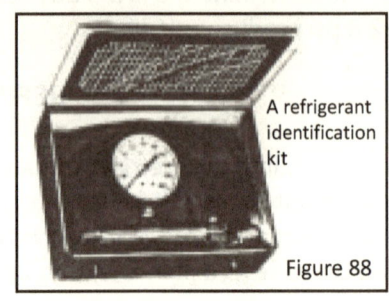

A refrigerant identification kit

Figure 88

*R-502.* This is a combination of R-12 and R-22 with a boiling point of -50.1°F, used basically for low temperature operation between -60°F to 0°F in ice machines, frozen-food display cases, frozen-food storage cabinets, and industrial refrigeration in frozen-food processing plants. It is nontoxic and nonflammable. It creates relatively low condensing pressure (head pressure) at low temperature. It is used only with the reciprocating type of compressor. The cylinders are color-coded orchid.

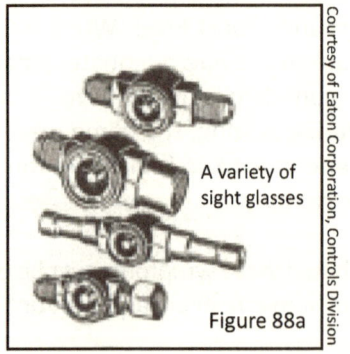

A variety of sight glasses

Figure 88a

Courtesy of Eaton Corporation, Controls Division

*R-503.* This refrigerant boils at -126°F at sea level. Because of this, it is used where a very low temperature range is needed (-125°F to -100°F), such as in the low state of cascade systems.

Because of such low temperatures, the oil in the compressor tends to freeze if oil separators are not used to return the lubricant to the compressor. This refrigerant comes in aquamarine cylinders.

*R-504.* This nontoxic, nonflammable, noncorrosive liquid boils at -70°F. It is normally used in industrial refrigeration applications where temperatures of -80°F to -40°F. are needed. It comes in tan cylinders.

Figure 89

A *sight glass* is brazed or flared into the liquid line. The unit should be charged until the bubbles in the sight glass disappear. Presence of bubbles indicate a shortage of refrigerant in the system.

Printing on the top of the sight glass rim shows that if the Freon is pink in color, there is moisture in it, and blue indicates no moisture.

Courtesy of Henry Valve Co.

There is an instrument that can be used to identify the type of refrigerant used in a system (see fig. 88). Follow the instructions supplied with it.

A *sight glass* is used in commercial units to aid in determining the status of the refrigerant (see fig. 89) They are really liquid indicators that show the condition of the refrigerant in the system. They show bubbles for a system low on refrigerant (while the compressor is running). Some of them indicate the presence of moisture in the system by a color change. That information is printed on the sight glass.

## SURGE TANK

When two evaporators with two different temperatures are used in a system, a two-temperature valve is installed at the outlet of the warmer evaporator to prevent the compressor suction power from bringing the pressure in the warmer evaporator lower than the desired pressure.

Look at figure 96 to see where the surge tank is installed. When a two-temperature valve is used, the compressor short-cycles, unless a surge tank is installed in the suction line. The reason for this is that immediately after the pressure in the suction lines drops to a predetermined cut-out setting, a switch in the low-pressure control opens, shutting off power to the compressor causing the two-temperature valve that controls the pressure inside the warmer evaporator to open, causing a rapid rise in the suction line. At this point, the switch in the low-pressure control (sensing the increase in pressures) closes its circuit and starts the compressor and causes the short-cycling. The surge tank prevents this by creating more low-pressure volume to absorb the sudden pressure rise that might affect the low-pressure control. It could be said that it is a type of shock absorber that smoothes out the fluctuations of the suction-line pressure changes.

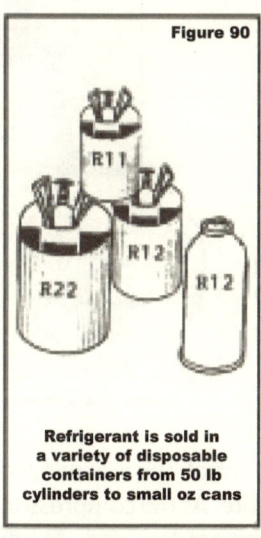

Figure 90

Refrigerant is sold in a variety of disposable containers from 50 lb cylinders to small oz cans

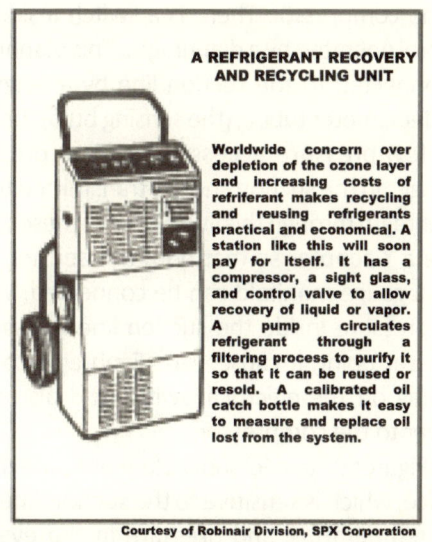

A REFRIGERANT RECOVERY AND RECYCLING UNIT

Worldwide concern over depletion of the ozone layer and increasing costs of refriferant makes recycling and reusing refrigerants practical and economical. A station like this will soon pay for itself. It has a compressor, a sight glass, and control valve to allow recovery of liquid or vapor. A pump circulates refrigerant through a filtering process to purify it so that it can be reused or resold. A calibrated oil catch bottle makes it easy to measure and replace oil lost from the system.

Courtesy of Robinair Division, SPX Corporation

# PRESSURE CONTROLS IN COMMERCIAL REFRIGERATION

In residential refrigeration units, a thermostat controls the temperature by connecting and disconnecting the power to the compressor. This type of cold control reacts to the temperature changes.

There is a correlation between the pressures created in the sealed system and the temperatures produced by the vaporizing refrigerant. Commercial units take advantage of this relationship by employing pressure controls that govern the operation of the compressor to regulate the temperature in the unit by controlling the pressures inside the sealed system. Serving a two-fold purpose, they regulate the temperature and, at the same time, protect the system from pressures that become too high or too low. They do this by disconnecting and reconnecting the power to the compressor motor. There are three types of pressure controls utilized in commercial refrigeration. They regulate either the high or low pressures or act as a safety device for the oil pressure in the compressor. Some units employ a low-pressure control, some employ a high-pressure, and some employ (including the oil-pressure control) all three. All of the pressure controls are safety devices.

I.   **LOW-PRESSURE CONTROL.**

The temperature in a commercial refrigeration unit is regulated by the low-pressure control.

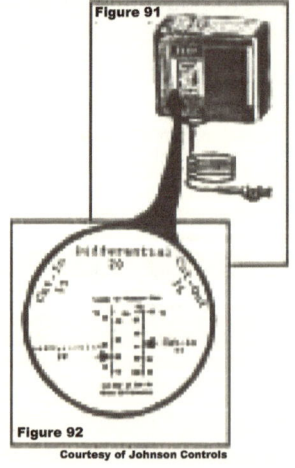

Figure 91

Figure 92
Courtesy of Johnson Controls

The low-pressure control intersects the wire going into either the run or common terminal of the compressor. There is a switch inside the control operated by a diaphragm. This diaphragm is connected to the suction line by a length of small-diameter tubing (the sensing bulb) through the low-pressure access valve. The pressure changes in the suction line are transmitted to the diaphragm through this line. Some of these valves have a Y-adapter so that both this sensing bulb and a gauge manifold can be connected. When the pressure inside the suction line falls below a preset point, it causes the diaphragm to flex, turning off the electrical switch and disrupting power to the compressor.

Figures 91 and 92 show a low-pressure motor control. This is a spring-loaded device, which is sensitive to the suction-line pressure. As the compressor runs, the temperature and pressure in the evaporator are lowered. As soon as pressure drops to a given point, it cuts off power to the compressor. When the pressure in the evaporator rises to a predetermined setting, the control reconnects power to the compressor, the compressor resumes running, and the cycle is repeated. This control device is often used in drinking fountains and other units where a constant temperature is needed.

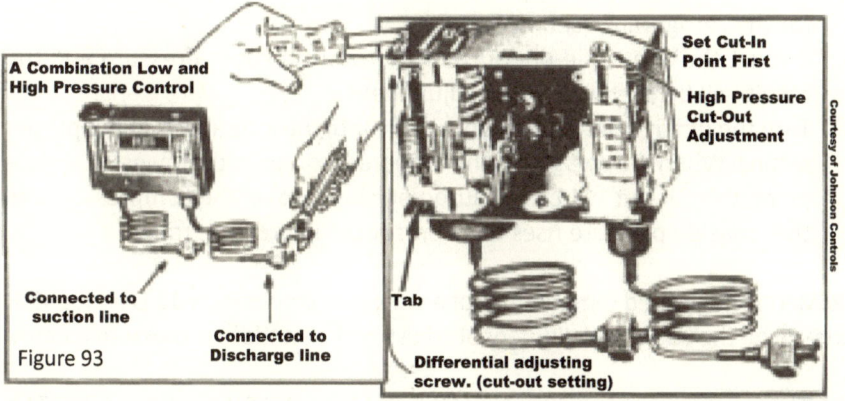

Figure 93

Until the pressure in the suction line gets high enough to flex the diaphragm, the unit remains in the off cycle.

NOTE: It is normal for a refrigeration unit to fail to start after opening the sealed system. The low-pressure control reacts to the abnormal pressures (created by opening the system) by having the bellows within the control keep the circuit to the compressor open. It can be restarted by manual operation of the tripping switch. Lift the bellows tab with the blade of a screwdriver (at arrow in fig. 93) for a few seconds (time depends on the size of the unit). Do NOT operate the control at any other point or it will be damaged.

There are two adjusting screws on top of the control. One for the *cut-in* adjustment, and one for the *cut-out*. As the adjustment screws are turned, an indicator moves up or down a scale calibrated in pounds. (See fig. 93) *Cut-in* refers to the pressure at which the compressor resumes operating. (Not warmer than that point.) And *cut-out* refers to pressure at which the compressor stops operating. (Not colder than that point.)

NOTE: After replacing the low-pressure control, check the pressures in the unit if it is suspected that any refrigerant has escaped. Watch it cycle two or three times after restarting it. Remember that pressure controls are affected by the pressures in the sealed system and not by the temperature. Use the chart on page 124 to convert any temperature to its corresponding pressure. To calculate cut-in and cut-out, do the following:

1. Using the chart on page 133, determine the evaporator pressure for the warmest allowable temperature. Then turn the cut-in screw until the needle indicates the correct pressure.
2. Determine the lowest allowable evaporator temperature. (Assume a 20°F drop in evaporator temperature is allowable.)
3. Find the corresponding pressure for the allowable temperature drop for the refrigerant used in the unit.
4. Deduct the allowable pressure drop from the cut-in pressure and the result will be the *differential* adjustment.
5. Turn the differential adjustment screw until the indicator is at the proper setting. When the pressure in the evaporator drops to the lowest allowable point, the cut-out switch will automatically turn off the compressor until the low-side pressure rises to its preset cut-in pressure range.

EXAMPLE: The desired temperature of a walk-in cooler using R-12 is about 35°F. Determine the cut-in and differential adjustments on the low-pressure control.

1. The corresponding pressure for R-12 at 35°F (on page 133, The Temperature-Pressure Chart) is 32.6 psi. This will be the cut-in setting.
2. Since the temperature of the evaporator should always be kept 20°F below the desired refrigerated ambient temperature, to determine the cut-out setting on the low-pressure control: 35°F - 20°F = 15°F.
3. Refer to the temperature-pressure chart to find the corresponding pressure at 15°F for R-12. Convert the 15°F to pressure, which will be 17.7 psi. According to step 4 above, 32.6 - 17.7 = 14.9 psi will be the *differential adjustment.* By setting the differential adjustment, the pressure control will cut in at 35°F and cut out at 15°F.
4. Using a screwdriver, turn the cut-in screw until the cut-in needle indicates 32.6 psi.
5. Turn the cut-out screw until the cut-out needle indicates 14.9 psi. Now the walk-in cooler temperature is set to be kept at about 35°F. The pressure control will cut in at 35°F and cut out at 15°F.

## HOW TO CHECK THE LOW-PRESSURE CONTROL

First, shut off the unit and remove the control cover to expose the switch. Then connect a test light with alligator clips to the two wire terminals in the low-pressure control. Now, connect the compound gauge to the suction-line service valve and restart the unit as the unit must be running when performing this check. Watch the test light and compound gauge while closing the liquid-line service valve and let the system pump down. Watch for the low switch opening, then open the liquid-line valve and watch for the high switch closing. (It must close as soon as the compound gauge rises to the cut-in pressure.)

CAUTION: Perform these checks for a brief time only to avoid damage to the unit.

**LOW-SIDE PRESSURE MOTOR CONTROL SETTINGS FOR TYPICAL REFRIGERATION APPLICATIONS**

| ★Inches Vacuum | R-12 | | R-22 | | R-502 | |
|---|---|---|---|---|---|---|
| | Out | In | Out | In | Out | In |
| Refrigeration applications | | | | | | |
| Florist box | 26 | 42 | 51 | 77 | 61 | 88 |
| Vegetable display | 11 | 35 | 27 | 66 | 35 | 77 |
| Walk-in cooler | 12 | 35 | 29 | 66 | 37 | 77 |
| Beer cooler | 15 | 34 | 33 | 64 | 42 | 76 |
| Show case | 18 | 34 | 39 | 64 | 49 | 76 |
| Reach-in cooler | 18 | 36 | 39 | 68 | 47 | 79 |
| Frozen-food open display case | 7★ | 5 | 4 | 17 | 9 | 23 |
| Frozen-food closed display case | 2 | 8 | 11 | 22 | 17 | 30 |
| Beer, milk, water cooler | 18 | 30 | 40 | 55 | 48 | 67 |
| Soda fountain | 20 | 28 | 42 | 56 | 61 | 66 |
| Ice machine | 4 | 17 | 15 | 36 | 22 | 47 |

Figure 93a

## II.  HIGH-PRESSURE CONTROL (HIGH-PRESSURE CUT-OUT)

It is primarily a safety device which differs from the low-pressure control. Its sensing bulb is connected to a service valve on the discharge line by a flare nut. When the head pressure in the system approaches a dangerous level, it will automatically shut off the electrical power to prevent a compressor burnout through overheating.

In many commercial units, depending on the type of refrigerant being used, the cut-out pressure is never set higher than 300 psi to prevent damage to the system. As a rule of thumb, the high-pressure control is usually set to cut out the power at about 20% above normal head pressure, which is 165 psi for units using R-12, 275 psi for units using R-22, 200 psi for units using R-500, and 295 psi for units using R-502. Some units use a combination high- and low-pressure control. (See fig. 93)

### HOW TO CHECK THE HIGH-PRESSURE CONTROL

Connect the high-pressure gauge to the high-side service valve. Open the valve. Disconnect the condenser fan or block the airflow, or if it is a water-cooled system, shut off the flow of water. As the temperature rises, so does the head pressure. If the gauge goes over 10% above the correct head pressure and the control fails to shut off the compressor, disconnect the power and adjust the high-pressure control to a correct setting or replace the control if necessary.

CAUTION: Perform this check for a brief time only to prevent damage to the unit.

### III.   OIL-PRESSURE SAFETY CONTROL

This type of control is used in heavy commercial and industrial refrigeration. They have two lines with flare nut connections: one is connected to the compressor oil line (usually located directly on the compressor), and the other one connects to the suction line.

Because the discharge line pressure causes some amount of oil to leave the compressor and circulate with the refrigerant, this oil must be returned to the compressor from the compressor suction line to maintain proper lubrication. If the low-side pressure goes higher than the oil pressure, the oil will not be circulated back to the compressor.

OIL-PRESSURE SAFETY CONTROLS

These controls are made for commercial systems and operate on the basis of the oil and suction-line pressure difference.

Figure 94

(Courtesy of Johnson Controls)

If, for any reason, the oil pressure drops below the low-side pressure, the pressure switch in the control will shut off the power to the compressor.

This type of control operates the same way as the low-pressure control in figure 94. There are two bellows (or diaphragms) within the pressure control which are set to sense a safe pressure difference between the suction-line pressure and the oil pressure. If this pressure difference drops below a predetermined point, the mechanism will open the circuit to the compressor.

## CHECK VALVES

*Check valves* are used in large commercial systems having multiple evaporators with different temperatures. The purpose of the check valve is to keep the liquid or vapor refrigerant flowing in one direction. Normally installed at the outlet of the colder evaporator, they only permit vapor to leave to prevent flooding of the colder evaporator during the off cycle. In some units, a check valve is placed near the suction line of the compressor to prevent a possible back flow of liquid refrigerant into the compressor during the off cycle. (See fig. 95a.)

## DOUBLE-TEMPERATURE VALVES
### (Pressure-regulating or Pressure-reducing Valves)

In some of the larger commercial units, the suction line is connected to a multiple evaporator system requiring different temperatures, such as one evaporator in a frozen-food display with an average temperature of -20°F, and another in a walk-in cooler set at 40°F. Since, in this system, the two evaporators cannot be the same temperature, a reducer valve (two-temperature valve) is installed at the outlet of the warmer evaporator to increase the pressure inside the warmer evaporator by reducing the constant suction force that the compressor applies to it. Also, a check valve is placed at the outlet of the colder evaporator to stop the back flow of refrigerant, thus preventing the buildup of more pressure (and consequent warm-up) of the colder evaporator during the off cycle. (See figs. 96 and 95.) They are manufactured with different physical appearances.

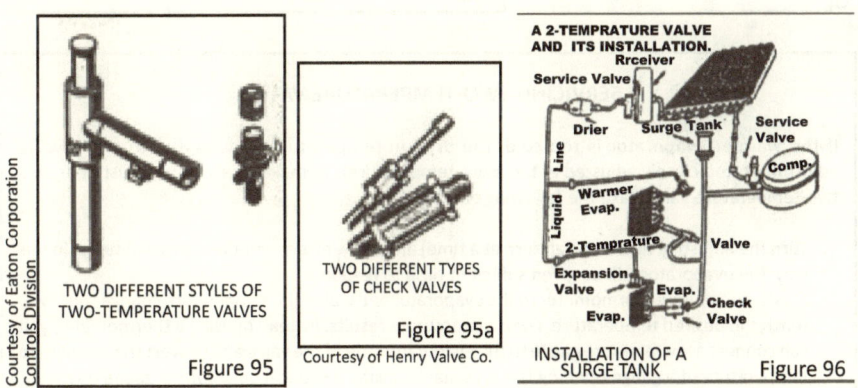

Figure 95 — TWO DIFFERENT STYLES OF TWO-TEMPERATURE VALVES (Courtesy of Eaton Corporation Controls Division)

Figure 95a — TWO DIFFERENT TYPES OF CHECK VALVES (Courtesy of Henry Valve Co.)

Figure 96 — A 2-TEMPRATURE VALVE AND ITS INSTALLATION. INSTALLATION OF A SURGE TANK

There are two general categories of two-temperature valves: those using a spring inside their mechanism (including spring valves with thermostats or solenoids) and valves employing a sensing bulb, very much like TEVs.

Operation of the spring-type valve is quite simple: When the pressure inside the evaporator goes up (due to evaporating refrigerant), the spring opens the valve to allow refrigerant to flow out of the evaporator (drawn by the compressor suction power).

The thermostatic-type double-temperature valve is very similar in operation to the thermostatic expansion valve with the difference that it only opens to let vaporized refrigerant be drawn by the compressor suction power when its sensing bulb perceives the need for cooling. When the valve opens, the pressure built up in the evaporator drops, causing the temperature to go down.

In the solenoid type, as in figure 97, an independent thermostat is placed in line before the expansion valve to operate the solenoid. When the thermostat is satisfied, the solenoid valve closes, allowing no more refrigerant from the liquid line into the TEV valve.

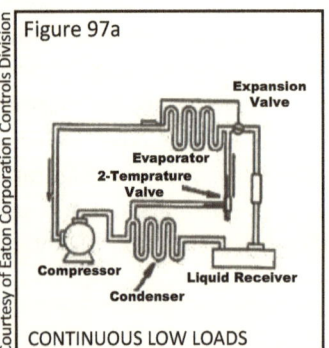

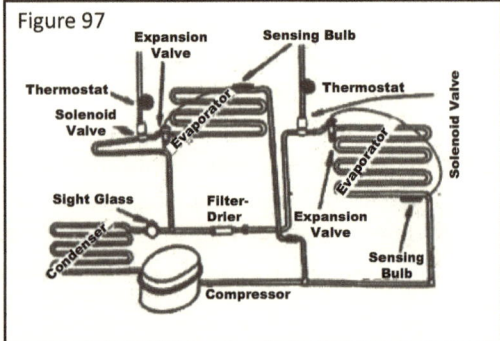

### SERVICING TWO-TEMPERATURE VALVES

If the warmer evaporator is too cold, the drop in temperature may be due to a bad valve adjustment (too closely adjusted at too low a temperature). To make an accurate adjustment on a two-temperature valve, take the following two steps:

1. Turn the adjusting nut (one-half turn at a time) and allow fifteen minutes between turns. In this way, the evaporator will be given sufficient time to respond.
2. Attach an accurate thermometer to the evaporator and watch the temperature changes until you reach the desired temperature. For more accurate results, instead of using a thermometer, you can connect a low-pressure gauge to the evaporator side of the valve and convert the pressure to temperature. If a gauge opening is not available, install a shut-off valve with a gauge opening.

If the valve is leaky, the evaporator temperature will not rise, in which case the valve must be replaced.
If the valve is stuck shut, the warmer evaporator will never cool and the valve must be replaced.

If the two-temperature valve is installed in or near the freezer compartment, frost will accumulate on the bellows. Remove the valve from the freezer compartment or cover the bellows with a light grease.

A double-temperature valve can be used as a high- to low-side bypass to assure continuous compressor operation at low loads. A low-load condition means that the evaporator has little cooling to do as the selected temperature has already been attained. This type of connection is normally used in commercial applications to limit evaporator temperature. (See fig. 97a.)

A constant-pressure valve can also be used to limit the compressor inlet pressure when the system starts after defrosting (at which time the suction pressure has greatly increased). By installing a constant-pressure valve in the suction line, the compressor is protected from overload or burnout. (See fig. 97b.)

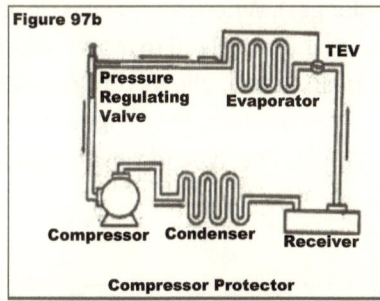

Figure 97b — Compressor Protector

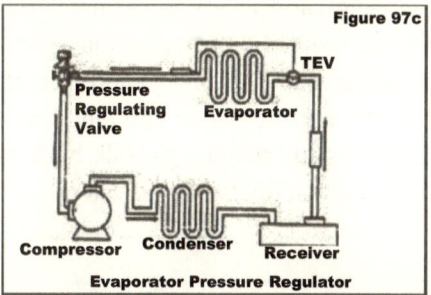

Figure 97c — Evaporator Pressure Regulator

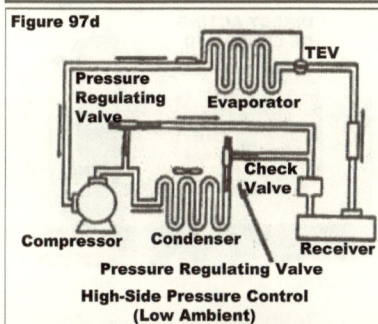

Figure 97d — Pressure Regulating Valve, High-Side Pressure Control (Low Ambient)

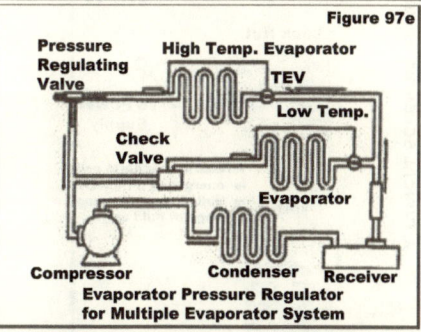

Figure 97e — Evaporator Pressure Regulator for Multiple Evaporator System

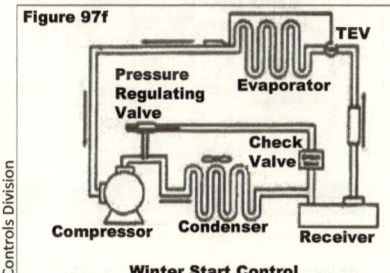

Figure 97f — Winter Start Control

*Courtesy of Eaton Corporation, Controls Division*

**INSTALLATION OF PRESSURE-REGULATING VALVE TO SERVE DIFFERENT PURPOSES**

97b. A valve installed in the system to protect the compressor against burnouts.

97c: Valve installed in evaporator suction line to stop evaporator pressures from dropping below a desired minimum.

97d. A valve installed in the liquid line at the condenser outlet to close at low ambient temperatures.

97e: A valve installed at the outlet of the warmest evaporator in a multiple evaporator system. It prevents pressure drops below a desired point in the warmest evaporator.

97f: A valve installed in a bypass line between the compressor and the receiver to quickly admit discharge gas into the receiver at the start of the refrigeration cycle to rapidly build operating receiver pressure. The valve closes when operating pressure is reached.

Figure 97c shows a pressure-reducing valve installed in the evaporator suction line to prevent evaporator pressures from dropping below a desired minimum. Installation of these valves, on the suction line for example, can prevent liquid in chillers from freezing.

## SOLENOID VALVES

A solenoid valve is an electromechanical device, which operates by creating and eliminating a magnetic field. It is primarily composed of a coil of wire (windings) and an armature (metal rod or plunger). When the circuit is closed, the coil creates a magnetic field which causes the metal rod to move upward its valve seat. The seat, when free of its obstruction, allows refrigerant to flow through it. When the electrical circuit is opened, the magnetic field disperses and the plunger falls back into its seat. This type of valve is known as a direct-acting solenoid valve.

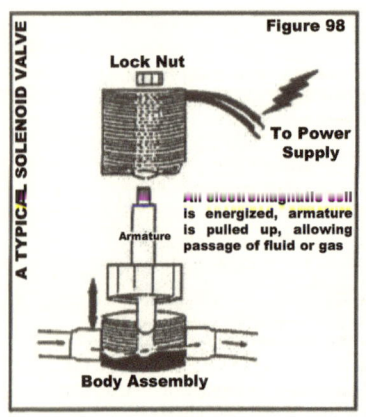

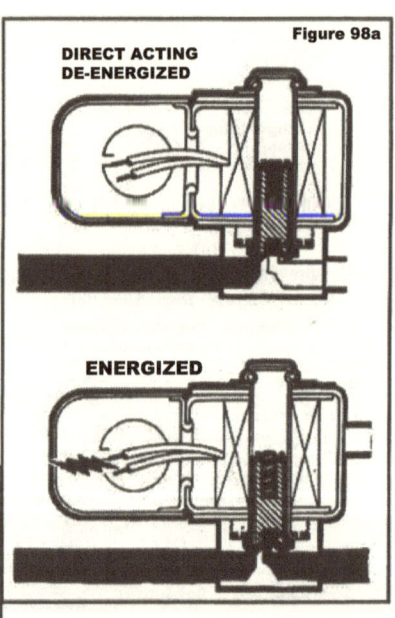

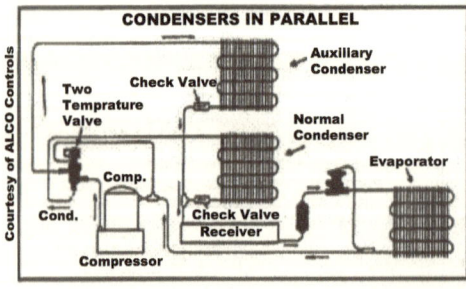

An illustration of a typical refrigerating system with condensers connected in parallel using a three-port solenoid valve for the heat-recaiiming application. Note the placement of the valve. The refrigerant circulates through the auxiliary condenser when the temprature exceeds a certain point.

Pilot-operated solenoid valves use a combination of the solenoid coil and the line pressure to operate. In this type of valve, the plunger is attached to a needle valve covering a pilot orifice rather than the main port. The line pressure holds an independent piston or diaphragm closed against the main port. (See fig. 98b.)

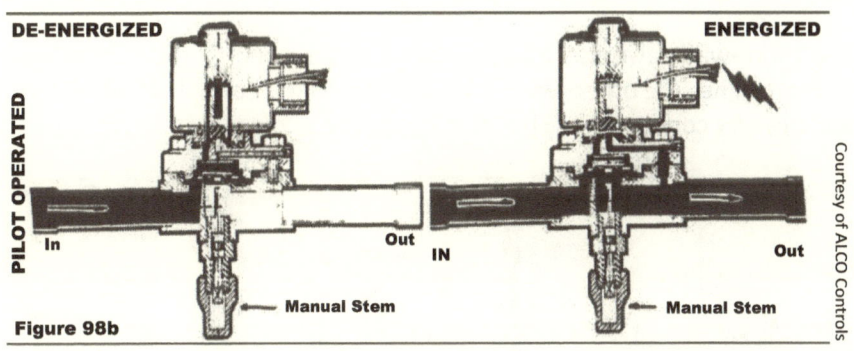

Figure 98b

When the coil is energized, the plunger is pulled into the center of the coil, opening the pilot orifice. Once the pilot port is opened, the line pressure above the diaphragm is allowed to bleed off to the low side or outlet of the valve, thus, relieving the pressure on the top of the diaphragm. The inlet pressure then pushes the diaphragm up and off the main valve port and holds it there, allowing full flow of the fluid. When the coil is de-energized, the plunger drops and closes the pilot orifice. Pressure begins to build up above the diaphragm by means of a bleed hole in the piston diaphragm until it, plus the weight of the diaphragm and spring, cause it to close on the main valve port. This type of solenoid valve requires a minimum pressure difference between inlet and outlet in order to operate.

The manual stem as shown in figure 98b is used to manually open the valve if the line current is not available or for flushing in cleanup or other service maintenance functions.

The two-way valve, which is the most common type of solenoid valve, controls fluid flow in one line. It has an inlet and an outlet connection. This valve can be of the direct-acting or pilot-operated type, depending on the need. When the coil is de-energized, the two-way valve is normally closed. Although normally closed is the most widely used, two-way valves are manufactured to be normally open when the coil is de-energized. See figure 98c for an example of a two-way valve.

The three-way valve has a connection, which is common to either of two different outlets and controls refrigerant flow in two different lines. They are used chiefly in commercial refrigeration units for heat-reclaiming applications, hot gas defrost, and discharge-gas unloading applications. See figure 98d for an example of a three-way valve.

The three-way valve has its common inlet attached to the compressor discharge line. The other two outlets connect to the normal condenser and the auxiliary condenser as shown in figure 98e.

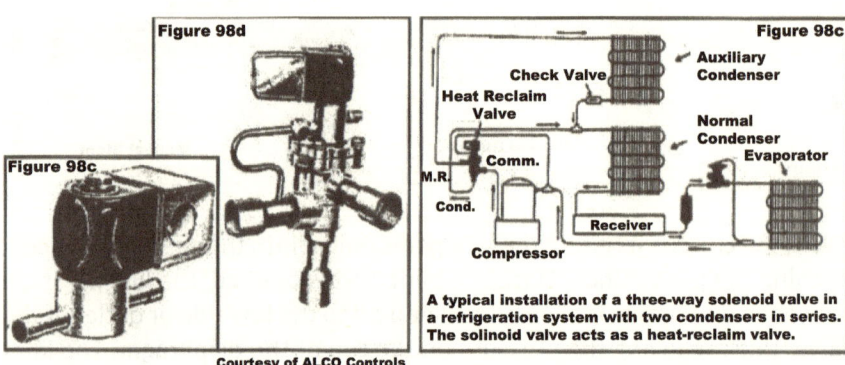

Courtesy of ALCO Controls

When the solenoid coil is de-energized, the pilot line to the suction side of the compressor is closed. Discharge gas pressure escapes through the bleed port into the top of the piston and drives it downward, closing the bottom seat. This allows discharge gas to flow to the normal or outside condenser.

Energizing the solenoid opens the pilot line to the suction side of the compressor and permits the discharge gas pressure on top of the piston to escape to the suction line. The discharge gas pressure below the piston now causes the piston to be driven upward, closing the outlet to the condenser and diverting the discharge gas to the auxiliary condenser. Some valves are available with an internal bleed, which drains the reclaim oil during normal operation.

This type of three-way solenoid valve is designed to meet the requirements of high temperatures and pressures existing in compressor discharge gas applications. It is specifically designed for discharge gas diverting in compressor unloading. Valves for compressor-unloading applications are usually designed to provide mounting directly on the compressor head. In this application, the valve is used for suction line use and is shown in a schematic diagram in figures 98h and 98i.

When the solenoid is de-energized, the pilot line to the suction side of the compressor is closed. This allows the suction gas to flow in the normal direction to the compressor as shown in figure 98g. When the solenoid is energized, the pilot port is opened. This enables the piston to be driven upward, closing off the suction line connection and permitting a reverse flow of the hot gas through the suction line to the evaporator for hot gas defrost.

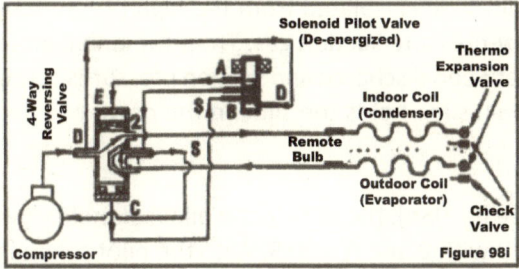

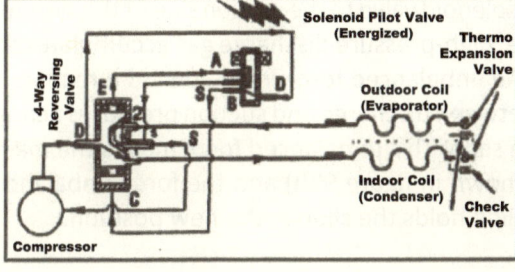

Four-way solenoid valves (often called reversing valves) are used almost exclusively on reverse-cycle heat pumps to select either the heating or cooling mode depending upon requirements. These valves have one common inlet and three outlets. Illustration 103 is a picture of a four-way valve.

Heat pumps and reversing valves should be increasing in volume in years to come since they conserve energy. A heat pump is a central air conditioner (or window unit) with a reverse cycle for heating. In the summer, the refrigerant absorbs heat from the house and exhausts it outdoors. In winter, the cycle is reversed with the refrigerant absorbing heat from outdoors and releasing it inside the house. The following is a detail of how the reversing valve operates.

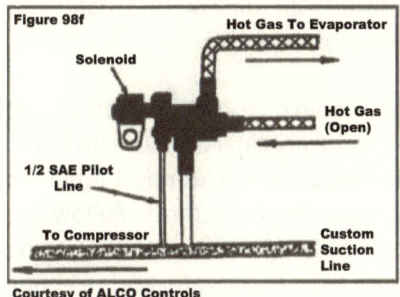

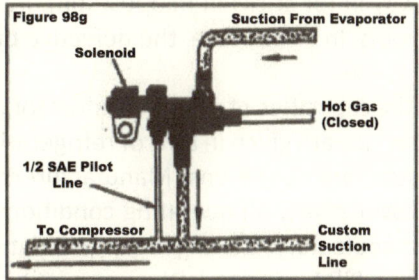

Figures 98h and i show a schematic diagram of a four-way valve on a typical reverse-cycle heat-pump system. In figure 98i, the system is on the heating cycle with discharge gas flowing through reversing valve ports D to 2, making the indoor coil the condenser. The suction gas flows from the outdoor coil (evaporator) through reversing valve ports 1 to S and back to the compressor. With the four-way solenoid pilot de-energized, the slide is positioned so as to connect ports D with A, and B with S. When the pilot is de-energized, high-pressure discharge gas builds up on top of the main slide. The area below the main slide is isolated from the high pressure by C-cup seal and exposed to low-pressure suction gas. Thus, the unbalanced force, due to the difference between discharge and suction pressures acting on the full end area of the main slide holds the slide in the *down* position as shown in figure 98i.

When the coil is energized (see fig. 98h), the slide in the pilot solenoid valve raises, now connecting ports D with B and A with S. With the pilot solenoid so positioned, the discharge pressure imposed on the top of the main slide area E flows through the pilot solenoid valve to the suction side of the system. At the C end of the main slide, high-pressure discharge gas accumulates so as to increase the pressure. An unbalanced force in an upward direction is again due to the difference between discharge and suction pressures acting on opposite ends of the main slide. This unbalanced force moves the main slide to the *up* position (as shown in figure 98h) and the force unbalance across the area of the main slide holds the slide in the new position.

Depending on the design requirements of the manufacturer, some reversing valves produce the cooling cycle when they are energized, and some produce it when they are de-energized.

When the solenoid in a reversing valve fails, the valve will hold the unit only in the cycle it takes the unit to upon being energized. Sometimes, due to an internal short in the winding, the solenoid loses the ability to pull all the way in (or sticks in midposition), in which case the unit neither heats nor cools. In either case, the defective part must be replaced.

The capacities of solenoid valves for normal liquid or suction gas refrigerant service are given in tons of refrigeration at some nominal pressure drop and standard conditions. Manufacturers' catalogs provide extended tables to cover nearly all operating conditions for common refrigerants. Follow the manufacturer's sizing recommendations. Do not select a valve based on line size. Pilot-operated valves require a pressure drop to operate, and selecting an oversize valve will result in the valve failing to open. Undersized valves result in excessive pressure drops.

The solenoid valve selected must have a MOPD (maximum operating pressure differential) rating equal to or in excess of the maximum possible differential against which the valve must open. The MOPD takes into consideration both the inlet and outlet valve pressures. If a valve has a 500 psi inlet pressure and a 250 psi outlet pressure and an MOPD rating of 300 psi, it will operate since the difference (or 500 psi -250 psi) is less than the 300 psi MOPD rating. If the pressure difference is larger than the MOPD, the valve will not open.

Consideration of the safe working pressure (SWP) required is also important for proper and safe operation. A solenoid valve should not be used for an application when the pressure is higher than the safe working pressure. Solenoid valves are designed for a given type of fluid so that the materials of construction will be compatible with that fluid. Steel or ferrous metals and aluminum are used in solenoid valves for ammonia service. Special seat materials and synthetics may be used for high-temperature or ultra-low-temperature service. Special materials are required for corrosive fluids.

Special attention to electrical characteristics is also important. Required voltage and hertz must be specified to ensure proper selection. Valves for DC service often have different internal construction than valves for AC applications, so it is important to study the manufacturer's brochure carefully.

Solenoid valves having a spring-loaded plunger or diaphragm may be installed and operated in any position; however, the older style conventional solenoid valve with a plunger, which depends on gravity to close, must always be installed with the plunger in an upright, vertical position with the pipe horizontal. An adequate strainer or filter-drier should be installed ahead of each solenoid valve to keep scale, pipe dope, solder, and other foreign matter out of the valve.

When installing a solenoid valve, be sure the arrow on the valve body points in the direction of refrigerant flow. When brazing solder-type connections, do not use a torch that is too hot and point the flame away from the valve. Allow the valve body to cool before replacing the valve inner parts to ensure that the seat material and gaskets are not damaged by the heat. Wet rags and/or chill blocks are recommended during brazing. They are necessary to keep the valve body cool so that body warpage and close coupled valves will not occur. When reassembling, do not overtorque.

## OIL SEPARATORS

The discharge of refrigerant from the compressor causes some small amount of compressor oil to be circulated with it. In heavier commercial units, the amount may be great enough to cause damage to the compressor if it runs with insufficient lubrication. This oil must be collected and returned to the compressor to maintain proper lubrication. An oil separator is installed between the compressor and the condenser. The separator screens out and collects the oil, and when it reaches a certain level, the float mechanism permits the oil to return to the compressor. This happens because the pressure in the oil separator is considerably higher than the pressure in the compressor crankcase. It is returned to the compressor through the oil return line connected directly to the compressor crankcase (see fig. 99). Some types of separators are serviceable in which case screens and/or elements can be replaced without having to replace the unit.

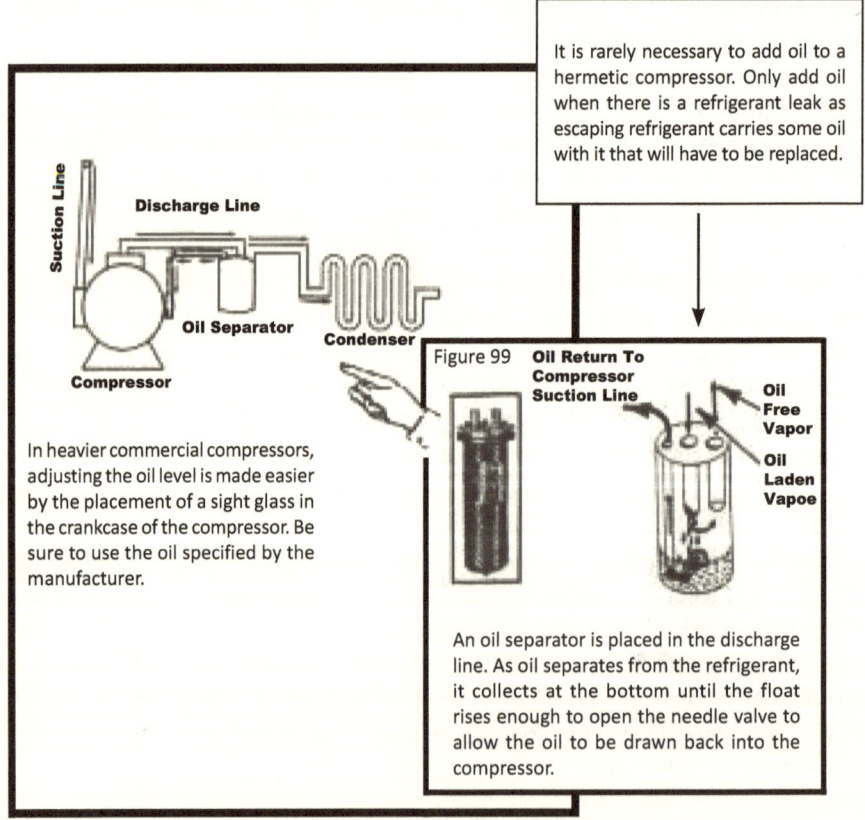

It is rarely necessary to add oil to a hermetic compressor. Only add oil when there is a refrigerant leak as escaping refrigerant carries some oil with it that will have to be replaced.

In heavier commercial compressors, adjusting the oil level is made easier by the placement of a sight glass in the crankcase of the compressor. Be sure to use the oil specified by the manufacturer.

Figure 99

An oil separator is placed in the discharge line. As oil separates from the refrigerant, it collects at the bottom until the float rises enough to open the needle valve to allow the oil to be drawn back into the compressor.

## COMPRESSOR SERVICE VALVES

Commercial refrigeration units are equipped with permanent service (access) valves, which make the service technician's job much easier. With these valves already in place, installation of piercing valves becomes unnecessary. Some units come with a valve installed directly on the compressor suction opening (see figs. 100 and 118) and one connected to the receiver (see figs. 101 and 118). Many bolted-type compressors are equipped with suction and discharge valves connected to the compressor housing (fig. 120). The valves seal off the system (by turning the valve stems all the way clockwise) and trap the refrigerant in the sealed system. By removing the bolts connecting the valves to the compressor housing, they can be removed and reconnected to the new compressor in a few minutes without breaking into the sealed system and having to do the regular procedures of evacuating, recharging, and installing a new filter-drier. In most belt-driven compressors (see figs. 31 and 32), these valves are also mounted directly on the high- and low-side ports of the compressor.

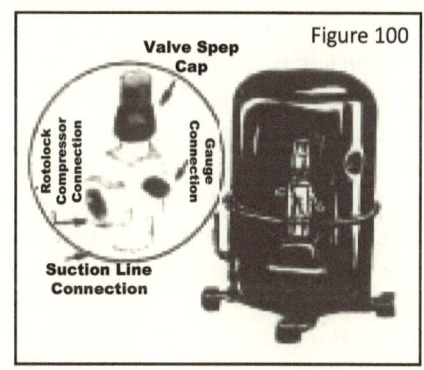

Figure 100

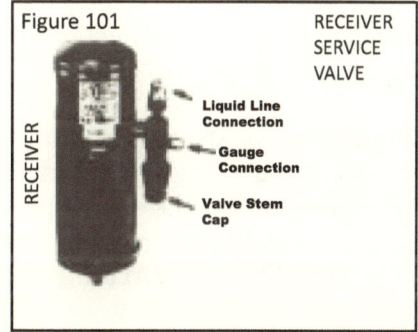

Figure 101 — RECEIVER SERVICE VALVE

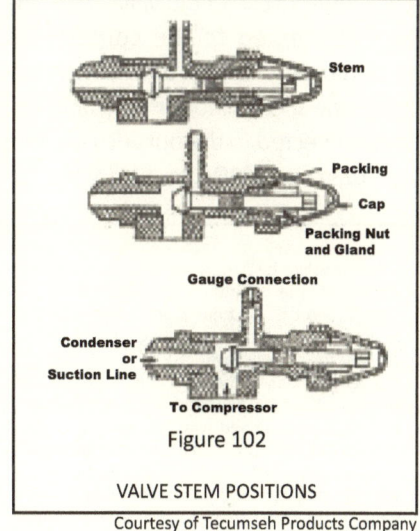

Figure 102

VALVE STEM POSITIONS

Courtesy of Tecumseh Products Company

To gain access to the system,

1. remove each valve stem cap and each gauge line access cap;
2. connect the compound gauge to the low-side valve and the high-pressure gauge to the high-side valve; and
3. using a service wrench, turn the valve stems to open the lines to the gauges.

In figure 102, you can see the different positions of the stem. When the stem is all the way in, the condenser or suction line is blocked, depending on whether the valve is installed on the suction or discharge line. When the stem is turned counterclockwise all the way out, the gauge connection is blocked. This is the position it must be in when disconnecting the gauges.

When the stem is in its midposition, the path is opened to the suction or discharge line and to the gauge manifold as well as the compressor. So then, every port is open.

When the stem is all the way in, it closes the suction or discharge line and leaves the compressor and gauge ports open.

## REVERSING VALVE

Reversing valves are used in heat pumps. It is a four-port solenoid valve with one port connected to the compressor suction tube, one port to the discharge tube, one to the indoor coil, and one port connected to the outside coil. (See figs. 103a and b, 104, and 105.)

Solenoid valves become inoperative due to either electrical or mechanical malfunction. If due to an internal short or disconnection a valve becomes incapable of creating enough magnetism to lift the plunger, the coil will have to be replaced. (Remove the screws holding the coil.) Any disconnection in the coil can be detected when the two coil terminals are touched with the two ohmmeter probes. (Set scale on RX1.) If the valve develops a leaky seat causing the needle to stick or chatter, the valve must be replaced. Most solenoid valves will not operate unless they are in a vertical position and right side up. The voltage rating must be compatible with the replacement valve.

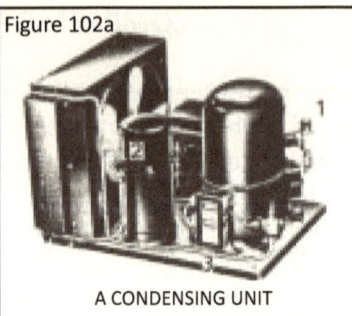

Figure 102a

A CONDENSING UNIT

1. Compressor suction-line access valve
2. Liquid receiver access valve (high side)
3. Pressure control

*Courtesy of Tecumseh Products Company*

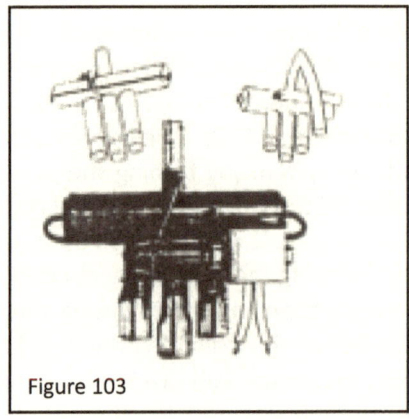

Figure 103

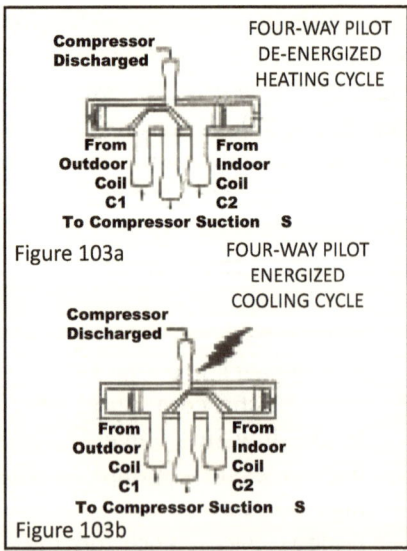

Figure 103a

Compressor Discharged

FOUR-WAY PILOT DE-ENERGIZED HEATING CYCLE

From Outdoor Coil C1
From Indoor Coil C2
To Compressor Suction  S

Figure 103b

Compressor Discharged

FOUR-WAY PILOT ENERGIZED COOLING CYCLE

From Outdoor Coil C1
From Indoor Coil C2
To Compressor Suction  S

*Courtesy of ALCO Controls*

## REVERSING VALVE

When this solenoid valve is de-energized, the unit is in the heating cycle which means that the indoor coil acts as a condenser and the outside coil acts as an evaporator. (See fig. 103a.)

The discharge gas flows through ports D and C2, making the indoor coil the condenser.

The suction gas flows from the outdoor coil (evaporator) through reversing ports C1 to S and back to the compressor. When the coil is energized (see fig. 103b), the unit changes over to the cooling cycle with the discharge gas flowing through reversing valve ports D to C1, making the outdoor coil the condenser, and the suction gas flows through port C2 to S, thus making the indoor coil the evaporator. Figure 105a shows a typical wiring of a reversing valve in a heat pump that uses line voltage. Figure 105b shows a typical heat pump circuit using a 24 V contactor, reversing valve, and thermostat.

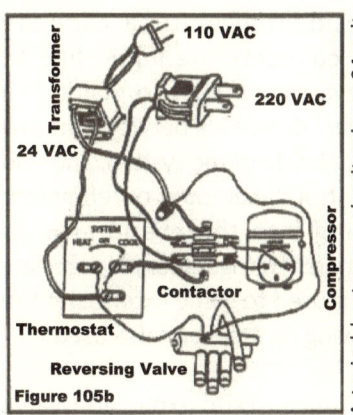

Figure 105b

A typical heat pump circuit using a 24-volt contactor coil, reversing valve, and thermostat.

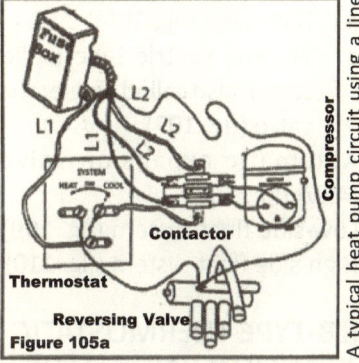

Figure 105a

A typical heat pump circuit using a line voltage contactor coil, reversing valve, and thermostat.

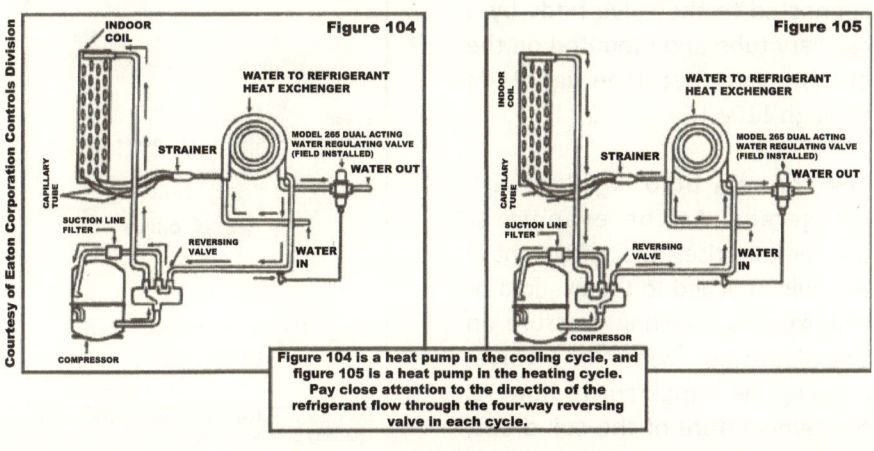

Figure 104 is a heat pump in the cooling cycle, and figure 105 is a heat pump in the heating cycle. Pay close attention to the direction of the refrigerant flow through the four-way reversing valve in each cycle.

## REFRIGERANT CONTROLS USED IN COMMERCIAL REFRIGERATION

In residential units, a capillary tube connects the liquid line to the evaporator inlet to maintain a pressure difference between the low and high side of the system, and also to control the amount of refrigerant flowing into the evaporator. In commercial refrigeration, the capillary tube is used to some extent, but the following controls are also used:

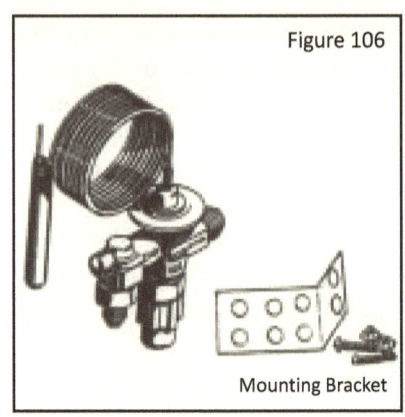

Figure 106

Mounting Bracket

1. Thermostatic expansion valve (TEV)
    a. Bulb type (fig. 106)
    b. Thermoelectric type (solid state), also called an electric valve (fig. 107)
2. Automatic expansion valve (AEV) (fig. 108)
3. Low-side float system (fig. 109)
4. High-side float system (fig. 110)

### BULB-TYPE THERMOSTATIC EXPANSION VALVE (TEV)

It is installed on the inlet of the evaporator with a sensing bulb connected to the valve body by a capillary tube and mounted on the evaporator outlet. (See figs. 110a through 110e.)

The thermal bulb is filled with refrigerant. As the evaporator temperature rises, the refrigerant in the bulb attached to the evaporator coil expands, exerting pressure on the diaphragm within the valve, allowing the refrigerant to flow. As the temperature of the coil drops,

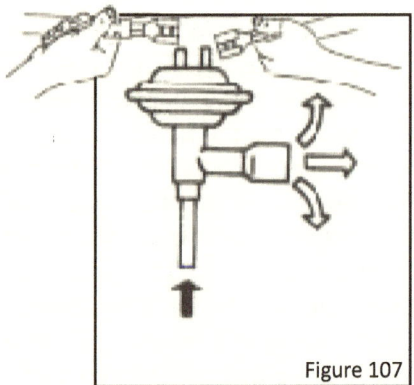

Figure 107

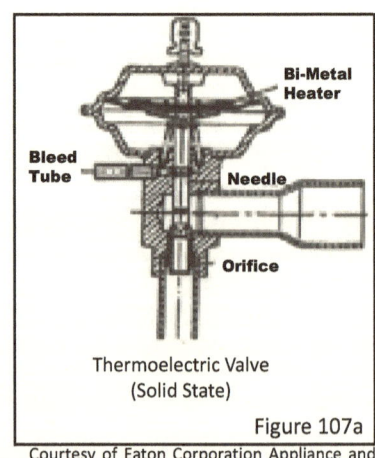

Thermoelectric Valve
(Solid State)

Figure 107a

Courtesy of Eaton Corporation Appliance and Specialty Controls Division the refrigerant in the bulb contracts and reduces pressure on the diaphragm, which closes the valve. The valve is spring loaded, so the spring tension and the pressure in the evaporator also work to shut off the valve. These three forces control the performance of the thermostatic expansion valves. (See figs. 110a and 110b).

Force F1 is applied to the valve diaphragm by the expansion of the refrigerant in the sensing bulb which tends to open the valve. F2 is the force applied by the evaporator pressure which tends to close the valve. F3 is the spring tension which acts to close the valve.

If force F1 is greater than forces F2 and F3, the valve will open. If forces F2 and F3 are greater than F1, the valve will close.

Both the temperature perceived by the sensing bulb, as well as the pressure inside the evaporator, work together to control the flow of refrigerant and, consequently, the temperature of the unit.

On the bottom of most TEV valves, there is an adjusting stem. To gain access to the stem, first remove the seal cap covering the stem. Use a service wrench to turn the stem clockwise to decrease the flow of refrigerant into the evaporator, and counterclockwise to increase the flow. Turn this valve stem only one-fourth turn at a time, then wait for about five minutes for a partial frost pattern to appear on the evaporator coil. A "hissing" sound from the evaporator while the

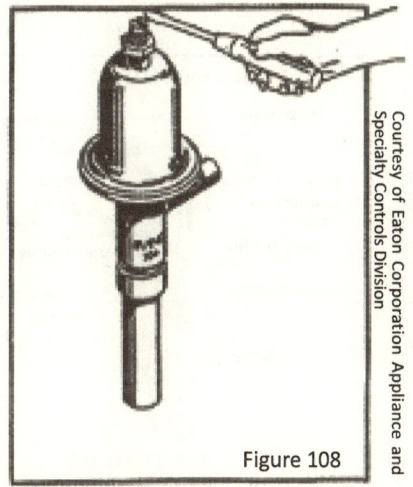

Figure 108

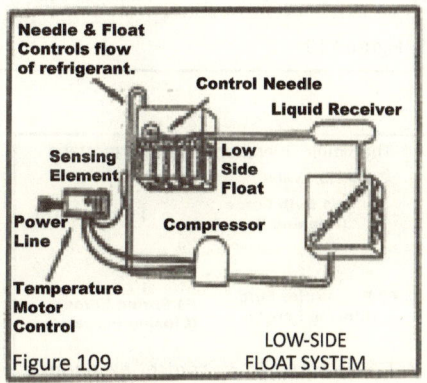

Figure 109 — LOW-SIDE FLOAT SYSTEM

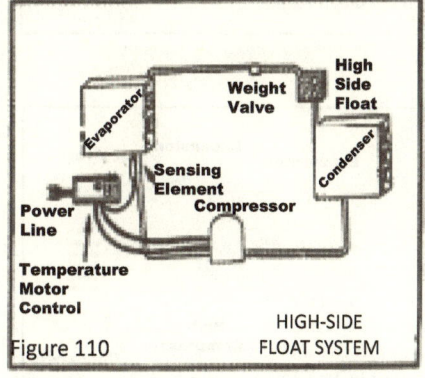

Figure 110 — HIGH-SIDE FLOAT SYSTEM

unit is turned on indicates a starved evaporator. In this case, turn the stem counterclockwise as described above.

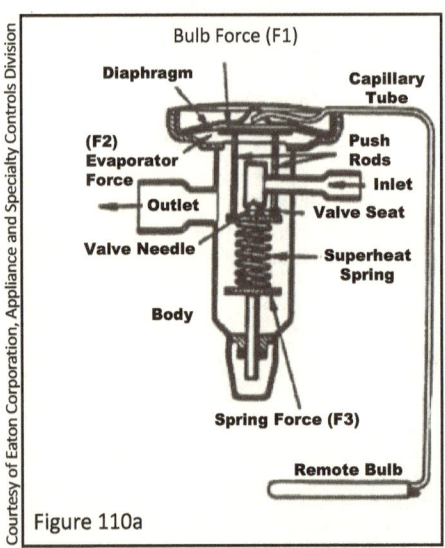

Figure 110a

The Simple Operation of the Thermostatic Expansion Valve

Figure 110b

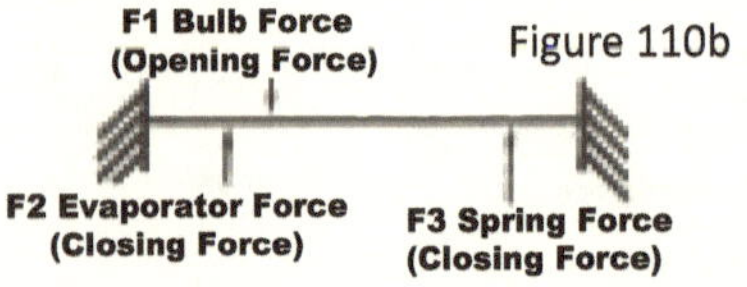

F1 Bulb Force (Opening Force)

F2 Evaporator Force (Closing Force)   F3 Spring Force (Closing Force)

A TEV is designed to control evaporator superheat by the reaction of the three forces on the diaphragm.

On replacement of any valve, it is most important to obtain an exact duplicate, (one with the same specifications). Just as containers for different types of refrigerant are color-coded, the majority of TEVs are color-coded indicating the type of refrigerant they can handle. If they are not color-coded, they have a label or tag giving the specifications of the unit for which they are suited. The label shows the type of refrigerant and the tonnage of the unit.

Valves can be obtained with flared or brazed connections. For those with flared connections, be sure they are fastened tight enough to prevent Freon leakage, but not too tight to damage the flare. Use Teflon tape on these connections.

The bulb is attached to the evaporator outlet by a couple of screws. It is important not to change the location of this bulb when a new one is installed. (See figs. 110c through 110e.)

Be sure not to bend the bulb line sharply during installation; otherwise, it may inhibit the free flow of refrigerant.

The bulb should be seated on a straight section of the last portion of the evaporator line, not on a curve.

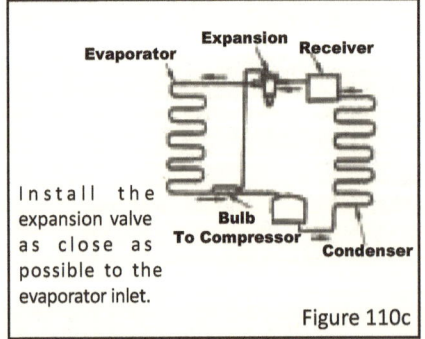

Figure 110c

The temperature on a curve is slightly different because the pressure produced at that particular point is different and produces a false temperature.

With the unit running, liquid refrigerant passes through the TEV and gets into the evaporator where it immediately absorbs heat and boils off (becomes superheated vapor). Superheated refrigerant, due to the absorption of heat, loses a great deal of its cooling capacity. A properly adjusted valve admits the correct amount of liquid refrigerant into the evaporator to produce the desired evaporator temperature (by absorbing heat and becoming superheated), and the least amount of refrigerant leave the evaporator in liquid state.

The sensing bulb is installed at the outlet of the evaporator. If all of the refrigerant becomes superheated before leaving the evaporator, the bulb senses the higher temperatures and transmits signals to the valve for more refrigerant. The valve opens wider and admits more refrigerant to a point where all of the refrigerant within the evaporator coil will not get a chance to boil off and become superheated, and a portion of the refrigerant remains in the liquid state while leaving the evaporator at which time the bulb senses the lowered temperatures due to the passage of this liquid refrigerant and transmits a signal to the valve to limit the flow of refrigerant into the evaporator. This cycle continues as long as the unit runs.

Since only superheated refrigerant (with very little heat-absorbing capacity) should reach the TEV sensing bulb, the temperature of the bulb is always above the actual evaporator temperature. This temperature difference is called superheat. The best way to adjust this type of valve is to do it by superheat adjustment.

Here is a general rule of thumb for superheat adjustment:

**A GENERAL RULE OF THUMB FOR SUPERHEAT ADJUSTMENT:**

1. 10°F to 12°F: Superheat for high-temperature evaporators. (30°F+ evaporators.)
2. 5°F to 10°F: Superheat for medium-temperature evaporators. (0°F to 30°F evaporators.)
3. 2°F to 5°F: Superheat for low-temperature evaporators. (0°F and below evaporators.)

Turn the valve stem *counterclockwise* to *decrease* the superheat (increasing the refrigerant flow to the evaporator).

Turn the valve stem *clockwise* to *increase* the superheat (decreasing the flow of refrigerant to the evaporator).

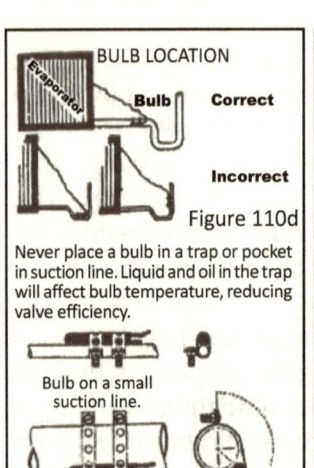

**BULB LOCATION**

Bulb — Correct

Incorrect

Figure 110d

Never place a bulb in a trap or pocket in suction line. Liquid and oil in the trap will affect bulb temperature, reducing valve efficiency.

Bulb on a small suction line.

Bulb on a large suction line.   Figure 110e

On suction lines 7/8" OD and larger, install the bulb at about a four o'clock position on the suction line.

Figure 110f

○ Superheated Vapor Refrigerant
● Liquid Refrigerant
◐ Superheated Vapor Refrigerant
◑ Liquid Refrigerant

**REFRIGERANT PHYSICAL CHANGES IN THE EVAPORATOR COIL**

At A, hot, high-pressure liquid refrigerant enters the TEV. At B, cold, low-pressure liquid plus *flash gas* enters the evaporator. At C, all of the liquid refrigerant is boiled off, or vaporized by the heat load (latent heat).

Between C and D, the vapor temperature increases dramatically as further heat load is applied (sensible heat). At this point, the gas is superheated above its saturation temperature. At D, suction-line temperature of the superheated gas is monitored by the sensing bulb which signals the TEV to open or close accordingly. Good TEV performance depends on the accurate adjustment of the superheat, so accurate measurement is vital.

*Courtesy of ALCO Controls*

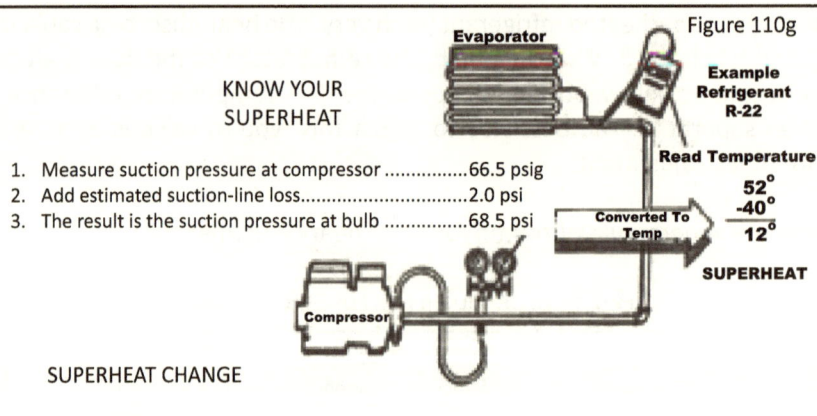

Figure 110g

**KNOW YOUR SUPERHEAT**

Example Refrigerant R-22

1. Measure suction pressure at compressor ................. 66.5 psig
2. Add estimated suction-line loss................................ 2.0 psi
3. The result is the suction pressure at bulb ................. 68.5 psi

Read Temperature
52°
Converted To Temp   -40°
12°
**SUPERHEAT**

**SUPERHEAT CHANGE**

If superheat on the TEV is readjusted, the result will be the following:

A. Increasing superheat will reduce system capacity.
B. Decreasing superheat will add to system capacity.

To increase valve superheat, turn adjusting stem clockwise. To decrease valve superheat, turn adjusting stem counterclockwise.

## HOW TO ACCURATELY MEASURE SUPERHEAT IN A SYSTEM USING A TEV WITH AN EQUALIZER

> Step A. Determine the suction pressure at the evaporator outlet with an accurate gauge. If there is no gauge connection, a tee installed in the valve external equalizer line can be used.
>
> Step B. Refer to the temperature-pressure chart (p. 133) for the refrigerant used in the system and determine the saturation temperature at the observed suction pressure.
>
> Step C. Measure the temperature of the suction line at the remote sensing bulb location. This can be accomplished by a "strap on" thermometer or an electric device similar to an "Annie" or "Simpson" meter. Be certain the spot chosen for measurement is clean to ensure accurate readings.
>
> Step D. Subtract the saturation temperature determined in Step B from the suction gas temperature measured in Step C. The difference is the operating superheat. (See fig. 110h.)

For best results, there are two types of TEV on the market today: (1) internally equalized TEV used in regular evaporators where pressure drop inside the evaporator is not significant and (2) externally equalized TEV (see fig. 110h) used on the evaporators in which pressure drop is considerably high.

The pressure inside the sensing bulb is the only opening force in the valve working against the closing forces of the spring and evaporator pressure. When the pressure drop at the evaporator outlet (where the bulb is installed) is substantial, this reduced pressure (low temperature) reduces the opening force applied to the valve diaphragm, and the valve tends to close and starve the evaporator.

To normalize this condition, it only makes sense to reduce the closing force (evaporator pressure) applied to the diaphragm. This is done by connecting the evaporator outlet reduced pressures to the valve by a tube. This allows the low pressures from the evaporator outlet (where the opening force is also affected) to reduce the excessively high pressures that tend to close the valve.

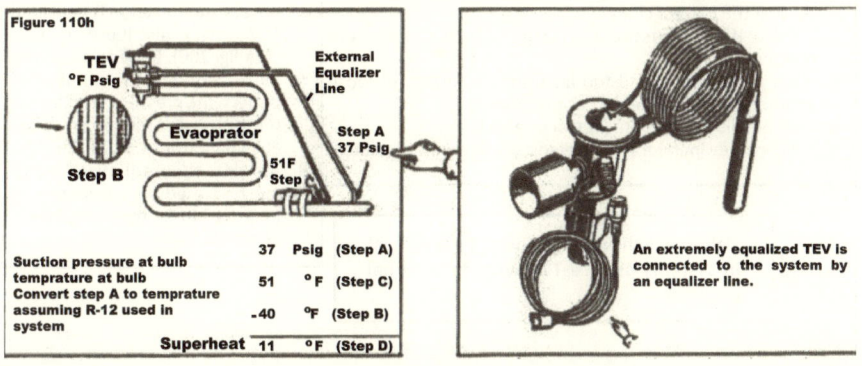

Figure 110h

Figures 110d and 110e show the ideal placement (horizontal) of the bulb in relation to suction-line size. Never put the bulb at 6 o'clock because it may sense the temperature of the oil flowing through the pipe rather than the temperature of the refrigerant. And be sure the bulb location is on a free-draining suction line.

A good way to check a suspected punctured sensing bulb is to close one hand around the bulb and its connecting line. The heat transferred to it should cause the refrigerant in it to expand and open the valve. Otherwise, the valve has to be replaced as the refrigerant in the bulb must have leaked out.

**DIAGNOSING THERMOSTATIC EXPANSION VALVES**

A STARVED EVAPORATOR WITH AN UNEVEN FROST PATTERN resulting in poor cabinet refrigeration may be due to the following:

1. Loss of refrigerant from the thermal bulb (if undependable, erratic refrigeration is evidenced).
2. Valve needle stuck shut. (This very seldom happens. The evaporator no longer cools.)
3. Clogged expansion valve screen. (Evaporator loses its cooling ability.)
4. Moisture in the system. (The evaporator cools sometimes and sometimes does not).
5. Under-capacity valve installed. (Evaporator temperature never drops to the desired point.)
6. Inside of valve covered with wax. (Poor or no refrigeration). This occurs when the wrong type of oil is used. (Different types of lubricant are used for different temperatures.)

A FLOODED EVAPORATOR AS THE RESULT OF TOO MUCH REFRIGERANT FLOW IS EVIDENCED BY A FROSTED OR SWEATING SUCTION LINE

The six most usual causes of this condition are the following:

1. Pressure drop too great in the evaporator coil. (Replace valve with one using an equalizer.)
2. Thermal bulb with wrong charge. (Replace valve with one having the correct charge.)
3. TEV orifice adjusted too large. (Turn valve stem clockwise).
4. Thermal bulb installed too far from evaporator or loose from suction line. (Correct as necessary.)
5. TEV needle stuck open. (Replace valve.)
6. Undersize evaporator. (This rarely occurs. Replace evaporator.)

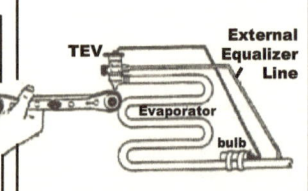

Turn valve stem 1/4 turn at a time. Then wait for about five minutes for a partial frost pattern to appear on the evaporator coil.

As a general rule of thumb, externally equalized TEVs are used where pressure drop in the evaporator is more than:

a. 3 psi through an evaporator in high temperature application,
b. 2 psi through an evaporator in a medium-temperature application,
c. 1 psi through an evaporator in a low temperature application
d. Always use an externally equalized TEV when a distributor is used (See fig. 200). Depending on the make, size and the number of outlets, the pressure drop across the distributor alone can range anywhere from 5 to 30 psi.
e. An externally equalized valve must be used on any system in excess of three tons (regardless of application).

Figure 110h shows how to measure superheat in systems using a TEV with an external equalizer.

## SOLID-STATE THERMOSTATIC EXPANSION VALVES
(Electric Valves)

Figures 107, 107a, and 111 show a solid-state type of TEV. This type of valve can be used on any system. Unlike the bulb-type TEV, the electric valve eliminates the problem of placement of the sensing bulb.

The electric valve is operated by and responds to low-voltage electricity. Its operation is simple and easy to understand.

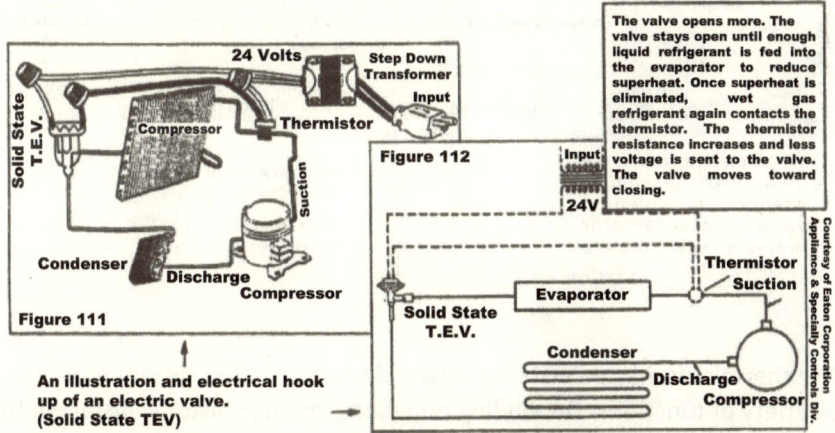

Figure 111
An illustration and electrical hook up of an electric valve.
(Solid State TEV)

A liquid-sensing thermistor (see fig. 112) is installed in the suction line at the outlet of the evaporator where a complete change of refrigerant state from liquid to gas occurs. It is wired in series with the electric valve. Here, the thermistor reacts, increasing or decreasing voltage in the valve circuit, depending on the state of the refrigerant passing through that part of the line. When it is exposed to hot refrigerant gas, the thermistor is heated to a high temperature by the voltage applied to it. The thermistor resistance drops as it self-heats. This negative-coefficient thermistor causes an increase in voltage to the bimetal heater inside the valve head.

The amount of low voltage applied to the heater bimetal within the valve controls the degree of valve opening (see figs. 208 and 209). At zero voltage, the valve is closed. As voltage is applied, the bimetal heater is deflected. The needle follows the bimetal deflection and opens the valve. The more voltage applied, the greater the valve opening.

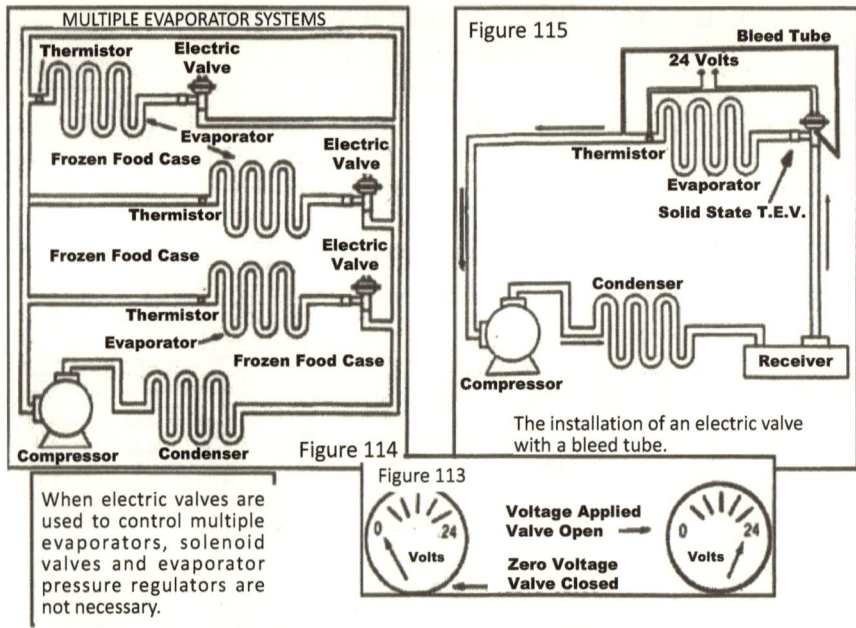

The thermistor may be installed nearly anywhere in the system to perform a variety of functions. Depending on where the thermistor is installed, the electric valve can control

* head pressure,
* maximum evaporator pressure,
* minimum evaporator pressure,
* flooded evaporator (in low-side float system),
* flooded condenser (in high-side float system) (see figs. 114 and 115).

The electric valve can make system analysis and system troubleshooting fast and easy. Service personnel need only attach a voltmeter to the electric valve. The readings obtained from the voltmeter will tell how the valve is operating at a glance.

A single check of system conditions will

* indicate valve reaction,
* identify problems elsewhere in the system.

Complete servicing details are given under the thermal electric valve troubleshooting guide in the Refrigeration Fluid Flow Controls and System Troubleshooting.

## AUTOMATIC EXPANSION VALVES

These are also referred to as constant-pressure or pressure-reducer valves (see fig. 108).

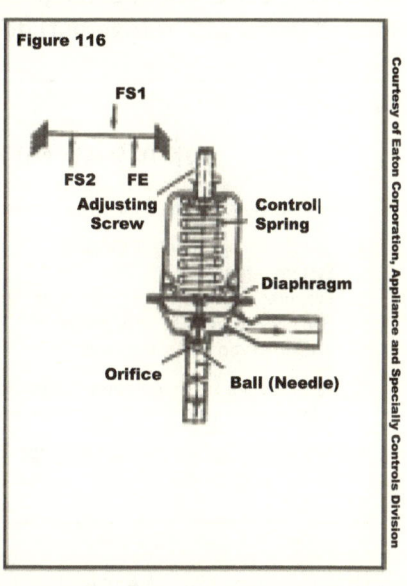

Figure 116

Because of their versatility, these valves are produced in a variety of types and sizes, and can be installed in different parts of the sealed system, then adjusted to the pressures desired. They can be used as high- to low-side bypass regulators, evaporator-pressure regulators, compressor overload protection, evaporator freeze-up protection, and also as an expansion valve for one-fourth- to three-ton-capacity room air conditioners, ice cream units, ice makers, water coolers, dehumidifiers, and central air conditioners.

These valves contain a diaphragm, control spring (FS1), seat, and the basic valve needle or ball. The control spring above the diaphragm moves the diaphragm down, this opens the valve. The opposing force is provided by low-side evaporator pressure (FE) and a constant body spring force (FS2), this moves the valve to close.

During the off cycle, evaporator pressure builds up and overcomes the spring pressure. This keeps the valve closed until the next cycle. Then the compressor quickly reduces evaporator pressure. When this pressure equals the control-spring pressure, the valve begins to open. The valve opens when evaporator pressure falls below the control-spring setting. This is the point, or setting, at which the valve opens. (See fig. 116.)

When it is used to control the evaporator temperature, it is mounted on the liquid line at the inlet of the evaporator to control the amount of refrigerant entering the evaporator. By operating only on the low-side pressure, it is activated when the evaporator pressure drops. At that time, the needle valve automatically opens and sprays refrigerant into the evaporator until the evaporator pressure rises to a predetermined point. This can only occur when the compressor is running and creating low pressure in the evaporator. When the compressor is stopped, and no more suction is applied to the low side, the pressure built up in the evaporator prevents the needle valve from opening. This prevents the evaporator from being flooded (filled with refrigerant) during the off cycle.

---

**HOW TO SELECT AUTOMATIC EXPANSION VALVES**

Determine the following data: (They can be found on the unit nameplate).

1. Btu or tonnage rating of the unit
2. System refrigerant
3. Evaporator temperature or pressure (Determine the expected evaporator temperature.)
4. High-side temperature (or pressure)
5. Pressure drop across the valve

These valves are set at a predetermined pressure and sealed with a lock nut. Loosen nut to readjust. Heat the screw with a heat gun to loosen for readjustment.

For servicing AEVs, see the double-temperature valve pages.

---

In this system, the thermostat sensing bulb is attached to the outlet of the evaporator. When sufficient drop in temperature is sensed by the bulb, the contacts within the thermostat mechanism open, causing power to the compressor to be disrupted, and the cooling cycle stops (see fig. 117).

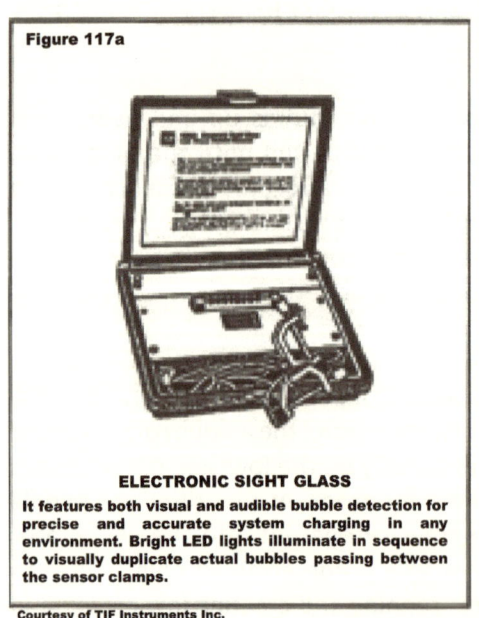

**Figure 117a**

**ELECTRONIC SIGHT GLASS**

It features both visual and audible bubble detection for precise and accurate system charging in any environment. Bright LED lights illuminate in sequence to visually duplicate actual bubbles passing between the sensor clamps.

Courtesy of TIF Instruments Inc.

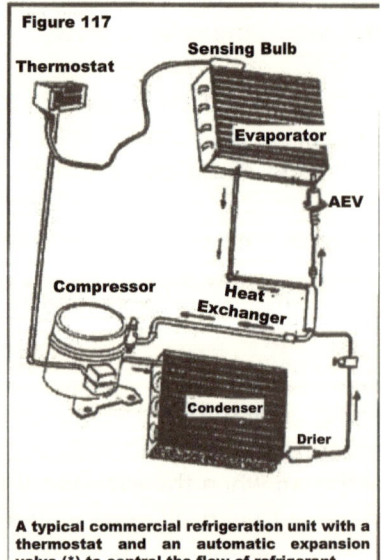

**Figure 117**

A typical commercial refrigeration unit with a thermostat and an automatic expansion valve (*) to control the flow of refrigerant

## LOW-SIDE FLOAT SYSTEM (Fig. 109)

Although the technical schools and many refrigeration manuals refer to high- and low-side float systems, units using these systems are very seldom encountered. The low-side float system was more popular in the early years of mechanical refrigeration. It is also called a "flooded system." This low side float mechanism is a part of the evaporator. The vapor refrigerant is drawn through the suction line to the compressor then discharged as a high-pressure gas into the condenser where it is cooled and changed to its liquid state. Liquid refrigerant flows into the evaporator through a valve that operates in conjunction with a float. (The lower portion of the evaporator is occupied by liquid refrigerant, while the portion above the float is filled with vapor.) When the desired low temperature is reached, the compressor is de-energized and this operation stops.

Flooded systems are easy to service and are very efficient because the cold liquid refrigerant wets the surface of the evaporator, causing a rapid heat transfer. These systems require a motor that can start under heavy load because the pressures do not equalize during the off cycle. They also take a larger than average refrigerant charge because liquid refrigerant occupies both the liquid receiver tank and the evaporator.

About the only service these units need is the replacement of the needle valve.

## HIGH-SIDE FLOAT SYSTEM (Fig. 110)

Since in this system the evaporator is always filled with liquid refrigerant, this is also considered a flooded system.

During the run cycle, liquid refrigerant is forced into the high-side float chamber through the condenser. As soon as enough liquid refrigerant enters the float chamber, the float ball moves up allowing the refrigerant to flow to the evaporator. Because the evaporator is under low pressure, the line entering it should be insulated. Often a capillary tube is used on the line connecting the high-side float chamber and the evaporator. If not, a weight valve should be installed to prevent the liquid from evaporating in that connecting line.

Refrigerant enters the evaporator under low pressure, evaporates and absorbs heat from the evaporator. It then flows through the suction line into the compressor where it is forced out under high pressure into the condenser again. The condenser then removes the heat absorbed in the evaporator, changing the vapor refrigerant back into its liquid state before it is forced into the float chamber to repeat the cycle. NOTE: The amount of the refrigerant charge in a high-side float system must be measured very accurately for the proper operation of the unit.

Liquid receiver tanks are used on all units with a high-side float system. Units with capillary tubes use an accumulator (installed at the outlet of the evaporator). The motor-control, temperature-sensing element is installed on the coil just before the accumulator on the suction line.

Liquid receiver tanks are used on all units with a high-side float system or systems using an expansion valve(s). Systems with capillary tubes use an accumulator, which is installed at the outlet of the evaporator. The motor-control, temperature-sensing element is installed on the coil just before the accumulator on the suction line.

Figure 118

A condensing unit typical of those used in commercial walk-in coolers or freezers, open or closed display cases, salad bars, beverage coolers, ice machines, large water coolers, ice cream and frozen yogurt machines, and food preparation bars. These are placed inside self-contained units or in back of the building or on the roof for split units. A pressure control (not shown) is normally located on the unit. (1) Compressor (2) Start Capacitor (3) Control Box where the starting relay and line wire connections are located. (4) Receiver (5) High-side Access Valve (6) Low-side Access Valve (7) Condenser Coil (8) Condenser Fan (9) Compressor Terminal Cover (10) Compressor Nameplate. (11) the nameplate for the condensing unit.

### CAUSES OF COMPRESSOR SHORT-CYCLING

Short-cycling means that a refrigeration unit starts and stops much more frequently than it should.

The most common causes of compressor short-cycling are (1) a weak overload protector, (2) low line voltage, (3) a dirty or linted condenser, (4) a defective condenser fan, (5) a bad pressure control (in commercial units), (6) a shorted start capacitor, (7) pressure control reaction to a restriction in the sealed system of commercial units, (8) the cut-in/cut-out set too close on the pressure control or thermostat, (9) a defective compressor, (10) loose connections—particularly at the compressor terminals, (11) an overcharged system, (12) an undercharged system (particularly one that uses a pressure control), (13) extension cord too long and/or too light for the electrical requirements of the unit, (14) a refrigerant leak in systems using pressure controls, and (15) a defective starting relay.

## SHORT-CYCLING AND A BAD STARTING RELAY

If a starting relay does not disconnect the power from the compressor start winding, the compressor will keep running on its start winding until it overheats and the overload protector shuts off the power (within four seconds). Once the compressor cools, the contacts within the overload protector close, causing the compressor to restart at short intervals. If short-cycling (cycling on overload) continues, the compressor start winding may overheat and burn. If the contacts within the starting relay don't close, the start winding will not become energized and the compressor will never start. In either case, the starting relay will have to be replaced.

## SHORT-CYCLING AND PRESSURE CONTROLS

A high-pressure control is wired in series with the compressor common terminal, and it is also connected to the discharge line. Its main function is to shut down the unit when the pressure in the discharge line rises to a dangerous level. The cut-in and cut-out settings on the control are manually adjustable. When the pressure goes higher than a preset cut-out point, a switch within the control snaps and shuts off the power to the compressor.

If the cut-in and cut-out settings are adjusted too close, the unit will short-cycle. If the unit is overcharged or if a restriction occurs in the sealed system, the discharge line pressure will rapidly rise higher than normal, causing the high-pressure control to react and shut down the system.

During the off cycle, the pressures in the sealed system tend to equalize (the head pressure decreases and the low-side pressure increases). As soon as the discharge line pressure drops to a cut-in point, the switch within the pressure control snaps closed and restores power to the compressor, causing the unit to resume operation. As long as this happens at short intervals, the short-cycling continues. (See the pages concerning pressure controls for more details.)

In a particularly restricted system, the unit will short-cycle as a result of the reaction of the low-pressure control to the lower-than-normal pressure in the suction line. The compressor keeps drawing from the suction line. It creates partial vacuum in the system from the restriction and keeps pumping into the discharge line against the restriction creating above normal pressure in the discharge line and the condenser.

Another reason for short-cycling in a system equipped with a low-pressure control is the occurrence of general low pressures in the suction line as a result of an undercharged system (or a leak in the system causing a loss of refrigerant). In this case, as soon as power is restored to the compressor, the lower-than-normal pressure in the suction line drops to the cut-out point causing the low-pressure control to react and shut down the system. Thus, short-cycling continues as long as the unit operates under this condition.

A low-pressure control is connected to the suction line to shut off power to the compressor as soon as the pressure in the suction line drops below its predetermined setting. As soon as the compressor is de-energized, the evaporator warms and pressure in the suction line increases, causing the bellows within the control to expand and restore power to the compressor. Sometimes the cut-in and the cut-out settings on the control are adjusted too close, In which case, as soon as the compressor is activated, the pressure in the system drops to the cut-out point and the low-pressure control shuts the unit off in short intervals. In this case, the settings must be recalibrated. (See fig. 93).

## SHORT-CYCLING AND A LINTED CONDENSER OR DEFECTIVE CONDENSER FAN MOTOR

When the condenser gets linted or when the condenser fan becomes sluggish or totally inoperative, the head pressure (high-side pressure) goes up because the condenser is no longer able to dissipate heat from the refrigerant into the surrounding air. The unit loses its cooling capacity and runs continuously as the temperature control never becomes satisfied. The compressor overheats to the point where the overload protector, sensing the abnormal heat, shuts it down until it cools enough to allow a restart. A dirty condenser is a common cause of compressor short-cycling (cycling on overload).

## SHORT-CYCLING AND A BAD START CAPACITOR

A start capacitor increases the voltage to the compressor start winding causing the compressor to run fast enough to reach the 75% of its normal rpm during the start-up. Compressors operating with a start capacitor will cycle on overload if the capacitor becomes inoperative. When this happens, the capacitor is no longer able to provide the compressor with the necessary voltage to speed up its initial rpm. The compressor start winding will never bring the unit to its initial start-up rpm required for its run winding to take over. The start winding remains in operation until it overheats causing the overload protector to shut off the compressor power. When it cools enough, power to the compressor is restored by the overload protector bimetal, the compressor start winding becomes energized without being able to provide the motor with adequate initial speed. It then heats up and the short-cycling (cycling on overload) continues.

## SHORT-CYCLING AND A BAD COMPRESSOR

Motors draw up to 600% more current at the instant of starting. This increased starting current tends to overheat the start winding if the motor cannot reach its operating rpm within 3½ seconds to permit the motor run winding to take over. If a short or disconnection in the motor run winding occurs, the compressor motor will continue running on the start winding and very rapidly overheat, causing the overload protector to open the circuit. When the overload protector bimetal cools, it will close the electrical circuit to the compressor until it senses the high temperatures again, and the cycling on overload continues. Also, a binding or stuck compressor rotor causes too much current draw and consequently cycles on overload.

## SHORT-CYCLING AND A BAD OVERLOAD PROTECTOR

When an overload protector cracks or gets weak, it loses its ability to transmit power long enough for the compressor to reach its initial speed. As soon as the overload protector warms, it disrupts the power, causing the unit to short-cycle.

Under normal conditions, a motor compressor operates at 125°F (52°C). When the temperature rises to about 225°F, the overload protector opens the circuit to de-energize the compressor motor. When the motor cools to about 160°F, the bimetal in the overload protector flexes and closes the circuit to energize the compressor motor again. In many commercial compressor motors, a solid-state overload protector is placed within the compressor and connected in the circuit. This type of overload protector is referred to as the thermistor type. Hermetic compressor motors that are equipped with internal overload protectors are replaced when the overload protector becomes defective.

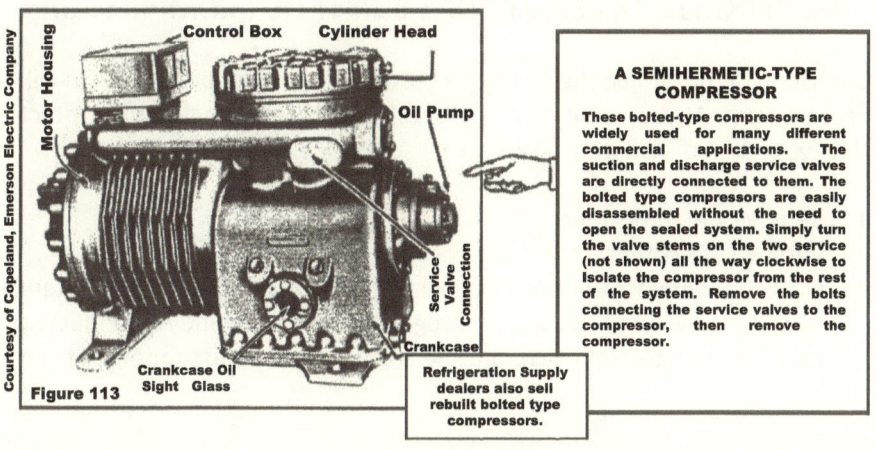

Figure 113

**A SEMIHERMETIC-TYPE COMPRESSOR**

These bolted-type compressors are widely used for many different commercial applications. The suction and discharge service valves are directly connected to them. The bolted type compressors are easily disassembled without the need to open the sealed system. Simply turn the valve stems on the two service (not shown) all the way clockwise to isolate the compressor from the rest of the system. Remove the bolts connecting the service valves to the compressor, then remove the compressor.

Refrigeration Supply dealers also sell rebuilt bolted type compressors.

## COMPRESSOR BURNOUT

Compressors often burn out by overheating caused by above-normal head pressures. Laboratory tests have proven that presence of air, dirt, and moisture in the sealed system contribute to this condition. Compressors are cooled by bypassing cooled oil mixed with vapor refrigerant returning from the evaporator over the windings.

When the compressor operates while the system is undercharged (or out of charge), insufficient cooling of the compressor motor can cause a compressor burnout. Compressor motors can also overheat and burn out by excessive current flow through the windings caused by low voltage supply or a binding compressor.

If the temperature of the circulating oil (mixed with refrigerant) leaving the compressor discharge valve rises to 350°F (177°C), it will break down and form hydrochloric and hydrofluoric acid, sludge, and varnish. The acid will corrode the insulation on the motor windings and increase the operating temperature of the compressor motor. Shortly after this happens, the motor windings will short-circuit and burn out. A compressor burnout is indicated by oil color changing from clear to black (depending on the severity of burnout) and by a very unpleasant, pungent odor. Oil test kits available from refrigeration supply houses can be used to determine the degree of contamination. The test is simple, and instructions are supplied with the kit.

> Because compressors are burnt out more frequently by above-normal temperatures caused by high head pressure, condensers should therefore be checked and cleaned on a regular basis. Condenser fins can be cleaned with a long bristle brush and/or high-pressure gas such as nitrogen, air, or carbon dioxide.

## PROCEDURE FOR REPLACING A BURNT OUT COMPRESSOR

CAUTION: Do not touch the oil from a burnt out compressor as the acid will cause a burn. Always wear rubber gloves. Also, safety goggles will protect the eyes when discharging a system.

1. Remove the filter-drier.
2. Remove the compressor.
3. Flush the condenser with Nitrogen or $CO_2$ gas. To do this, purge the system by forcing Carbon Dioxide or Nitrogen gas from the line disconnected from the compressor discharge line and out the line disconnected from

---

**REPLACING A BOLTED-TYPE COMPRESSOR**

1. Turn the valve stems on both the high- and low-pressure sides all the way clockwise to seal off the refrigerant in the commercial system.
2. Remove the bolts connecting both service valves to the compressor and unbolt the compressor from the chassis. Disconnect power, detach wires, and remove the compressor.
3. Install the new compressor by reconnecting both service valves and the wires to the proper terminals in the new unit. The new compressor comes with new gaskets to be placed on the compressor high and low side connections after you have moistened them with a little compressor oil. (In these cases, the refrigerant does not have to be discharged from the system.)
4. Turn the high- and low-pressure service valve stems all the way counterclockwise to restore refrigerant circulation (see figs. 102, 120, and 121.)

---

the filter-drier. Do this for at least twenty seconds for average size units to clean the high side of the system by forcing the contaminated oil out of the system.

Many condensers have a strainer in their last pass. This strainer must be removed before flushing. The new drier has a strainer which will take its place. (See fig. 142.)

★4. Disconnect the capillary tube from the evaporator, and flush the evaporator with $CO_2$ or Nitrogen gas for at least eight seconds.
5. Install a new compressor, a new filter-drier, and then reconnect the tubing. (See p. 108.)
6. A new suction-line filter-drier will have to be installed on the suction line after a compressor burnout too.
7. For residential units, charge the unit with about five ounces of the same type of refrigerant used in the system. For commercial units, charge until the head pressure rises above atmospheric pressure. Stop charging, turn on the unit and let the unit run for five minutes, then purge the system. (Refer to pp. 100 through 110 for procedures for evacuation and charging a sealed system.)
8. Repeat step 7 two more times.
9. Using a vacuum pump, evacuate the system and allow the pump to run for thirty minutes. (See page 117.)
10. Recharge the system with the proper refrigerant as indicated on the nameplate of the unit as instructed in the section on charging the system.

A service valve for a suction or discharge line is located on each side.

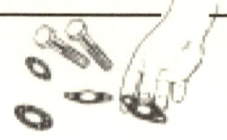

Figure 121

1. Put some compressor oil on the new gaskets supplied with replacement compressors.
2. Gaskets are placed between the compressor housing and the service valves.

\* It is good practice to replace the existing expansion valve or capillary tube.

Figure 120

A commercial type Copelametic compressor. Note the attached service valve and location of the nameplate. The compressor terminals are located inside the box.

Courtesy of Copeland, Emerson Electric Company

### SIGNS OF A DIRTY CONDENSER

Dirty condensers cause (a) higher evaporator temperatures and consequently higher temperatures in the cabinet, (b) higher operating temperatures in compressors and condensers, (c) longer running time, and (d) slow ice production and poor refrigeration. This is usually misdiagnosed as a low charge or bad compressor due to the overheated compressor short-cycling on overload.

Figure 122

### SIGNS OF LOSS OF REFRIGERANT

1. Low-side pressure reads lower than normal.
2. High-side pressure reads lower than normal.
3. Low amperage reading. Read the amperage the compressor draws. Do it by placing the ammeter around the wire to the compressor run or common terminal.

(See fig. 124) The nameplate on the unit will indicate the correct amperage the unit should draw while running, shown as FLA, as opposed to the amperage it draws at the instant of starting, shown as LRA on the nameplate (or in the reference book mentioned on page 105).

4. Little or no frost on the evaporator coil.
5. Unit runs continuously (nonstop).
6. Unit starts immediately after it is turned off. (Instead of the usual two- to-three-minute delay for the head and back pressures to equalize through the capillary tube during the off cycle.)
7. In cycle-defrost units, the temperature in the ice compartment drops below normal while the temperature in the fresh-food compartment rises above normal because an insufficient amount of vaporizing refrigerant circulating through the evaporator becomes superheated by the time it reaches the last passes of the coil in the fresh-food compartment.

Since the thermostat sensing bulb is attached to the lower part of the coil in the fresh-food compartment, it never allows the thermostat contacts to open. This causes the unit to run continuously and the temperature in the freezer compartment to drop below normal.

Figure 123

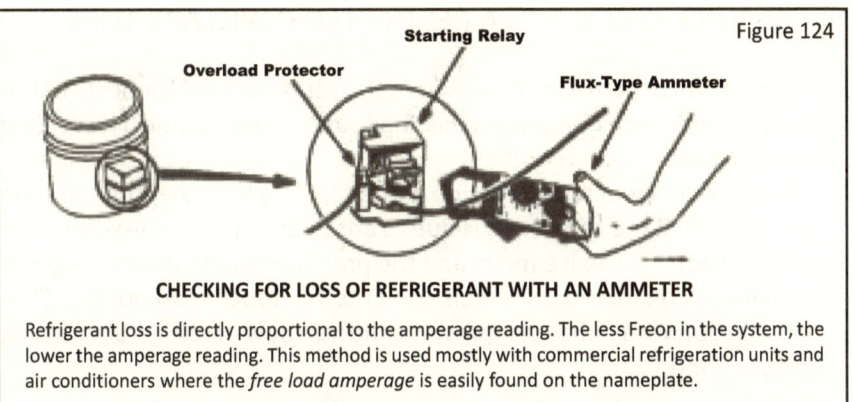

**CHECKING FOR LOSS OF REFRIGERANT WITH AN AMMETER**

Refrigerant loss is directly proportional to the amperage reading. The less Freon in the system, the lower the amperage reading. This method is used mostly with commercial refrigeration units and air conditioners where the *free load amperage* is easily found on the nameplate.

## SIGNS OF RESTRICTION OF REFRIGERANT FLOW

A restriction of refrigerant flow is caused by the formation of ice from moist air in the system. Air can penetrate into the system from a very tiny hole in the tubing, perhaps caused by an improperly sealed joint, flared connection, or failure to properly vacuum the system if previously repaired.

This restriction occurs in the capillary tube, TEV or AEV, because they are the narrowest passages through which refrigerant flows in a sealed system. When the capillary tube becomes restricted, the most common complaint is that "the unit sometimes seems to cool and sometimes it doesn't cool at all. It acts crazy!" This is due to the fact that the formation of ice in the capillary tube temporarily disrupts the flow of refrigerant to the evaporator causing the unit to stop cooling. As the evaporator temperature goes up, the ice melts, causing the circulation of refrigerant to be restored and the evaporator to cool again until the temperature drops low enough to refreeze the circulating moisture in the capillary tube.

**THE MOST COMMON INDICATIONS OF A RESTRICTION**

1. Unit runs at warmer than normal temperature with little or no frost on evaporator coil.
2. Low-side pressure reads partial vacuum.
3. High-side pressure reads higher than normal.
4. Unit runs continuously.
5. Higher than normal wattage draw.
6. The condenser, capillary tube, or drier feels cool.
7. Low-torque compressors used in residential and light commercial applications cycle on overload and take longer than three or four minutes to restart after being shut off. This is due to high-pressure refrigerant being trapped and separated in the system by the restriction. Equalization of high- and low-side pressures through the capillary tube is difficult to achieve during off cycles.

## A QUICK CHECK FOR A RESTRICTED CAPILLARY TUBE

1. Turn off the system and listen where the capillary tube connects to the evaporator. If you don't hear a hissing sound, there may be a restriction at that point.
2. Apply a warm, wet rag to the side of the capillary tube that runs into the evaporator. If the restriction is due to an ice formation, a hissing sound will be heard as the ice melts and the pressures in the system begin to equalize. A restricted drier feels cooler to the touch than normal. (The obstruction within the drier prevents the free flow of refrigerant causing evaporation.)

---

**A QUICK TEST FOR A RESTRICTION OR LEAK IN RESIDENTIAL UNITS**

When a residential unit runs warmer than normal (or does not cool), the problem can be due to a leak or a restriction in the system or several other reasons. Check the following before conducting this test: (a) condenser fan, (b) evaporator fan, (c) door gasket, (d) linted condenser, (e) cabinet light (Does it stay on while door switch is depressed?), (f) cold-control (thermostat) settings, (g) high ambient and frequency of cabinet door openings by customer, and (h) defrost system (timer, termination switch, heater).

If items a through h are working correctly, begin the test:

1. Connect power to unit.
2. Install a piercing valve on the suction line and connect it to the compound gauge hose.
3. While the unit operates, check the pressure in the suction line (with the valve on the gauge closed.)
4. If the suction pressure reads above zero, stop the test. The problem is an inefficient compressor which must be replaced.
5. If the suction pressure reads below zero (a vacuum), continue the test to step 6.
6. Look to see if the last pass of the capillary tube, the drier, or the condenser is sweating or cold. If so, the problem is a restriction in the system.
7. If the entire condenser feels cool to the touch, the problem can be a restriction or a complete loss of refrigerant. Add refrigerant to the system. If the temperature rises throughout the condenser, the problem is a leak. Otherwise, there is a restriction in the system.
8. If the last pass of the capillary tube, drier, or the condenser feels warm, the problem is a leak. If there is a restriction in the system, replace the drier, evacuate, and recharge the system. If there is a leak, locate and seal the leak, replace the filter-drier, evacuate, and recharge the system.

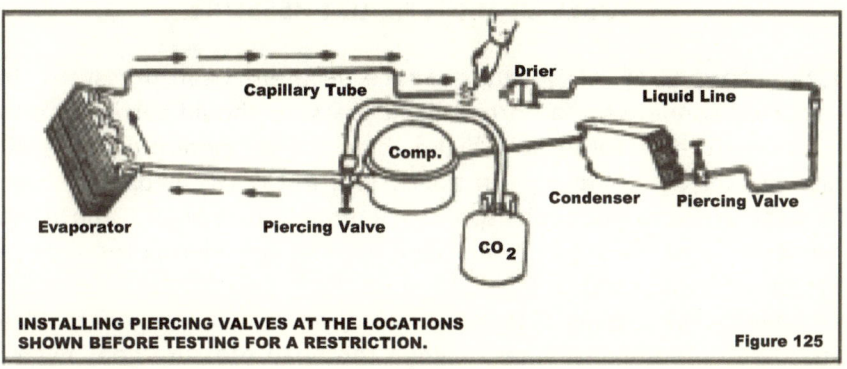

**INSTALLING PIERCING VALVES AT THE LOCATIONS SHOWN BEFORE TESTING FOR A RESTRICTION.**

Figure 125

## CAUSES OF HIGH HEAD PRESSURE

Different types of refrigerant produce different head pressures. In air-cooled condensers, a normal head pressure should correspond to a temperature of about 30°F higher than the temperature of the air passing over the condenser. In the case of a water-cooled condenser, the head pressure must correspond to a temperature of 20°F above the temperature of the return water (see fig. 87), considering the type of refrigerant being used.

Keeping these rules of thumb in mind, you will have a general idea what head pressure to expect. When the head pressure goes higher than normal, the cooling efficiency of the unit will drop as most of the heat does not dissipate from the refrigerant by the time it leaves the condenser.

Higher-than-normal head pressures are caused by the following:

1. Restriction in the sealed system. (As the compressor discharge line keeps pumping against the restriction, it creates high pressures in the condenser and the liquid line.
2. In the case of water-cooled condensers, poor water circulation or presence of air bubbles in the water.
3. Too much refrigerant in the system. Some of the refrigerant must be released by using the manifold gauge.

    a. Connect the high-pressure gauge hose to the access valve on the discharge (high) side of the compressor while the valves on the manifold gauge are closed.
    b. Turn on the unit.
    c. Turn the high-pressure gauge valve on and off in short intervals and check the gauge.
    d. By repeating this procedure a few times, enough refrigerant is released to bring the reading within normal range.

4. Ambient temperature above 85°F. It is normal to have a higher head pressure in the summer with higher ambient air temperatures.
5. Dirty condenser. When a condenser gets covered with grease, lint, and dirt, the necessary heat transfer to change the refrigerant from vapor to its liquid state does not take place. Consequently, head pressure goes higher than normal, and the unit no longer cools.

In places where pets such as dogs and cats are kept indoors, the condenser requires cleaning more often as hair shed by the pets is drawn into the fins and coil and restricts airflow.
6. Inefficient condenser fan. A fan that does not run (or runs too slowly) due to worn bearings or an internal short reduces (or stops) air circulation over the condenser fins and prevents the refrigerant from losing its latent heat and changing to its liquid state. This causes the head pressure to rise, and if not corrected, the compressor will burn out.

## CLEANING THE CONDENSER

There are several cleaning solutions in the market suitable for condensers. They are primarily alkaline or acid, about the same price and about equally effective. The alkaline is preferred since it is less destructive. Nevertheless, it should be cleaned up promptly if it runs off onto the flooring.

Additional care must be taken when applying these chemicals to the cooling fins of a condenser. The condenser fan(s) may have to be covered to prevent moisture from entering the motor(s). Once the solution is applied, the dirt, grease, and lint begin to boil and steam. This indicates it is working.

After the solution is given a chance to work for a while, it can then be easily removed with a high-pressure stream of air. Most technicians carry a cylinder of $CO_2$ (carbon dioxide) or nitrogen in their trucks. It is an excellent tool to clean a linted condenser very quickly.

For cleaning condensers on residential units, all you need is a good long-bristled brush and a vacuum cleaner.

Very often, especially in commercial units, a failure to cool properly can be due to a dirty or linked condenser. Even if the condenser appears clean on the outside, dirt accumulated at the base of the fins and coil can prevent proper heat exchange and keep the unit from cooling.

Furthermore, particularly in the case of roof-mounted condensers, make certain they are protected from direct exposure to the sun and have adequate shade to assist in the cooling process. A unit left exposed to the sun absorbs so much heat that it becomes incapable of transferring heat necessary to change the hot vapor refrigerant back to its liquid state. When this happens, the head pressure will rise above normal. The temperature will never drop low enough to satisfy the cold control causing the compressor to run continuously. Eventually, the overheated compressor will cycle on overload. Do not rely on natural shading (taller buildings, trees, etc.). It will be necessary to build a structure over the condensing unit if one is not already provided.

The symptoms are often misdiagnosed as a bad compressor or a weak overload protector.

## SYMPTOMS OF AN OVERCHARGED SYSTEM

There are several symptoms that indicate the presence of an excessive amount of refrigerant in the system.

1. Long running time of the unit. When there is too much refrigerant in the system, the unit runs for an exceedingly long period of time before the temperature drops to the point where the thermostat becomes satisfied and shuts off the compressor. This increase in compressor running time is due to the increased pressure in the evaporator from overcharging. The more refrigerant charged into the system, the higher the head and back pressures go. (See the temperature-pressure chart on page 133).
2. Head and back pressures read higher than normal.
3. High temperatures in the freezer and fresh-food compartments. The customer complaint here is that the unit does not get as cold as it used to. This is true because the evaporator plates do not get as cold as they should.
4. Suction line near the compressor sweats or frosts up.
5. Compressor-operating amperage goes too high. Note the FLA (run amperage) on the nameplate (example: FLA 12). With the unit running, close the jaws of the flux-type ammeter around the wire going to the compressor run or common terminal. If the meter reads 18 A for example, the unit can be overcharged with refrigerant.

Look again at the section on "Signs of Loss of Refrigerant" (item number 3). By using an ammeter, it can be determined if the unit is overcharged with refrigerant (fig. 124). With the unit running and the ammeter placed around the run or common wire, a high amperage reading indicates an overcharge (too much refrigerant).

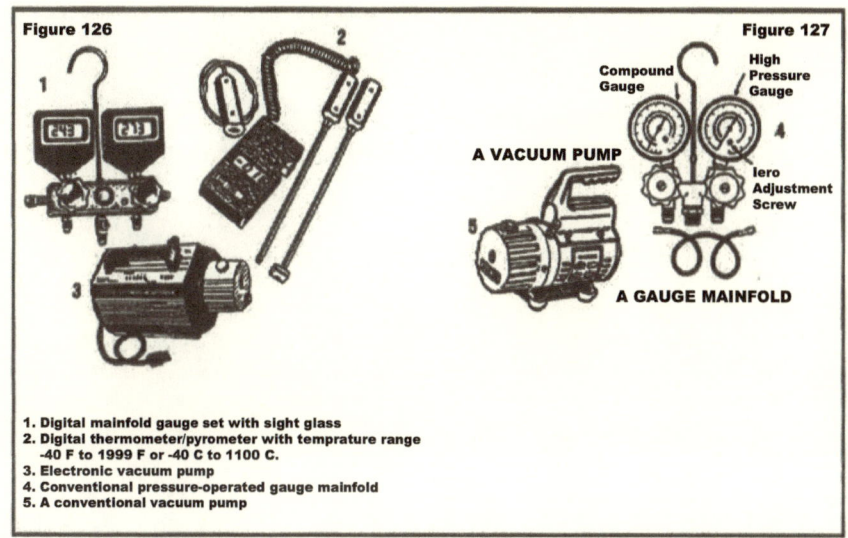

**Figure 126**

1. Digital mainfold gauge set with sight glass
2. Digital thermometer/pyrometer with temprature range -40 F to 1999 F or -40 C to 1100 C.
3. Electronic vacuum pump
4. Conventional pressure-operated gauge mainfold
5. A conventional vacuum pump

Courtesy of TIF Instruments Inc.

**Figure 127**

Courtesy of Robinair Div, SPX Corporation

# ICE MACHINES

This section covers the operation of various types of ice machines. It includes an easy-to-follow diagnostic chart and provides an electrical diagram of a typical ice machine.

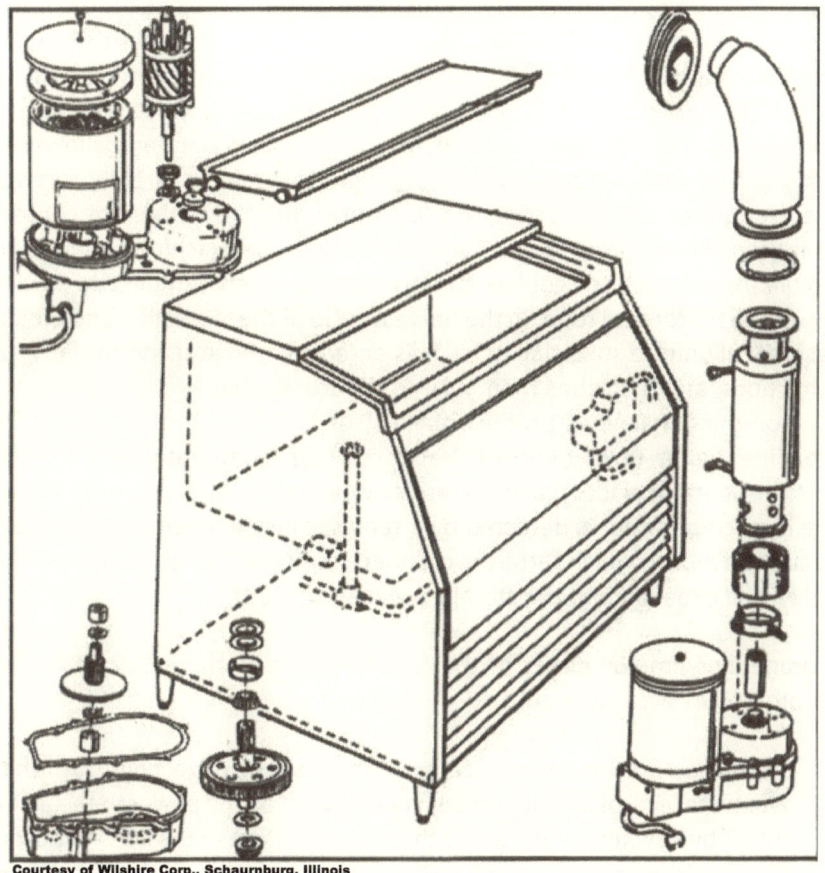

Courtesy of Wilshire Corp., Schaumburg, Illinois

# ICE MACHINES

Primarily, ice machines are nothing more than simple refrigeration units with a few differences. Only water and refrigeration are involved with a well-insulated, but unrefrigerated bin to store the ice chunks or flakes.

In the cube producing units, water enters a reservoir tank. It is then circulated by a water pump and then sprayed onto the extremely cold surface of the mold (evaporator) where a thin layer is frozen almost instantly (see fig. 128). Excess water is circulated back into the reservoir and the process is repeated. One thin layer forms at a time until the molds are filled. A thermostat (See fig. 130a) is installed close to the molds, when the ice formed in the molds become thick enough to touch the thermostat, power to the compressor is interrupted, and the circuit to a heating element is closed to loosen the ice from the molds, causing the cubes to fall into the storage bin (see fig. 128) This cycle is referred to as the *harvest cycle*.

In the harvest cycle of units that use a hot gas defrost system, the compressor is not stopped. Instead, a solenoid valve opens and allows the hot gas from the compressor discharge line to circulate directly through the evaporator coil (bypassing the condenser and the refrigerant control) to loosen the ice. (See figs. 16 and 16a). Other ice machines operate by circulating water over a cold evaporator plate or over the molds by flowing through a perforated tube. In the harvest cycle of these smaller, and mostly residential units, a solid slab of ice falls onto a heated grid and thereby cut into cubes, and the cubes then fall into the storage bin.

Some ice machines produce flakes (see fig. 130). In this type of ice machine, water flows over a freezing cylinder (evaporator) with sharp, spiraling, cutting blades, called an auger. When a thin layer of ice is formed, the low temperature is detected by a sensor, causing a switch to close the circuit to a motor, which turns the cylinder, causing the ice to be shaved into flakes and expelled into the storage bin (see fig. 130).

Storage bins employ either of the following two methods to control the amount of ice accumulated:

1. *Mechanical* (a lever-operated control). This is an arm projecting into the bin. When the level of ice rises enough to lift the lever, the power to the unit is cut off by a switch connected to the arm until some of the ice is removed.
2. *Thermostatic* (a sensing bulb is installed near the top of the bin). When the ice level in the bin reaches the bulb, the contraction of the refrigerant within the bulb transmits pressure changes to the thermostat, causing a switch within the thermostat to open the main circuit to the unit, shutting off the system.

In the lower-right system, water enters the reservoir pan. It is then pumped across the top of the evaporator. It then flows down the face of the evaporator. The unfrozen portion of water flows back to the reservoir to be pumped over the evaporator plate again.

1. Ice thickness thermostat
2. Evaporator plate
3. Condenser
4. Thermostatic expansion valve
5. Filter-drier
6. Pump and float mechanism in the reservoir pan.

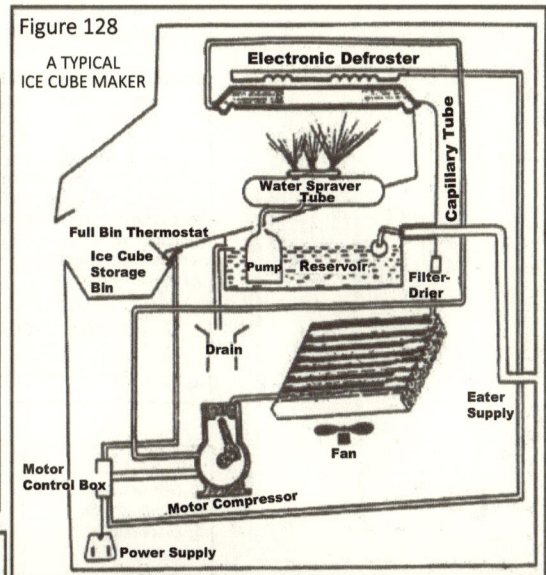

Figure 128
A TYPICAL ICE CUBE MAKER

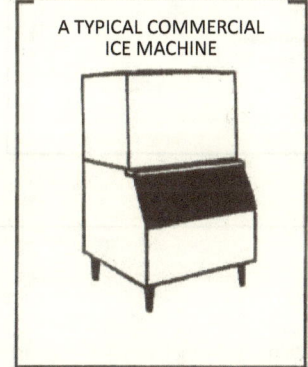

A TYPICAL COMMERCIAL ICE MACHINE

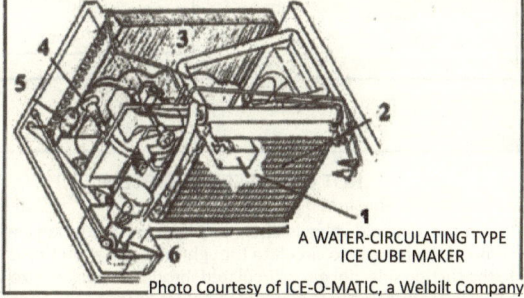

A WATER-CIRCULATING TYPE ICE CUBE MAKER

Photo Courtesy of ICE-O-MATIC, a Welbilt Company

As the water level in the reservoir lowers, the water inlet float opens a valve to the water supply to allow the reservoir to refill. Water from the reservoir is pumped to the sprayer tube where it is sprayed onto the mold continuously and instantly frozen in thin layers. Run-off water is returned to the reservoir. Ice continues to build until it becomes thick enough to touch a sensor, at which time the low temperatures cause the sensor to shut off power to the compressor and water pump and to energize the electric defroster. As the evaporator heats, the ice cubes loosen and fall into the ice bin. The cycle is repeated until the level in the bin goes high enough to touch the sensing bulb. The low temperatures are then transmitted to the thermostat-operating mechanism, causing a set of contacts to open, disrupting the flow of power to the unit. At this point, the ice-making operation is stopped until some of the ice is removed from the bin.

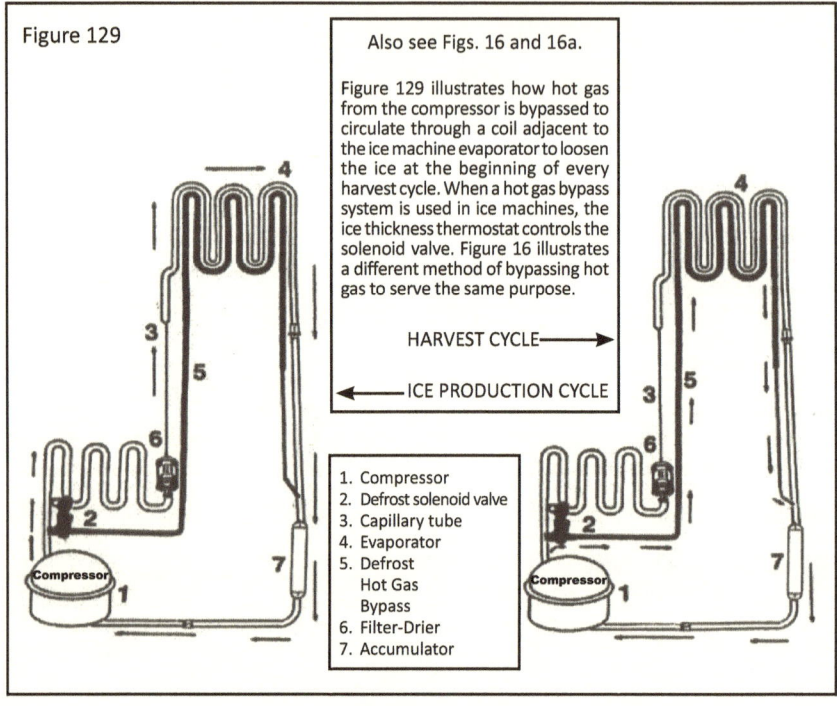

**Figure 129**

Also see Figs. 16 and 16a.

Figure 129 illustrates how hot gas from the compressor is bypassed to circulate through a coil adjacent to the ice machine evaporator to loosen the ice at the beginning of every harvest cycle. When a hot gas bypass system is used in ice machines, the ice thickness thermostat controls the solenoid valve. Figure 16 illustrates a different method of bypassing hot gas to serve the same purpose.

HARVEST CYCLE⟶

⟵ICE PRODUCTION CYCLE

1. Compressor
2. Defrost solenoid valve
3. Capillary tube
4. Evaporator
5. Defrost Hot Gas Bypass
6. Filter-Drier
7. Accumulator

## HOT GAS DEFROST SYSTEM

When the solenoid valve is energized by a timer, it opens, allowing some of the hot gas to circulate though the tubing. On the way back to the suction side, hot gas is circulated through a capillary-size section of tubing to become partially cooled before entering the compressor. When the frost is melted, the defrost bimetal, which is wired in series with the solenoid valve, cuts off power to the valve and closes off the supply of hot gas (if the frost buildup is melted before the timer completes its defrosting cycle). If this system is used in an ice machine, instead of a defrost timer, a thermostat controls the operation of the defrost solenoid. The defrost solenoid is energized in the beginning of every harvest cycle.

Usually, the two solenoid contacts are wired to a lamp that becomes energized when the unit is in the defrost cycle.

The solenoid valve is activated by a timer at a preset time, usually midnight or 1:00 AM in commercial units.

Using this very efficient type of system, makes it unnecessary to match hot gas tubing coil-for-coil with the evaporator.

AIR CONDITIONING AND REFRIGERATION REPAIR MADE EASY     189

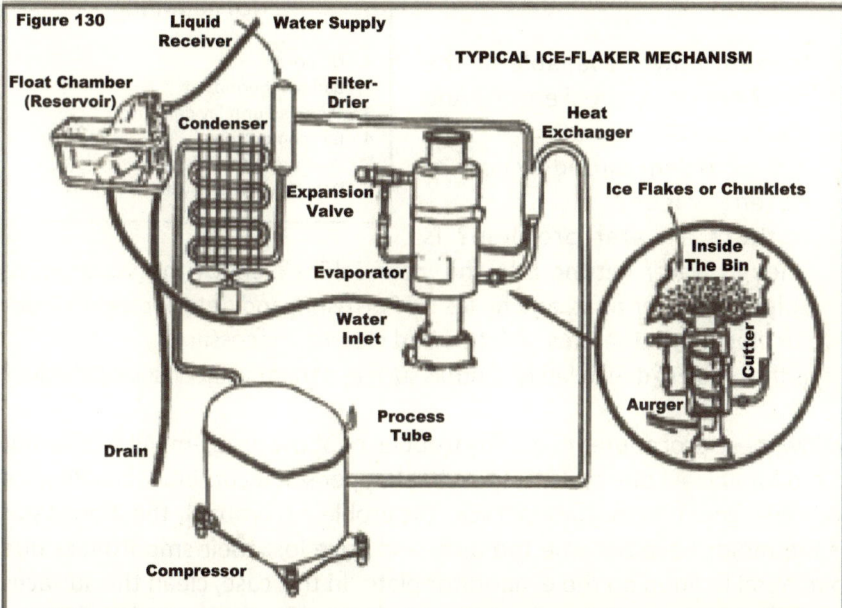

This is a typical ice machine that produces flakes. Water enters the float-controlled reservoir from the water inlet valve. It then flows (in some units, it is pumped) into the evaporator freezing chamber where the auger is located. Layers of ice are rapidly formed inside the freezing chamber and shaved into flakes then moved into the storage bin by the rotation of the auger.

Selected Parts Courtesy of Wilshire Corp., Schaumburg, Ill.

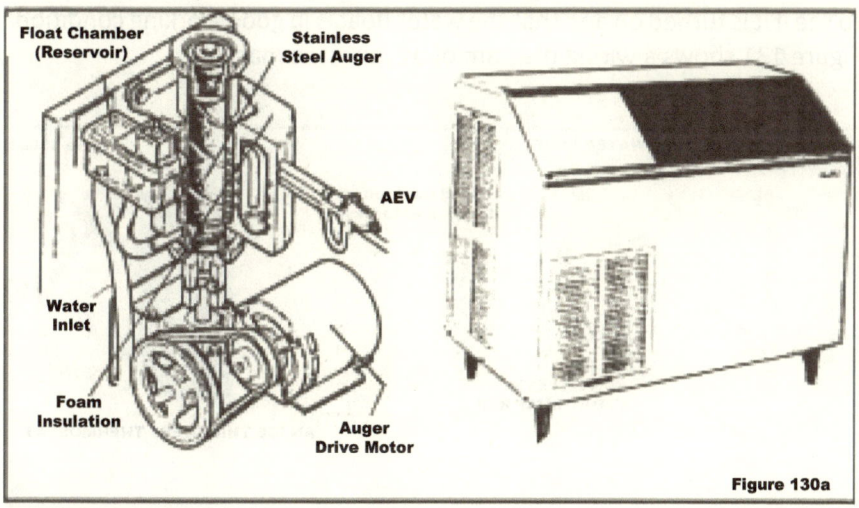

Photo Courtesy of Ice-O-Matic
A Welbilt Company

When there is a complaint of little or no ice production, check the following:

| REASONS FOR THE COMPRESSOR NOT RUNNING |
|---|
| 1. No power |
| 2. Bad bin thermostat |
| 3. Overload or relay defective |
| 4. Bad compressor |
| 5. Pressure control defective |
| 6. Defective master switch |

1. Is there power available at the unit? Are circuit breakers on? And are fuses OK?
2. Is the problem caused by poor or no refrigeration?
3. Is there a water problem? Is enough water getting into the ice machine? Check the water-valve solenoid, water float, and hoses for corrosion and deterioration. Check for clogged water lines or filters and clean as necessary.
4. Is the problem due to faulty controls such as a thermostat or solenoid valve?

Follow these procedures to quickly troubleshoot the problem: Allow the unit to run for at least one cycle to see what happens. If it continues building ice and never goes into the harvest cycle, the problem is probably the thermostat or the molds have become too dirty and have lost their smoothness due to mineral buildup on the evaporator plate. In this case, clean the surfaces carefully and thoroughly, using the manufacturer's recommended cleaning agents. If the smoothness cannot be restored, try spraying them with Teflon or replacing the evaporator plate if necessary.

Very often, the master water valve (sometimes located under a kitchen sink) may have been inadvertently shut off by the owner or an employee. Check to see if it is turned on and that the water float is in good working condition. Figure 131 shows a wiring diagram of a typical ice machine.

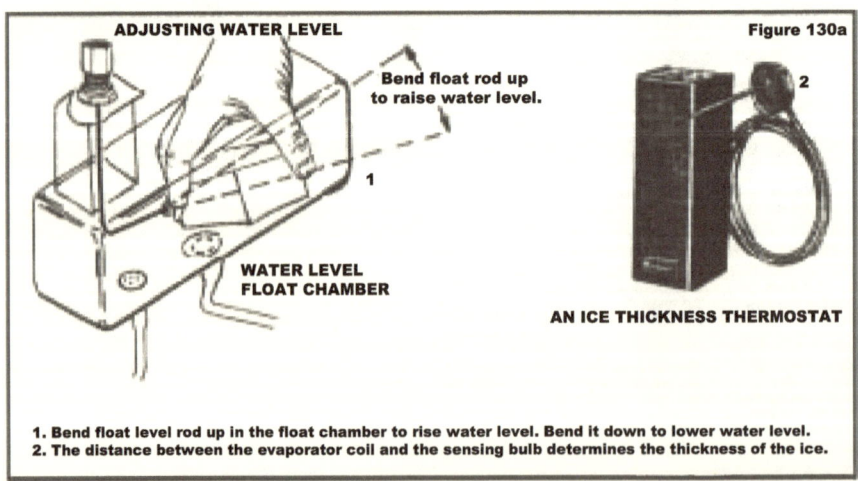

Figure 130a

1. Bend float level rod up in the float chamber to rise water level. Bend it down to lower water level.
2. The distance between the evaporator coil and the sensing bulb determines the thickness of the ice.

Courtesy of Honeywell

### REASONS FOR NO ICE PRODUCTION

1. Water valve closed
2. Evaporator thermostat inoperative or set wrong
3. Water pump inoperative or clogged
4. Cutter grid has open circuit
5. Refrigeration system inoperative
6. Mineral built up on the evaporator surface

7. Open bin thermostat
8. Open electrical circuit to the compressor

## COMMON SEALED SYSTEM PROBLEMS

1. Restriction
2. Compressor failure
3. Leaky hot gas valve
4. Refrigerant leak

## SIGNS OF AN UNDERCHARGED ICE MACHINE

1. Donut-shaped ice cubes
2. Hissing sound heard in the evaporator while running

If ice cubes come out too thin or too thick, the reason can be due to an out of adjustment or defective evaporator thermostat. The evaporator thermostat should have a cut-in temperature of between 31°F and 32°, and a cut-out temperature of -3°F to +10°F.

Sometimes the complaint is that the bin is full of ice and the machine will not shut off. The amount of ice in the bin is controlled by a bin thermostat or a lever-operated switch in the bin that will have to be replaced.

A linted condenser is a common reason for not enough ice production. Figure 131 is a typical wiring diagram of an ice machine. Please refer to the "Basic Electricity" section to get a better understanding of the illustration.

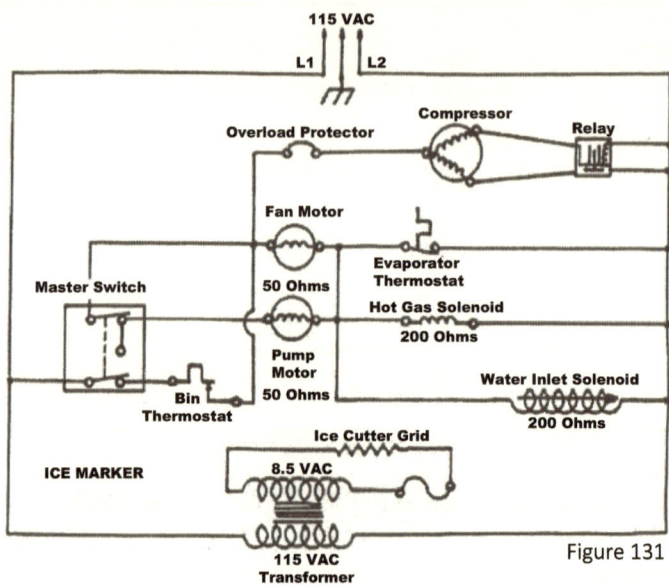

Figure 131

HOW AN ICE MACHINE WITH A CUTTER GRID WORKS WHEN SWITCH-OPERATING ROD IS MOVED UPWARD TO "ON" POSITION

Compressor runs
Condenser fan Runs
Water pump runs and circulates water
Cutter grid is warm

WHEN ICE SLAB REACHES PRESET THICKNESS, HARVEST CYCLE BEGINS, AND THE FOLLOWING HAPPENS:

Compressor keeps running
Evaporator thermostat is satisfied
Condenser fan stops or slows
Water pump stops
Hot gas solenoid opens
Cutter grid is warm
Harvest cycle lasts one to two minutes

UNIT RESTARTS FREEZING CYCLE WHEN SLAB IS RELEASED FROM EVAPORATOR AND CUTTING PROCESS BEGINS.

WHEN STORAGE BIN IS FULL,
BIN THERMOSTAT OPENS.
NOTHING WORKS.
Cutter Grid Remains On

To obtain a service manual for any particular ice machine, call the factory's 800 number hotline. The manufacturer's name and factory location can be found on the nameplate. The service manual contains photographs, line drawings, schematic wiring diagrams, and troubleshooting charts to further explain operation and repair of that specific unit.

## ADJUSTMENTS TO CHUNKLET / ICE FLAKERS

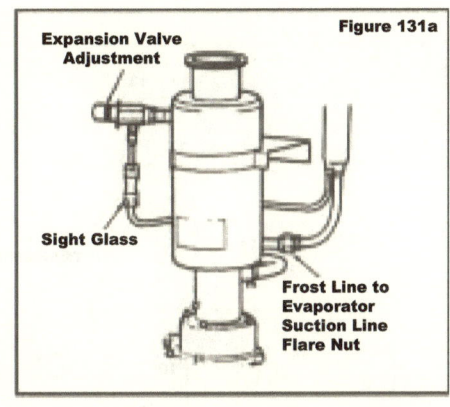

Figure 131a

The purpose of the water-level control is to automatically maintain proper water level in the ice flaker. Proper water level should be horizontal with the center of the AEV (See fig. 131b).

You must use high- and low-side pressure readings, water and air temperatures, the frost line, the sight glass, plus general conditions of cleanliness to assess the refrigeration system status when making any adjustments.

The location of the frost line can be very helpful in determining proper operation of the refrigeration system. The frost line is lowered by higher than normal water temperature or higher ambient air temperature and also poor refrigeration system efficiency. Ideally, the frost line should be seen on the flare nut attaching the evaporator to the heat exchanger, after the unit runs for at least fifty minutes, making ice flakes continually. (See fig. 131a.)

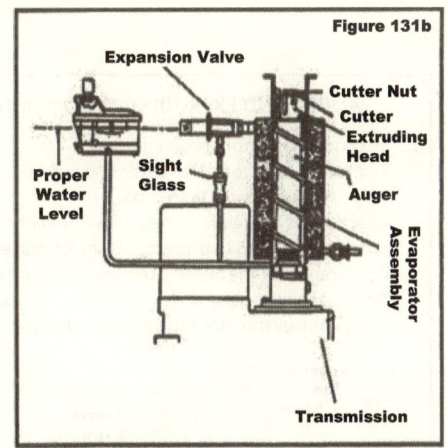

Figure 131b

Courtesy of Wilshire Corporation, Schaumburg, Illinois

To locate the frost line (1) remove the side panels, and (2) remove the black insulation over the flare nut and suction line by cutting it with a knife (save the insulation as it must be replaced). In dry areas where a frost line may not be visible, moisten the suction return line with a damp cloth to make it appear.

Turn the AEV adjustment screw *clockwise* to move the frost line toward the heat exchanger and raise the suction pressure. Turn the AEV adjustment screw *counterclockwise* to move the frost line toward the evaporator. (See fig. 131a.)

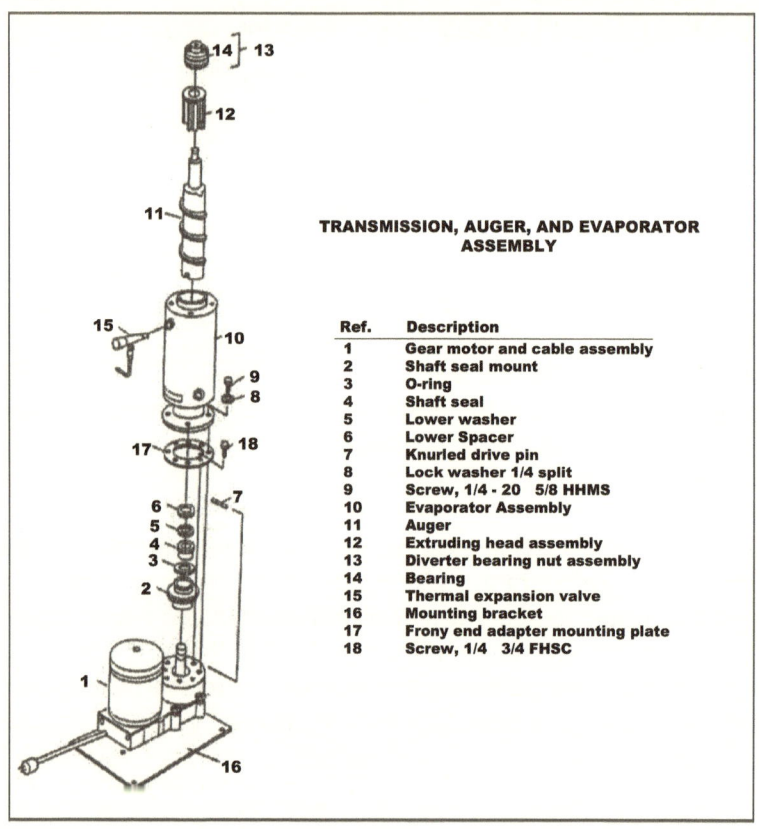

**TRANSMISSION, AUGER, AND EVAPORATOR ASSEMBLY**

| Ref. | Description |
|---|---|
| 1 | Gear motor and cable assembly |
| 2 | Shaft seal mount |
| 3 | O-ring |
| 4 | Shaft seal |
| 5 | Lower washer |
| 6 | Lower Spacer |
| 7 | Knurled drive pin |
| 8 | Lock washer 1/4 split |
| 9 | Screw, 1/4 - 20  5/8 HHMS |
| 10 | Evaporator Assembly |
| 11 | Auger |
| 12 | Extruding head assembly |
| 13 | Diverter bearing nut assembly |
| 14 | Bearing |
| 15 | Thermal expansion valve |
| 16 | Mounting bracket |
| 17 | Frony end adapter mounting plate |
| 18 | Screw, 1/4  3/4 FHSC |

**AUGER AND EXTRUDING HEAD REMOVAL**

1. Disconnect unit from power supply.
2. Remove storage container cover and put aside.
3. Turn off water supply to ice maker.
4. After ice has melted from head, take hold of the auger nut and lift straight up to disengage from ice maker.
5. When replacing the auger assembly, make certain that both the auger engages the drive pin and the extruding head ribs engage the evaporator tube (see figs. 131d and 131e).

Figure 131d

Figure 131e

Courtesy of Wilshire Corporation, Schaumburg, Illinois

## INSTALLATION AND SHAFT SEAL REPLACEMENT
(See fig. 131f)

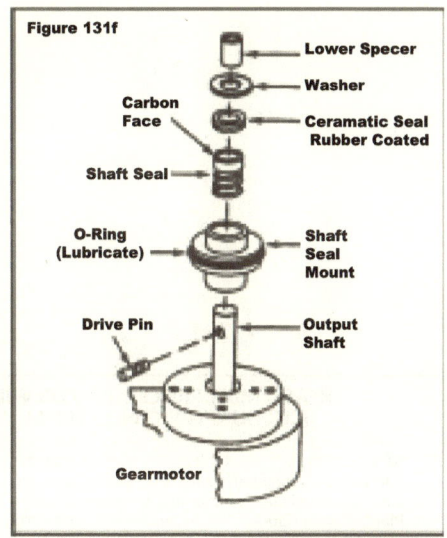

Figure 131f

*Courtesy of Wilshire Corp. Schaumburg, Illinois*

1. Place shaft seal mount-over gear motor output shaft and push down until shaft seal mount rests flush on top of gear motor.
2. Place shaft seal with carbon faceup (spring down) over output shaft and push (gently) downward until seal nests in recess of the shaft seal mount.
3. Place rubber-coated ceramic seal (important: ceramic face down) over output shaft and push down until seal rests on carbon face of the output shaft seal. (Lubricate rubber on ceramic seal with number 06195 rubber lubricant.)
4. Place flat washer over output shaft and let rest on the ceramic seal. Push down on the washer compressing the spring on the output seal and nesting ceramic seal inside recess of shaft seal mount. While holding the seals (down) in place, put lower spacer over output shaft. Insert drive pin into hole to hold assembly.
5. Place gear motor on its side supporting end of shaft with block and tap drive pin with hammer until fully engaged. (Pin must have approximately one-fourth inch protruding on either side of output shaft.) Do not use excessive force with hammer, resulting in damage to drive pin and/or gear-motor bearings.

## REASSEMBLY PROCEDURE FOR WILSHIRE ICE CHUNKLET/FLAKER FCS MODELS

A. After the unit is disassembled, check all the parts for wear and discard worn parts. Replace bad parts with new, authorized parts only. (See fig. 131d.)
B. Lubricate the inside of coupling (number 2) with a light coating of Lubri-Plate number 630-AA. Place coupling onto the output shaft of transmission (number 1).
C. Place lower bearing (number 3) into housing (number 4). (See installation lower bearing and housing.)
D. Place housing on transmission and secure with bolts (number 5).
E. Place gasket on top of bearing.
F. Locate ceramic seal (number 7) on top of housing.
G. Install shaft seal (number 8) onto bottom journal of the auger (number 9). For proper replacement, see Shaft Seal Replacement in figure 131f.
H. Carefully set the auger and seal assembly down onto the ceramic seal and lower housing, and load seal.
I. Slide evaporator assembly (number 10) down over auger and housing. Use P-80 or water on the O-ring that is around the ceramic seal.
J. Secure evaporator to housing with bolts (number 1).
K. Check threaded holes in the extruding head. Make sure they are clean and moisture free.
L. Place extruding head down into the evaporator tube.
M. Treat extruding head bolts with Grade "AA" Loctite. (Loctite is a gluelike substance that is used on bolt and screw threads when a lock washer is impractical or unsightly. It can be broken loose by a firm blow to the screw or bolt head. It is usually available at most hardware and sporting goods stores.) Make sure threads are clean and moisture free. NOTE: Loctite set up time is half an hour. Do not run ice maker until Loctite has time to set up.
N. Hold down on the auger and screw bolts into the extruding head through the neck of the hopper. Hand-tighten. Tighten extruding head bolts evenly and in sequence.
O. Install cutter, cutter nut, and transport tube to evaporator flange with clamp.
P. Connect ice maker to condensing unit and check for operation.

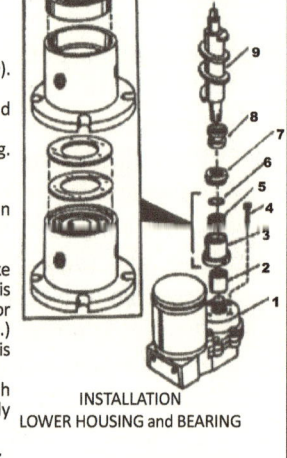

INSTALLATION
LOWER HOUSING and BEARING

Figure 131g

Courtesy of Wilshire Corp. Schaumburg, Illinois

## GEARMOTOR ASSEMBLY

Figure 131h

| Ref. | Description |
|---|---|
| 1 | Screw 8/32 × 3/8 |
| 2 | Motor shield |
| 3 | End bell |
| 4 | Motor screw |
| 5 | Lock washer |
| 6 | Gear motor stator |
| 7 | Spring washer |
| 8 | Washer |
| 9 | Motor bearing (upper) |
| 10 | Rotor assembly |
| 11 | Seal, motor |
| 12 | Shoulder washer |
| 13 | Motor bearing (front) |
| 14 | Screw T/O 6/32 × 1/4 PHS |
| 15 | Cable clamp, 3/16 |
| 16 | Manual T/O |
| 17 | Oiler assembly |
| 18 | Gear case screw |
| 19 | Cover |
| 20 | Junction box screw |
| 21 | Junction box cover |
| 22 | Woodruff key |
| 23 | Retainer ring |
| 24 | Seal output |
| 25 | Bearing shim |
| 26 | Bearing cup |
| 27 | Bearing cone |
| 28 | Output shaft |
| 29 | Output gear |
| 30 | Bearing, no. 3 pinion |
| 31 | Nylatron washer |
| 32 | number 4 gear and 3 pinion assembly |
| 33 | Bearing, no. 5 pinion |
| 34 | Nylatron washer |
| 35 | number 4 gear and 5 pinion assembly |
| 36 | Gasket |
| 37 | Case |
| 38 | Spacer, output shaft |
| ★ | Lube, "quarts" |
| 40 | Cable |
| 41 | Bushing |

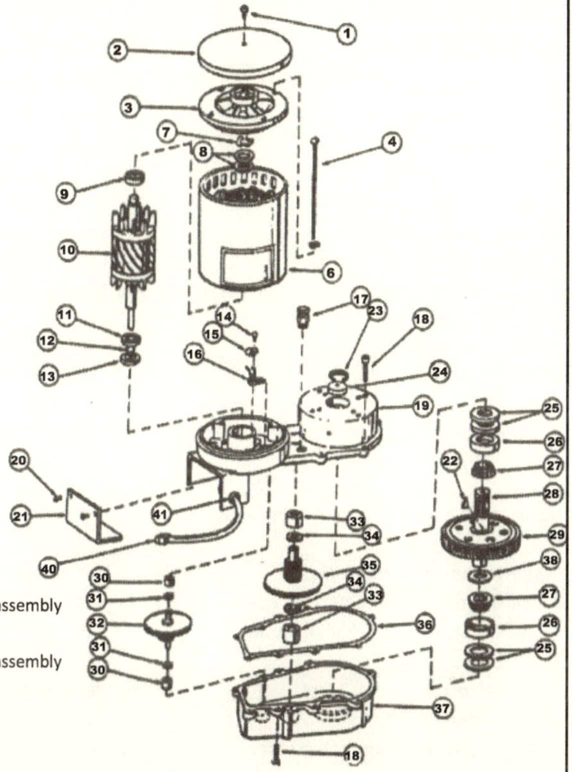

## WATER-LEVEL-CONTROL ASSEMBLY

Figure 131i

Some ice machines are equipped with a low-water safety control (reed switch) attached to the water reservoir to shut down unit if water level drops below bottom of reservoir. It can be adjusted by bending the magnet arm (number 3) as needed.

| No. | Description |
|---|---|
| 1 | Reservoir assembly (reservoir, cover, O-ring, and wing nut) |
| 2 | Water-level-control hardware (valve body, cotter pin, cap, spacer, rubber plunger, fiber washer, nut, sleeve, and compression nut) |
| 3 | Float-and-stem assembly (magnet, bracket assembly, and float) |
| 4 | Reed switch assembly |
| 5 | Water-level bracket |
| 6 | Reed switch clamp |
| 7 | Screw 6/32" × 1/4" long |

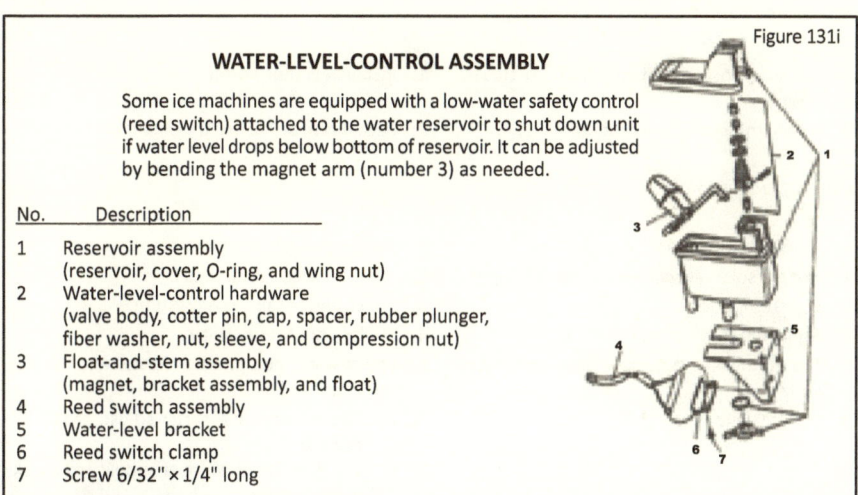

Courtesy of Wilshire Corporation, Schaumburg, Illinois

# TROUBLESHOOTING COMMERCIAL ICE MACHINES

| Problem and Possible Cause | Remedy |
|---|---|
| **No Ice in the Bin while Compressor Runs** | |
| 1. Hot gas bypass solenoid stuck open. | 1. Replace solenoid valve coil. |
| 2. Leak in the sealed system. | 2. Check refrigerant pressures; repair leak; change filter-drier and recharge. |
| 3. Water supply faucet turned off. | 3. Turn on faucet. |
| 4. Water float inoperative. | 4. Replace water float. |
| 5. Water solenoid valve bad. | 5. Check and replace if necessary. |
| 6. Evaporator thermostat out of calibration or inoperative. | 6. Install new thermostat or recalibrate. |
| 7. Excessive use of ice. | 7. Unit may be too small for needs. |
| 8. Cutter grid circuit open. | 8. Check for inoperative transformer, fuse, broken wire, etc., in grid circuit. |
| 9. Mineral buildup obstructing water-circulating system. | 9. Remove mineral deposits with recommended chemicals. |
| **No Ice in Bin and Compressor Does Not Run** | |
| 1. Power supply disconnected. | 1. Reconnect or turn on. |
| 2. Bin thermostat contacts stuck open. | 2. Replace bin thermostat. |
| 3. Master switch left in 'Clean' position. | 3. Turn switch to 'On' position. |
| 4. Defective relay. | 4. Replace relay. |
| 5. Defective compressor motor. | 5. Replace compressor motor. |
| 6. Master switch off. | 6. Turn master switch on. |
| 7. Loose connection or broken wire. | 7. Repair as necessary. |
| 8. Ambient air below 55°F. | 8. Relocate unit to warmer area. |
| **Bin Full of Ice and Compressor Runs Nonstop** | |
| 1. Bin thermostat contacts stuck in closed position. | 1. Install new thermostat. |
| 2. Bin thermostat out of calibration. | 2. Recalibrate or replace thermostat. |
| **Little Ice Production** | |
| 1. Ambient air below 55°F. | 1. Relocate unit to warmer area. |
| 2. Hot gas solenoid valve stuck partially open. | 2. Replace solenoid valve coil. |
| 3. Poor refrigeration. | 3. Check and repair sealed system. |
| 4. Faulty bin thermostat. | 4. Replace bin thermostat. |
| 5. Too little water circulated over evaporator surface. | 5. Check water pump, water float,0 and for any restriction in lines. |
| 6. Water falling in the bin. | 6. a. Check water line parts; adjust, repair, or replace.<br>b. Check water pressure at supply and check water faucets. |
| 7. Room temperature too high. | 7. Advise customer. |

| Problem and Possible Cause | Remedy |
|---|---|
| 8  Mineral buildup on evaporator plate. | 8. Clean evaporator plate. Recommend that a water softener be installed for the unit to end problems of mineral deposit buildup on plate and in lines and valves. |
| 9  Linted condenser. | 9. Clean condenser. |
| 10  Inoperative condenser fan. | 10. Repair or replace condenser fan. |
| 11  Water continuously fills reservoir and is siphoned off. | 11. Check for stuck water float; repair, readjust, or replace as necessary. |

### Ice Cubes Come Out "Milky"

| | |
|---|---|
| 1  Not enough water in water tank. | 1. Check water tank. |
| 2  Water hardness exceeds fifteen grains. | 2. Install water softener for unit. |

### Ice Cubes Come Out Too Thin

| | |
|---|---|
| 1  Faulty evaporator thermostat. | 1. Replace thermostat. |
| 2  Too little water circulated over evaporator plate. | 2. Check water pump and lines. Clean or repair as necessary. |
| 3  Evaporator thermostat set on "Thin Ice Cubes." | 3. Recalibrate thermostat by turning its adjusting screw clockwise to obtain desired thickness. |
| 4  Evaporator thermostat bulb installed in wrong location. | 4. Make sure there is a shim between the thermostat feeler tube and evaporator bracket. |

### Ice Cubes Come Out Too Thick

| | |
|---|---|
| 1  Faulty evaporator thermostat. | 1. Replace thermostat. |
| 2  Evaporator thermostat set to produce thick cubes. | 2. Turn adjusting screw to obtain desired thickness of cubes. |

### Ice Cubes Have Unpleasant Taste

| | |
|---|---|
| 1  Food stored in ice bin. | 1. Advise customer to remove food. |
| 2  Some packaging material left in unit. | 2. Be sure all packing material is removed from unit. |
| 3  Water supply has excessive mineral content. | 3. Refer customer to local water treatment authorities. |

### Ice Slab Won't Relate in Defrost Cycle

| | |
|---|---|
| 1  Mineral deposits built up on evaporator plate. | 1. Clean evaporator plate with correct cleaning chemicals. |
| 2  Faulty hot gas defrost solenoid valve. | 2. Check valve and replace if necessary. |
| 3  Faulty ice-thickness control. | 3. Replace ice-thickness control. |
| 4  Ice-thickness control out of adjustment. | 4. Readjust ice-thickness control. |

| Problem and Possible Cause | Remedy |
|---|---|
| **Ice Slab Builds Up Unevenly on Evaporator Plate** | |
| 1. Unit running on low charge. | 1. Locate leak, replace drier, evacuate, and recharge. |
| 2. Moisture in sealed system. | 2. Look for possible leak; repair as required. |
| 3. Faulty expansion valve. | 3. Replace expansion valve if necessary. |
| 4. Water distribution pipe does not distribute water evenly over evaporator plate. | 4. Clean holes in distribution pipe and repair or replace water float if water is too low in reservoir. |
| **Empty Water Tank** | |
| 1. Clogged water inlet screen. | 1. Remove and clean screen. |
| 2. Defective evaporator thermostat. | 2. Replace thermostat. |
| 3. No power to water solenoid. | 3. Check electrical diagram, locate disconnection, and repair. |
| 4. Faulty water solenoid valve. | 4. Replace solenoid valve coil. |
| 5. Restriction in water line. | 5. Clean restriction from lines and water shut-off valves. |
| 6. In newly installed smaller units (50 lbs.), water solenoid valves are not energized until the first defrost cycle. | 6. Wait for the first defrost cycle. |
| **No Ice Built Up on Evaporator Plate (in 100, 200 and 400 lb units)** | |
| 1. Defrost solenoid valve stuck open. | 1. Replace valve. (Check valve by lightly tapping on it. If ice builds up on evaporator plate, valve is faulty and should be replaced. |
| 2. Faulty water pump. | 2. If you notice a frost buildup on evaporator plate while water pump is not running, replace pump. |
| 3. Faulty float valve. | 3. Remove, clean, or replace. |
| 4. Master switch in "Clean" position. (Some models) | 4. Turn switch to ON position. |
| 5. Grid relay stuck on "Defrost." | 5. In some models, a faculty relay stays on "defrost." Replace relay. |
| 6. Faulty ice-thickness control. | 6. A faulty control does not open and the unit stays in defrost cycle. Replace control. |
| 7. No water circulation. | 7. Check for mineral buildup on water strainer and in water lines. |
| 8. No power or no water supply. | 8. Check power and water supply to unit. |
| 9. Faulty refrigeration system. | 9. Check system and repair as necessary. |
| **Ice Slab Released Slowly (All Models)** | |
| 1. Evaporator plate covered with mineral buildup. | 1. Clean evaporator plate thoroughly. |
| 2. Defrost solenoid valve stuck in open position. | 2. Check and replace if necessary. |
| 3. When ambient temperatures are too low, head pressures drop below 100 lb, refrigeration slows, and will slow ice formation. | 3. Relocate unit to where ambient temperatures are higher. |

| Problem and Possible Cause | Remedy |
|---|---|
| **Ice Build Up on Cutter Grids** | |
| 1. Poor connection between the power supply to the grids and the cutter grid connecting pin. | 1. Check and repair cutter grid connection. |
| 2. Broken wire in cutter grid. | 2. Check and replace grid. |
| **Short-Cycling of the Compressor** | |
| 1. Low voltage. | 1. a. Check for an extension cord that is too long or undersized. Replace.<br>b. Check power receptacle for proper voltage and inform owner of an overload problem in the house wiring. |
| 2. Loose electrical connections. | 2. Inspect and correct problem. |
| 3. Weak or faulty overload protector. | 3. Check and replace if necessary. |
| 4. Faulty thermostat. (May act erratic.) | 4. Bypass thermostat. If short-cycling stops, replace thermostat. |
| 5. Relay wired incorrectly. | 5. Check electrical diagram and correct. |
| 6. Faulty start capacitor or compressor. | 6. Check and replace if necessary. |
| 7. Faulty pressure control. | 7. Check and replace if necessary. |
| 8. Leak in sealed system. | 8. Check and repair as needed. |
| 9. Linted condenser or faulty condenser fan. | 9. Check and repair as necessary. |
| **Condenser Fan Will Not Run in Ice-Making Period (Machines of 50lbs or Less)** | |
| 1. Fan blade caught on shroud. | 1. Adjust shroud or blade as necessary. |
| 2. Faulty fan motor. | 2. Replace fan motor. |
| 3. Faulty evaporator thermostat. | 3. Replace thermostat. |
| 4. Open circuit in wiring to the fan motor. | 4. Disconnect wires to fan motor and check for voltage. If you have a correct reading, replace motor. If there's no power, check for broken wire or bad connection. |
| **Water Dripping into Ice Cube Storage Bin** | |
| 1. Water line leak at water valve. | 1. Check and tighten connections. |
| 2. Water dripping from ice slab remaining on warm grid. | 2. Check cutter grid, cutter grid circuit and minerals buildup on evaporator plate. |
| 3. Water return line moved out of position. | 3. Correct water line position. |
| 4. Water reservoir overflowing. | 4. Check for restriction in overflow tube; make sure overflow tube is properly inserted in liner drain and check for worn water valve washer. |

One of the better products on the market for cleaning evaporator plates is *nickel-safe ice machine cleaner* made by Calgon Vestal, a division of Calgon Corporation. This, or a similar product, is available at refrigeration supply businesses. It comes in a 16 oz bottle with the directions printed on the back, which are reprinted here to provide an outline of what is involved:

## DIRECTIONS

1. Turn off refrigeration, shut off water supply, and remove ice from bin.
2. Remove water trough, water curtain(s), water distribution tube(s), and other parts that may be scaled with deposits.
3. Mix 3 oz of nickel-safe ice machine cleaner per gallon of water in plastic container and place components in solution. Soak the components until they are free of deposits. For stubborn or thick deposits, use a soft brush to help the dissolving action.
4. Use above solution to clean storage bin top, bottom, and side extrusions and other components where deposits have collected, then rinse cleaned areas with fresh water.
5. Replace cleaned components and turn on water.
6. To clean evaporator as well as the remaining recirculating-water system, add nickel-safe to the water in ice maker according to the manufacturer's instructions. If none are available, use 5 fl oz of nickel-safe per gallon of water in the machine.
7. Allow cleaning solution to circulate for up to ten minutes. It may be necessary to recirculate the solution for longer than ten minutes to remove heavier and thicker scale deposits. Be sure all distribution and weep holes are clear. Drain cleaning solution and flush with water for a minimum of thirty seconds. After flushing, plug the drain.
8. Thoroughly rinse bin with clean water after all components are cleaned.
9. Return machine to service, discard first batch of ice.

Remember, these directions are for Calgon Vestal Laboratories' product, nickel-safe ice machine cleaner, only. Other products may have different instructions and precautions.

Also note the danger label on this and all other products of this nature. Since they contain caustic substances that are eye and skin irritants, always follow the manufacturer's recommendations for safe handling.

# WATER COOLERS AND FOUNTAINS

The framework of a water cooler is usually made of steel and covered with a sheet-metal housing. Some water coolers provide only chilled water, and others produce both chilled and hot water.

Normally, access to the mechanism is provided by removing a side panel. Basins in these units are usually made of porcelain-coated cast iron or porcelain-coated steel to prevent leakage and corrosion. The condensing units (condenser, condenser fan, and compressor) are located on the bottom of the units while the evaporators are positioned above them. Styrofoam is used as insulating material in most water fountains.

A heat exchanger is used to precool the water by using the chilled waste water. Temperature of the drinking water should generally be around 50°F.

Businesses employing large numbers of people use water-cooling units with several dispensers (bubblers) and one large condensing unit.

The smaller type office water coolers with bubblers for dispensing hot water as well as cold have a hot-water storage tank and separate electrical heating units with thermostats to control the temperature.

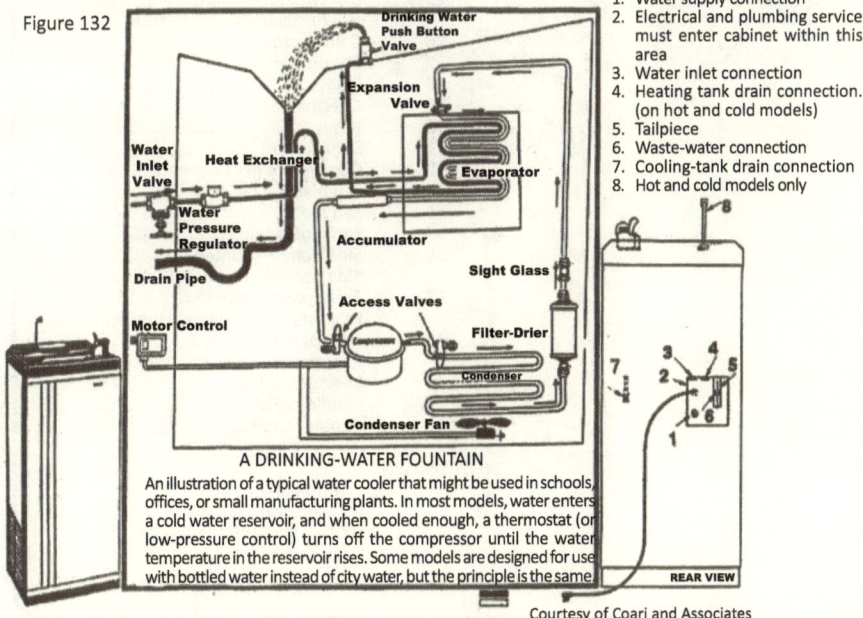

Figure 132

1. Water supply connection
2. Electrical and plumbing service must enter cabinet within this area
3. Water inlet connection
4. Heating tank drain connection. (on hot and cold models)
5. Tailpiece
6. Waste-water connection
7. Cooling-tank drain connection
8. Hot and cold models only

A DRINKING-WATER FOUNTAIN

An illustration of a typical water cooler that might be used in schools, offices, or small manufacturing plants. In most models, water enters a cold water reservoir, and when cooled enough, a thermostat (or low-pressure control) turns off the compressor until the water temperature in the reservoir rises. Some models are designed for use with bottled water instead of city water, but the principle is the same.

Courtesy of Coari and Associates

## A TYPICAL COMBINATION WATER FOUNTAIN AND HOT-WATER DISPENSER

A combination water fountain and hot-water dispenser typical of those available today.

This combination water fountain and hot water dispenser by OASIS has a 1/5 hp, 5 A compressor. It is charged with 5 oz of R-12 and produces cold water at the rate of 8 gph (gallons per hour).

| ITEM NO. | DESCRIPTION |
|---|---|
| 1 | Bubbler valve |
| 2 | Bubbler valve gasket |
| 3 | Friction washer |
| 4 | Gooseneck |
| 5 | Handle |
| 6 | Tee nut |
| 7 | Cover |
| 8 | Nut |
| 9 | Drain screw |
| 10 | Condenser shroud |
| 11 | Top |
| 12 | Top screw |
| 13 | Gasket (under top) |
| 14 | Waste gasket |
| 15 | Hot-valve gasket |
| 16 | Hot-valve body assembly |
| 17 | Check valve |
| 18 | Hot thermostat kit with clip |
| 19 | Heat limiter kit with clip |
| 20 | Hot tank |
| 21 | Precooler or waste assembly |
| 22 | Insulated cooling-tank assembly |
| 23 | Cooling-tank insulation (RH) |
| 24 | Cooling-tank insulation (LH) |
| 25 | Utility cover—painted |
| | Utility cover—stainless steel |
| 26 | Cold control |
| 27 | Condenser |
| 28 | Relay |
| 29 | Overload |
| 30 | Base |
| 31 | Service cord |
| 32 | Compressor (includes items 28 and 29) |
| 33 | Fan blade |
| 34 | Fan motor |
| 35 | Fan motor bracket |
| 36 | Side panel—painted |
| | Side panel—stainless steel |
| 37 | Front panel—painted |
| | Front panel—stainless steel |
| 38 | Rear panel—painted |
| | Rear panel—stainless steel |
| 39 | Shelf assembly |
| 40 | Paddle switch |
| 41 | Bubbler spud |
| NOT SHOWN | |
| | Terminal base |
| | Base cover |
| | Glass filler installation kit |
| | Foot pedal kit |

*Courtesy of EBCO Manufacturing Company*

# WATER FOUNTAIN TROUBLESHOOTING CHART

**Problem and Possible Cause**      **Remedy**

Water Does Not Come Out Cold while Compressor Is Running.

1. Freon leak in system.
2. Linted condenser.
3. Restriction.
4. Bad condenser fan.
5. Bad pressure control.
6. Faulty expansion valve.
7. Extensive use without giving water a chance to cool.
8. Leak in push-button valve.
9. Inefficient compressor.

1. Repair leak, change filter-drier, evacuate, and recharge.
2. Clean condenser.
3. Check by using touch method. Change filter-drier, evacuate, and recharge.
4. Repair or replace as required
5. Check and replace if necessary.
6. Check and replace TEV valve.
7. Capacity of unit may be too small for number of people using it.
8. Repair or replace valve.
9. Replace compressor.

Water Does Not Come Out Cool and Compressor Does Not Run.

1. No power to the unit.
2. Faulty motor-control pressure.
3. Incorrect voltage.
4. Bad thermostat. (In units that use a cold control in lieu of a pressure control.)
5. Bad starting relay.
6. Faulty starting capacitor.
7. Faulty overload protector.
8. Bad connection or broken wire.
9. Shorted compressor.

1. Check for blown fuse or tripped circuit breaker.
2. Observe safety instructions and bypass the two terminals in the control. If the compressor starts, it could be a bad pressure control or a leak. If manifold gauge shows vacuum in the black pressure (or close), there is a leak in the system. Repair leak, change filter-drier, and recharge. If pressures are normal, the pressure control needs to be replaced.
3. Check starting voltage drop in receptacle (as previously instructed) against required voltage on nameplate. Also check for undersized or a too long extension cord and replace it.
4. Observe safety rules and bypass the thermostat terminals. If compressor starts, thermostat is bad and must be replace.
5. Observing safety rules, check and replace relay if necessary.
6. Following the safety instructions, check and replace capacitor if necessary.
7. Observing safety precautions, check and replace the overload protector if necessary.
8. Check all wiring and connections and repair as necessary.
9. Test compressor and replace if necessary.

Water Does Not Come Out Cool while Unit Short-Cycles.

1. Faulty overload protector.
2. Faulty starting capacitor.

1. Check or replace if necessary.
2. Check or replace if necessary.

| Problem and Possible Cause | Remedy |
|---|---|
| 3. Faulty starting relay.<br><br>(This test has a potential shock hazard. Be sure to take all safety precautions before and during the procedure.) | 3. a. Unplug unit and allow compressor to cool.<br>b. Leave wire connected to compressor common terminal.<br>c. Disconnect the relay connection to the compressor run winding.<br>d. Connect a wire between compressor run terminal and the line that brings current to the run terminal of the relay.<br>e. Plug in the unit.<br>f. Momentarily, short between the compressor run and start terminals.<br>g. If compressor starts, relay needs to be replaced. |
| 4. Linted condenser. | 4. Clean condenser. |
| 5. Faulty condenser fan or stuck (or restricted) fan blade. | 5. Check condenser and fan blade. Clear restriction, or repair or replace fan blade. |
| 6. Loose connections. | 6. Check and tighten connections. |
| 7. Shorted or burned-out compressor. | 7. Replace compressor. |

**Water Does Not Come Out with Enough Pressure**

| | |
|---|---|
| 1. Low water pressure in building. | 1. Customer should contact local water company or plumber. |
| 2. Water push-button valve defective. | 2. Repair or replace valve. |
| 3. Inlet water valve (faucet) partially closed. | 3. Open faucet all the way. |
| 4. Bad water pressure regulator. | 4. If pressure is adequate, replace regulator. |
| 5. Clogged water valve inlet screen | 5. Clean or replace screen. |

**Water Does Not Come Out of Push-Button Water Valve**

| | |
|---|---|
| 1. Faulty water-dispensing valve. | 1. Repair or replace valve. |
| 2. No water supply in building. | 2. Advise customer to call local water company. |
| 3. Water inlet valve screen completely restricted by minerals. | 3. Check valve screen and clean or replace as necessary. |
| 4. Restricted water pressure regulator. | 4. Repair or replace regulator. |
| 5. Inlet water line kinked or restricted by minerals in water. | 5. Straighten or repair pipe or use a factory recommended chemical to clear line.* |

*Substantial experience has proven that Lime Away is an excellent product to do this job. If this, or any other similar product is used to dissolve waterborne minerals, the system must be thoroughly flushed with clear water prior to putting the fountain back in service. Even in small quantities, these chemicals can cause gastrointestinal injury or other problems if swallowed.

## TROUBLESHOOTING BY TOUCH

This method of troubleshooting refrigeration problems is very quick and easy and practiced by many technicians in the industry. As refrigerant circulates in the sealed system under different pressures, various temperatures are produced in its path. The hand-felt temperature of the components and tubing can be a good indication of the internal condition of the unit when compared with the standard expected temperatures when the unit operates in an ambient temperature between 75°F and 85°F.

*Compressor.* A warm to very warm compressor is normal. A hot or very hot compressor indicates a continuous excessive load or running time.

*Evaporator.* If it produces temperatures below 32°F, a frost pattern is observed. A restriction or a shortage of refrigerant is indicated by a below-average frost line. An inefficient compressor may show an above-average frost line. Suction-line temperature at the compressor should feel cool.

*Condenser.* The discharge line from the compressor should be very hot. The top of the condenser should feel very hot to the touch. It should gradually lose that heat as your hand passes down the surface to feel warm at the bottom. A clogged filter-drier or capillary tube can be indicated by a cool condenser because of a liquid buildup in the condenser.

*Filter-drier.* It should feel warm (about 12°F or 15°F above room temperature). A very warm or hot filter-drier can indicate poor air movement through the condenser coil or too much load on the evaporator (such as too much warm food to be cooled, etc.) A restriction in the sealed system is indicated by a cool or cold filter-drier.

*Capillary tube.* It should feel about room temperature coming from the filter-drier

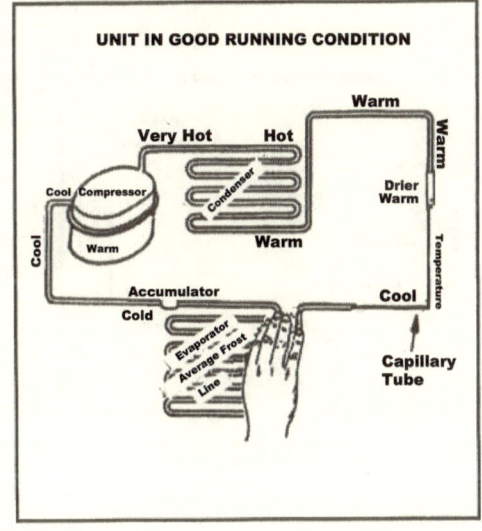

and cool upon entering the evaporator. A cold or frosted capillary tube is an indication of a restriction.

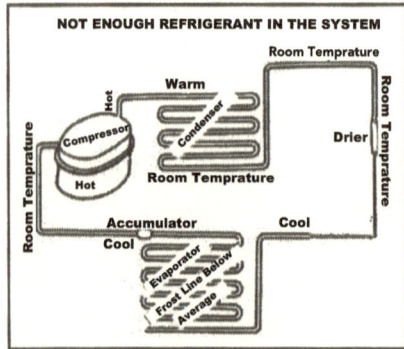
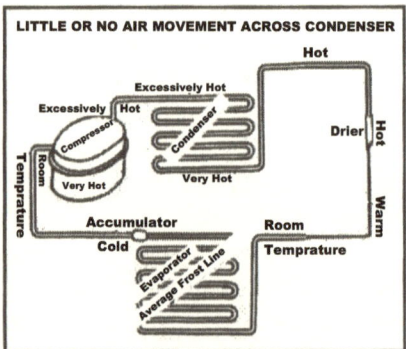
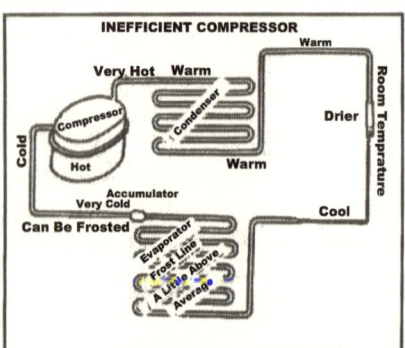
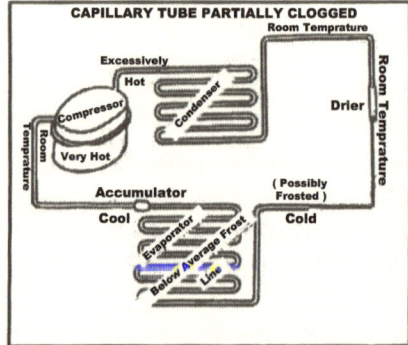
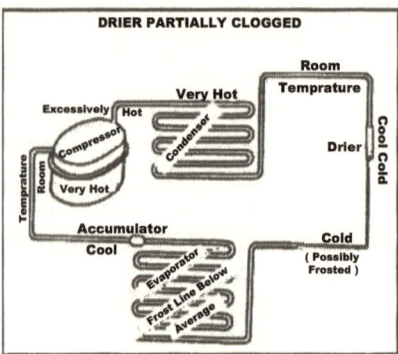
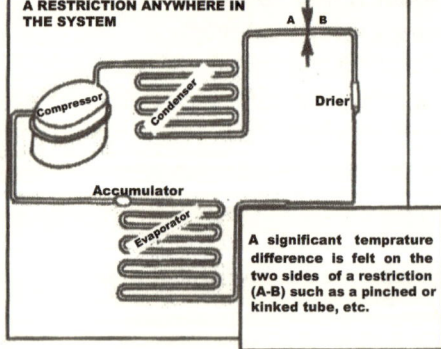

# REFRIGERATED DISPLAY CASES AND WALK-IN COOLERS

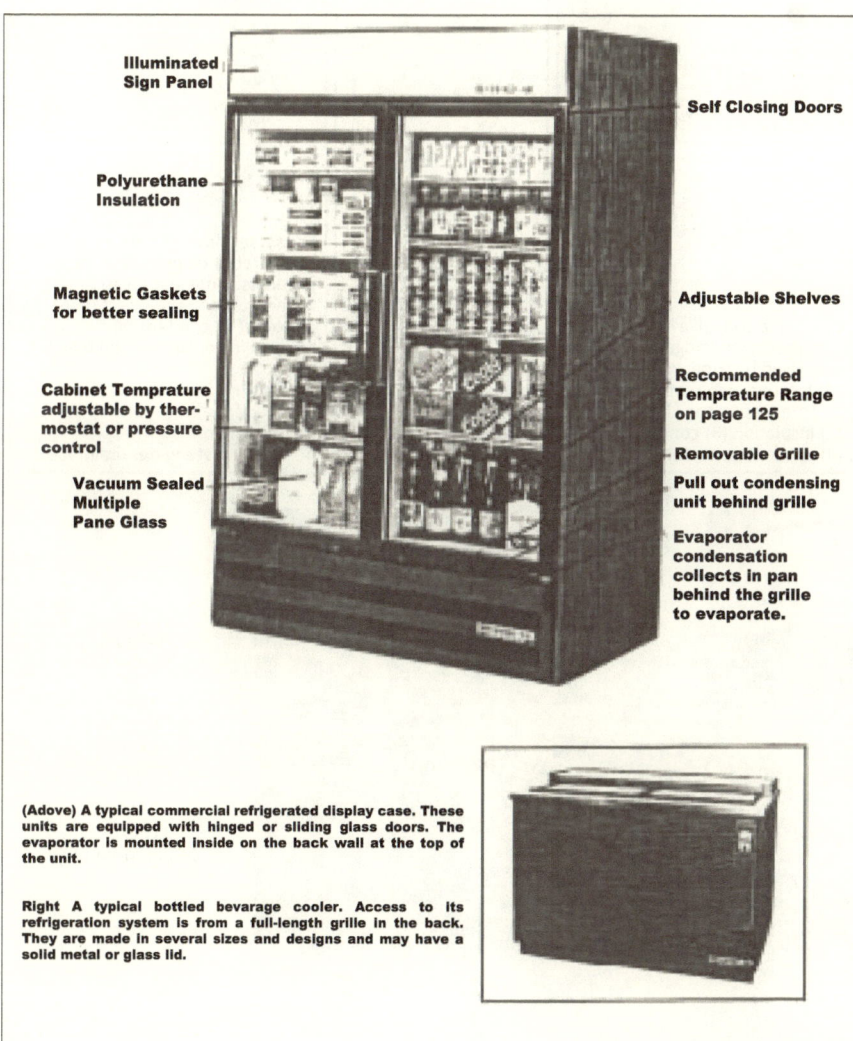

Illuminated Sign Panel

Polyurethane Insulation

Magnetic Gaskets for better sealing

Cabinet Temprature adjustable by thermostat or pressure control

Vacuum Sealed Multiple Pane Glass

Self Closing Doors

Adjustable Shelves

Recommended Temprature Range on page 125

Removable Grille

Pull out condensing unit behind grille

Evaporator condensation collects in pan behind the grille to evaporate.

(Above) A typical commercial refrigerated display case. These units are equipped with hinged or sliding glass doors. The evaporator is mounted inside on the back wall at the top of the unit.

Right A typical bottled bevarage cooler. Access to its refrigeration system is from a full-length grille in the back. They are made in several sizes and designs and may have a solid metal or glass lid.

Courtesy of Bevarage-Air

Figures a, b, and c are cross-sectional views of forced air convection evaporators as used in typical open refrigerated display cases and bottled beverage coolers. The fan circulates the air inside the cooler through the evaporator coil and back into the storage area again. The condensing unit could be inside the unit, enclosed in a metal housing located outside of the building or on the roof. (See figs. 102a and 118.)

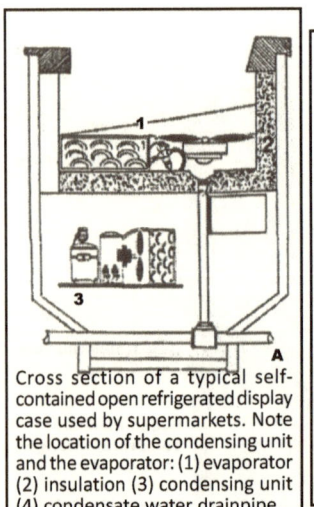

**A**
Cross section of a typical self-contained open refrigerated display case used by supermarkets. Note the location of the condensing unit and the evaporator: (1) evaporator (2) insulation (3) condensing unit (4) condensate water drainpipe.

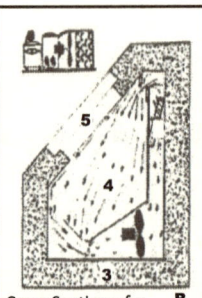

**B**
Cross Section of a FORCED AIR CONVECTION EVAPORATOR BOTTLED BEVERAGE COOLER
1. Fan
2. Evaporator
3. Foam insulation
4. Air circulation
5. Sliding glass door

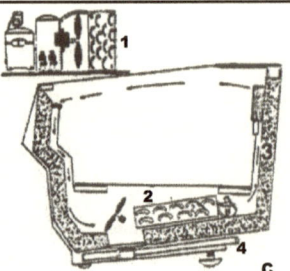

**C**
Cross section of a typical open refrigerated display case for meat products. Note the location of the evaporator and the direction of the airflow. The condensing unit could be installed outside of the building.

1. Condensing unit
2. Evaporator
3. Insulation
4. Condensate-water drainpipe

A TYPICAL APPLICATION OF OPEN REFRIGERATED CASES INSTALLED IN A LOCAL SUPERMARKET

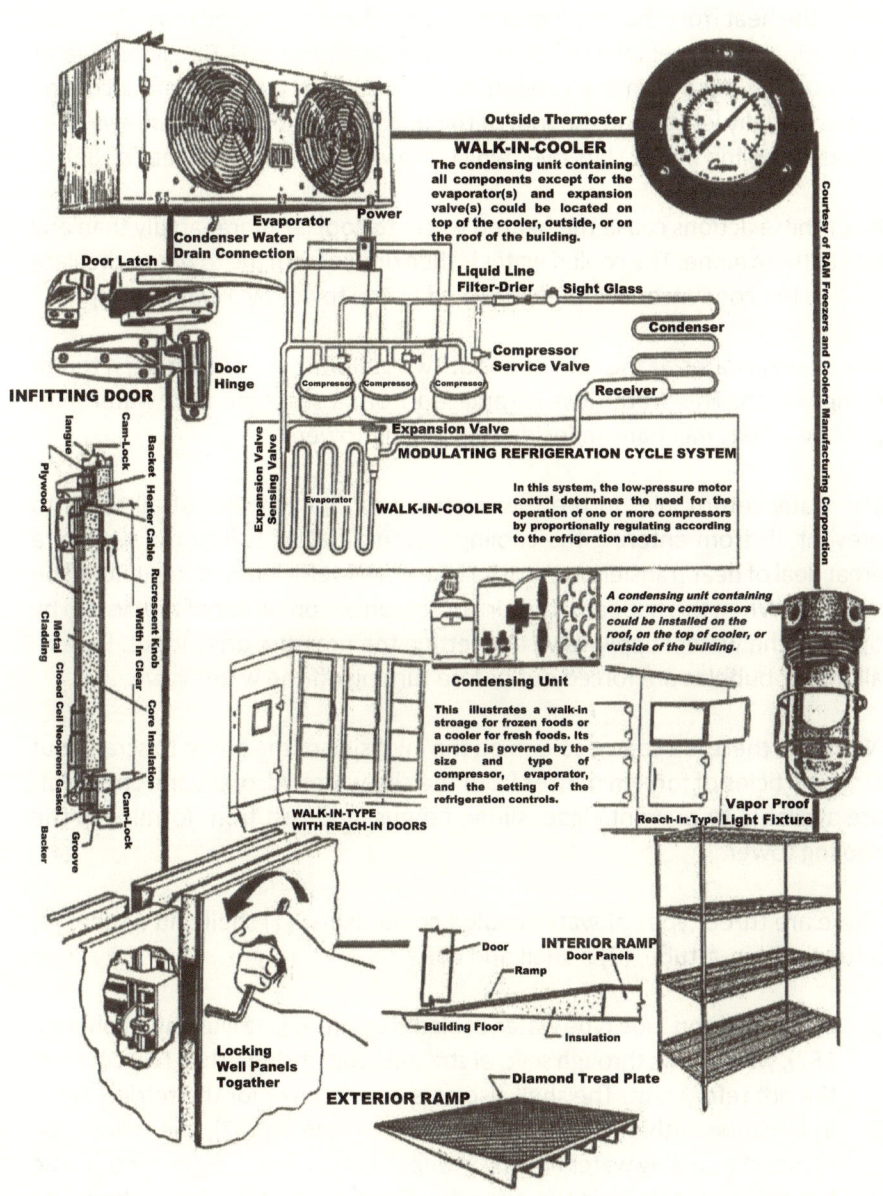

In many commercial refrigeration units that produce low temperatures, such as frozen yogurt machines, etc., the condenser is required to cool more rapidly than what can be accomplished by just circulating air through it. In these units, the heat from the condenser is absorbed and dissipated much faster by circulating cool water over or adjacent to the condenser coil. Circulating water picks up the heat from the condenser coil and then pumped into a cooling tower (usually located on the roof of the building), where a fan(s) draws air in through cooling fins or slats over which the water is sprayed. (See fig. 87)

All of these actions cause the heated water to cool far more rapidly than any one of them alone. The cooled water is then drawn from the tower to circulate next to the condenser coil then returned to the tower by a water pump.

Water evaporates at the rate of about two gallons per hour for each ton of refrigeration. This water loss is replaced from a reservoir or a water inlet pipe by a float mechanism inside the cooling tower.

The water-return inlet in the tower must stay under water at all times to prevent air from entering the cooling system. If air gets into the system, a great deal of heat transference will be lost and the efficiency of the unit will be reduced. When this happens, the air pocket can be forced out of the system by opening the water supply valve and letting the pressure drive it out. Be sure all the air bubbles are forced out before turning off the water valve.

Normally, there is a coarse screen on the inlet side of the pump to screen out large particles of foreign matter from the city water. Also, several chemicals are available to prevent algae, slime, fungus, and rust from forming in the cooling tower.

There are three types of water-cooled condensers: (1) shell and a tube, (2) a tube-within-a-tube, (3) a shell and coil.

1. In the shell and tube type water-cooled condenser (see illustration on page 187), water flows through several straight tubes inside a shell that contains the hot refrigerant. The shell also acts as a receiver for the refrigerant.
2. In the tube-within-a-tube type condenser (see fig. 87), the inner tube carries the cooling water flowing in one direction and in the outer tube the hot refrigerant flows in the opposite direction. Water velocity should be about seven to ten feet per second. If it is faster than that, it can remove the oxide coating and cause pitting. If it is circulated slower than three or four feet per second, it may cause scaling. The groove in the tube increases surface area and consequently provides a higher cooling efficiency.

3. In the shell and coil type condenser (see fig. 87), a coil of tubing is placed inside the condenser shell. In this arrangement, the cooling water flows through a coil rather than several straight tubes.

Sometimes the condenser is placed directly inside the cooling tower and the air-cooled water runs right over it. Continuous circulation of water is provided by a water pump installed in the piping. In smaller systems, city water is circulated through a tank that contains the condenser coil. Water flow through this tank is controlled by a pressure-actuated, water-circulation control valve, which has a tube connected to the compressor-discharge line. When the head pressure goes up, the pressure exerted from the discharge line to a diaphragm within the water valve causes the valve to open further and let more cool water flow into the condenser until the condenser cools and the head pressure drops. At this point, the pressure exerted on the valve diaphragm is decreased, causing the water valve to decrease or shut off the flow of water. (See picture at upper right.) For more details, refer to page 360.

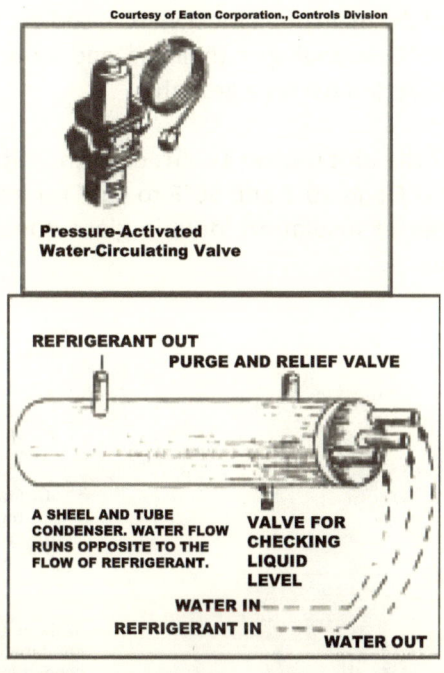

## WALK-IN CABINETS

They are often referred to as walk-in coolers and are in wide general use in restaurants, supermarkets, convenience stores, and floral shops. Their size varies according to the needs of the business, but their height is generally seven feet seven inches up to nine feet ten inches. Most units are prefabricated, making transportation and setup easier. They are usually constructed of galvanized aluminum or stainless steel for the exterior and have metal inside walls and foam for insulation. They are sometimes called knockdown boxes. For safety, they must have door latches that can be opened from the inside.

Evaporators in these units are usually wall mounted with two, three, or four fans depending on the size of the unit. Condensing units are placed outside or on the roof.

Some walk-in coolers have a frozen-food compartment with a separate evaporator and condensing unit. Many walk-in coolers installed in restaurants have reach-in compartments with separate doors for frequently used small items. This saves time for the food handlers and prevents frequent opening of the main door. Most walk-in coolers have a temperature range of 35°F to 45°F depending on their use, and some are equipped with ultraviolet lamps to retard bacteria growth.

If the unit is used as a freezer, the cabinet temperature ranges between 10°F and 30°F and 50°F to 55°F for a floral display and storage case. For better insulation, double glass doors and windows are used in these units.

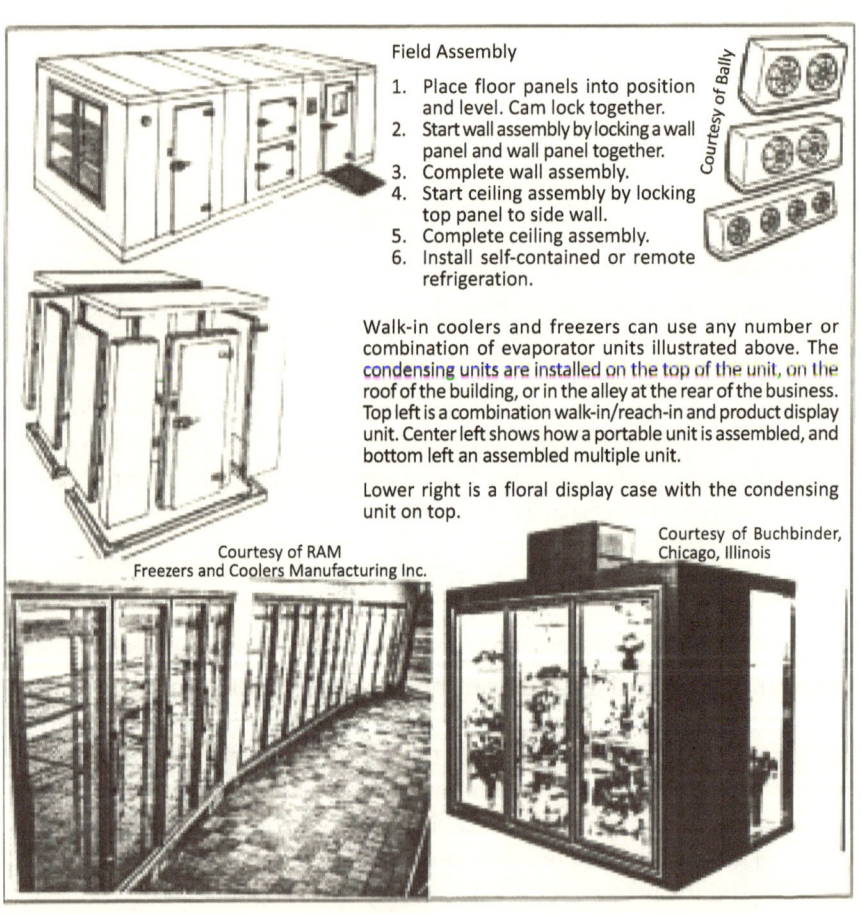

Field Assembly
1. Place floor panels into position and level. Cam lock together.
2. Start wall assembly by locking a wall panel and wall panel together.
3. Complete wall assembly.
4. Start ceiling assembly by locking top panel to side wall.
5. Complete ceiling assembly.
6. Install self-contained or remote refrigeration.

Courtesy of Bally

Walk-in coolers and freezers can use any number or combination of evaporator units illustrated above. The condensing units are installed on the top of the unit, on the roof of the building, or in the alley at the rear of the business. Top left is a combination walk-in/reach-in and product display unit. Center left shows how a portable unit is assembled, and bottom left an assembled multiple unit.

Lower right is a floral display case with the condensing unit on top.

Courtesy of RAM Freezers and Coolers Manufacturing Inc.

Courtesy of Buchbinder, Chicago, Illinois

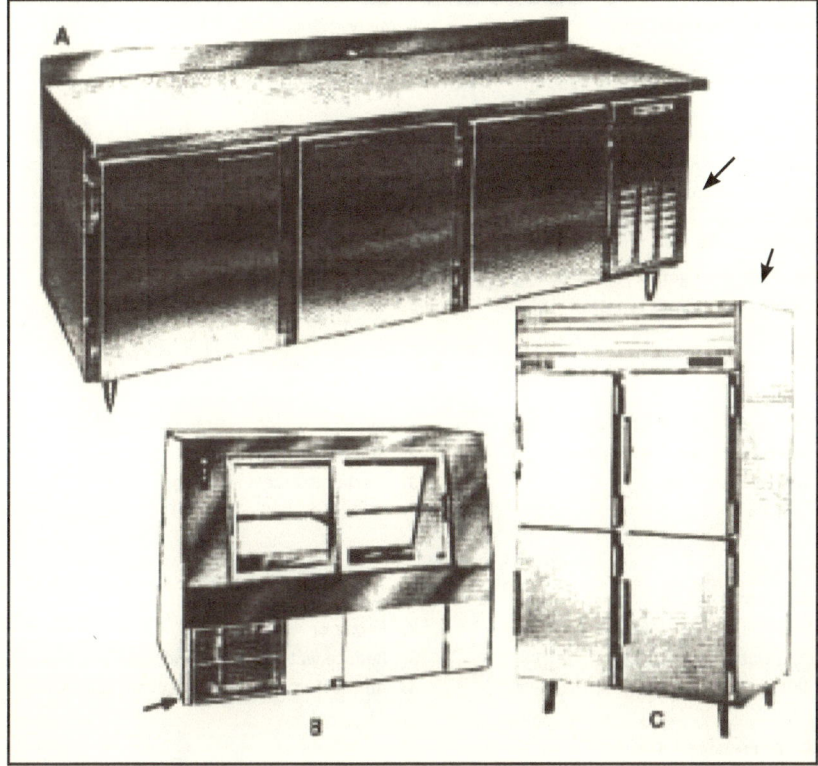

Courtesy of RAM Freezers and Coolers Manufacturing Inc.

A. Food-preparation refrigerator (Note location of condensing unit behind grille)
B. Rear view of a typical refrigerator case. (Condensing unit behind grille)
C. Commercial kitchen self-contained refrigerator/freezer (Condensing unit located on top behind grille)

*Bottom left*: Inside of a typical walk-in cooler. Arrow under evaporators points to condensate-water drain hose.

# COMPRESSOR TROUBLESHOOTING CHART
# (COMMERCIAL AND RESIDENTIAL UNITS)

**Problem and Possible Cause** | **Remedy**

Suction Pressure Too Low

1. Low refrigerant.
2. Restriction in liquid-line filter-drier.
3. Clogged expansion valve strainer.
4. Kinked tubing.
5. Insufficient refrigerant passing through valve.
6. Evaporator fan runs too slow.

1. Add refrigerant.
2. Replace filter-drier if it feels cold.
3. Remove and clean.
4. Inspect all tubing, particularly between expansion valve and compressor.
5. Turn valve stem for higher flow.
6. Replace evaporator fan.

Suction Pressure Too High

1. Refrigerant control valve open too far.
2. Low compressor suction power.
3. System overcharged.

1. Adjust valve for lower flow.
2. Replace compressor.
3. Discharge excess refrigerant.

Discharge pressure too low

1. Refrigerant charge too low.
2. Leaky discharge valve.
3. Cold water flow too high.**
4. Bad refrigerant control valve.

1. Add refrigerant.
2. Repair or replace valve.
3. Reduce water flow to condenser.
4. Adjust (turn stem clockwise) or replace valve.

Discharge Pressure Too High

1. Condenser linted or blocked.
2. Air or moisture in system.
3. Too much liquid in receiver.
4. Water flow too low.**
5. Water too warm at inlet.**

1. Clean condenser, remove obstruction.
2. Evacuate and recharge system, increase water flow, and check lines and pump.**
3. Discharge excess refrigerant.
4. Increase water flow to condenser; clean clogged water pump strainer or lines.
5. Increase water flow.

High-Pressure Cut-Out Causes Compressor to Stop

1. Too much refrigerant.
2. Cut-out setting incorrect.
3. Water-cooling failure.**
4. Water pump failure.**
5. Water line clogged.**

1. Discharge excess refrigerant.
2. Adjust high-pressure cut-out.
3. Check replace water failure switch.
4. Check pump and motor.
5. Check water strainer, valves, and lines.

Low Pressure Control Causes Compressor to Short-Cycle

1. Too much frost on evaporator.
2. Refrigerant control valve screen clogged.

1. Defrost evaporator coil.
2. Remove and clean screens.

| Problem and Possible Cause | Remedy |
|---|---|
| 3. Low refrigerant charge. | 3. Add refrigerant. |
| 4. Restriction in liquid-line filter-drier. | 4. Replace filter-drier if it feels cold. |
| 5. Clogged expansion valve strainer. | 5. Remove and clean if valve frosted. |
| 6. Kinked tubing. | 6. Inspect tubing, repair as necessary. |
| 7. Insufficient refrigerant passing through valve. | 7. Adjust valve stem for higher flow. |
| 8. Evaporator fan runs too slow. | 8. Replace evaporator fan. |

Compressor Will Not Start

| | |
|---|---|
| 1. Unit low on refrigerant. | 1. Check for leak, repair as necessary. |
| 2. Low-pressure cut-out open leak. | 2. Check head and back pressures for leaks and check low-pressure switch. |
| 3. High-pressure cutout open. | 3. Depress reset button. |
| 4. Water failure switch open.** | 4. Check for adequate water supply.** |
| 5. Overload has disconnected power. | 5. Depress reset button. |
| 6. Solenoid valve closed. | 6. Verify thermostat is on and power at the valve. |
| 7. Low line voltage. | 7. Check voltage; notify owner to call electrician for remedy. |
| 8. Defective start relay. | 8. Check and replace if necessary. |
| 9. Defective compressor. | 9. Check and replace if necessary. |

Compressor Never Stops Running

| | |
|---|---|
| 1. Compressor has low compression. | 1. Test, replace if necessary. |
| 2. Low refrigerant charge. | 2. Add refrigerant. |
| 3. Linted condenser coil. | 3. Clean condenser coil. |
| 4. System overcharged. | 4. Discharge an appropriate amount of refrigerant. |
| 5. Faulty low-pressure control. | 5. Check and replace if necessary. |

Overload Protector Causes Compressor to Short-Cycle

| | |
|---|---|
| 1. Faulty overload protector. | 1. Check, replace if necessary. |
| 2. Faulty start relay. | 2. Check, replace if necessary. |
| 3. Faulty start capacitor. | 3. Check, replace if necessary. |
| 4. Linted condenser. | 4. Clean condenser coils. |
| 5. Restriction or overcharge. | 5. Check for restriction and/or discharge refrigerant as necessary. |

Freezer Is Too Warm with Compressor Running

| | |
|---|---|
| 1. Cold control set wrong. | 1. Adjust thermostat and/or recalibrate cut-in and cut-out pressure control. |
| 2. Bad door gasket. | 2. Replace gasket; check for warped door. |
| 3. Faulty compressor. | 3. Check high- and low-side pressure. If head pressure reads too low, and back pressure reads too high, replace the compressor. |
| 4. Undercharged system. | 4. Check for leaks and charge system if necessary. |

| Problem and Possible Cause | Remedy |
|---|---|
| 5. Restriction in system. | 5. Check for bent tubing, a cold drier, blocked, or improperly set valves.<br><br>If a pressure increase is created behind a restriction point, it will feel warmer or hotter than usual, and a pressure drop will feel colder than usual. In other words, the temperature difference before and beyond a restriction point becomes drastic. |

Refrigeration Unit Is Noisy

| | |
|---|---|
| 1. Compressor mounts loose. | 1. Check rubber mounts for wear; replace, adjust, or tighten mounts and bolts as necessary. |
| 2. Vibrating tubing. | 2. Secure tubing with tape, wood, foam or rubber blocks. |
| 3. Unit not level or on weak floor. | 3. Residential refrigerators and freezers require leveling every time they are moved or if flooring is uneven. |

Compressor Loses Oil

| | |
|---|---|
| 1. Low Freon-oil ratio. | 1. (1 pt oil/10 lbs Freon) Keep ratio same if adding to factory charge. |
| 2. Oil trapped in lines. | 2. Position tubing to drain into compressor. |
| 3. Refrigerant low. | 3. Check for leaks; add refrigerant and oil. |
| 4. Unit short-cycles. | 4. See other remedies under this title. |

**Applies only to those units have water-cooled condensers.

# ELECTRICAL TROUBLESHOOTING CHART
# COMMERCIAL AND RESIDENTIAL UNITS

| Problem and Possible Cause | Remedy |
|---|---|
| **Compressor "Hums" But Will Not Start (Cycling on Overload)** | |
| 1. Contacts in relay not closing. | 1. Check and replace relay if necessary. |
| 2. Wired wrong. | 2. Check wiring with schematic diagram. |
| 3. Start winding open or shorted. | 3. Check winding; replace compressor if necessary. |
| 4. Low voltage. | 4. Check voltage at source to determine that it is not more than a 10% drop. |
| 5. Discharge pressure too high. | 5. Check for excessive amount of Freon, restriction, linted condenser, bad condenser fan, and that discharge service is fully open. |
| 6. Frozen compressor. | 6. Check, replace if necessary. |
| **Compressor Will Not Run** | |
| 1. Electrical control open (such as a pressure control, thermostat, etc.). | 1. Check, replace if necessary. |
| 2. No power. | 2. Check power source, fuses, or circuit breakers. |
| 3. Blown fuse or tripped breaker. | 3. Replace fuse or reset breaker. |
| 4. Motor or compressor "frozen." | 4. Replace compressor if all else is OK. |
| 5. Overload relay tripped. | 5. Reset overload switch. |
| 6. Control location too cold. | 6. Relocate to warmer place where it will function properly. |
| 7. Compressor piston stuck. | 7. On other than hermetic compressors, remove head and check for broken or jammed parts. Replace hermetic. |
| **Compressor Runs Only on Start Winding** | |
| 1. Run windings shorted. | 1. Check compressor terminals with ohmmeter, replace if necessary. |
| 2. Shorted run capacitor. | 2. Disconnect, discharge, and check. |
| 3. Weak start capacitor. | 3. Check and replace if necessary. |
| 4. Line voltage too low. | 4. Try switching to another circuit or call electrician. |
| 5. Discharge pressure too high. | 5. Check pressure and shut-off valve. |
| 6. Faulty relay. | 6. Test and replace if necessary. |
| 7. Wired wrong. | 7. Check wiring with schematic diagram. |
| **Run Capacitor Burnout** | |
| 1. High voltage or excessive running time. | Make sure line voltage is not over 10% above compressor rating; if so, call electrician to correct it. |
| **Start Capacitor Burnout** | |
| 1. Sticking relay contacts. | 1. Clean contacts or replace relay. |

| Problem and Possible Cause | Remedy |
|---|---|
| 2. Wrong capacitor | 2. Verify correct capacitor rating. |
| 3. Unit short-cycles. | 3. Reduce cycling frequency by in-increasing the differential on the pressure control or change capacitor. |
| 4. Runs too long on start winding | 4. Check for low voltage or reduce frequent starting by regulating valve installed on suction line (in multiple evaporator system). |

**Relay Malfunction or Burnout**

| | |
|---|---|
| 1. Relay vibrates. | 1. Secure mounts or move relay to a less shaky location. |
| 2. Unit short-cycles. | 2. Reduce cycling frequency. (See 3 above.) |
| 3. Wrong relay. | 3. Install a relay compatible with compressor motor specifications. |
| 4. Relay mounting or connections wrong. | 4. Install relay correctly and check for proper connections. |
| 5. Wrong run capacitor. | 5. Install proper capacitor. |
| 6. Line voltage too low. | 6. Voltage cannot be more than 10% lower than motor rating. |
| 7. Line voltage too high. | 7. Voltage cannot be more than 10% above motor rating. |

## OTHER PROBLEMS

Unit Runs Too Long

| | |
|---|---|
| 1. Dirty coil/heavily frosted coil. | 1. Clean condenser or de-ice evaporator. |
| 2. Inefficient compressor. | 2. Replace hermetic type/have others repaired. |
| 3. Low refrigerant charge. | 3. Repair leaks, add correct charge. |
| 4. Air or moisture in system. | 4. Purge system, replace filter-drier, and recharge. |
| 5. Evaporator coil too small. | 5. Replace with larger coil compatible with compressor horsepower. |
| 6. Tubing too small or restricted | 6. Replace with larger tubing, remove |
| 7. Pressure/temperature control contacts struck closed. | 7. Check, replace control if necessary. |
| 8. Expansion valve too small. | 8. Replace with larger valve. |
| 9. Expansion valve open too far. | 9. Reset valve. (Turn stem clockwise.) |
| 10. Ambient air too warm | 10. Provide better cooling or move condenser to cooler location. |

Unit Short-Cycles

| | |
|---|---|
| 1. Low on refrigerant. | 1. Add proper type and amount of Freon. |
| 2. Overload protector cuts out. | 2. Look for linted condenser, slowed water cooling, pressure too high, or inefficient compressor. |
| 3. System overcharged. | 3. Release Freon (may be necessary to purge system.) |
| 4. Cut-in/cut-out differential set too close. | 4. Increase differential. |

| Problem and Possible Cause | Remedy |
|---|---|
| 5. Unit with water-cooled condenser cycles on high-pressure cut-out. | 5. Look for low water supply or flow. |
| 6. Leaky valve on discharge line. | 6. Check, replace valve if necessary. |
| 7. Leaky expansion valve. | 7. Replace valve. |

**Suction Line Has Condensation or Frost**

| | |
|---|---|
| 1. Expansion valve open too far. | 1. Readjust valve. |
| 2. Inefficient compressor. | 2. Replace compressor. |

**Liquid Line Is Frosted**

| | |
|---|---|
| 1. Restriction in filter-drier or strainer. | 1. Replace drier and/or strainer. |
| 2. Valve on liquid receiver plugged or partially closed. | 2. Remove restriction or open valve. |

**Hot Liquid Line**

| | |
|---|---|
| 1. Expansion valve open too far. | 1. Readjust valve. |
| 2. System low on refrigerant. | 2. Look for leak and recharge system. |

**Upper Condenser Coil Cool When Compressor Runs**

| | |
|---|---|
| 1. System low on refrigerant. | 1. Look for leak and recharge system. |
| 2. Inefficient compressor. | 2. Check, replace if necessary. |

NOTE: If the expansion valve is frosted as unit runs on vacuum, check for a clogged strainer or an orifice plugged with ice. Remove the restriction, or replace valve, or wrap valve with hot, wet cloth until suction pressure reduces and replace drier.

# A TEST OF KNOWLEDGE ON CONTROLS AND SERVICE VALVES

1. How can a particular evaporator temperature by achieved? (p. 132)
2. What evaporator temperature is required for a reach-in refrigerator? (p. 134)
3. How is the high-side pressure determined? (p. 135)
4. Where is a sight glass installed? (p. 138)
5. What is the purpose of a check valve? (p. 144)
6. Where is a pressure-regulating valve installed? (p. 145)
7. How does an oil separator work? (p. 154)
8. What is the function of a service valve? (p. 156)
9. What is the purpose of a reversing valve? (p. 156)
10. How does a bulb-type expansion valve work? (pp. 159,160)
11. What is the function of a thermistor? (p. 165)
12. How does an automatic-expansion valve differ from a capillary tube? (p. 167)
13. What is a patented tube? (p. 65)
14. How would the length of a 0.042 inch capillary tube be determined for a 1/4 hp low-temperature compressor? (p. 66)
15. What symptoms indicate a loss of refrigerant? (pp. 176,177)
16. What symptoms indicate a restriction in the sealed system? (pp. 176,177)
17. What causes high head pressure? (p. 180)
18. Would the symptoms of an overcharged system be the same or different from a system with a dirty condenser? (pp. 181,182)
19. When is heat applied to an evaporator of an ice machine? (p. 187)
20. Do ice flakers use a defrosting method in the harvest cycle? (p. 189)
21. What are the symptoms of an open bin thermostat in an ice machine? (p. 191)
22. When is power to the ice cutter grid shut off? (p. 182)
23. What would happen in an ice maker if the bin thermostat contacts were to stick in the closed position? (p. 198)
24. What indicates a bad expansion valve in an ice machine? (pp. 199,200)
25. What problem would an incorrectly wired relay cause? (p. 201)
26. How is calcium removed from evaporator plates? (p. 202)
27. Describe the operation of a heat exchanger in a water fountain. (p. 203)
28. What would be a symptom of a dirty condenser in a water cooler? (pp.205,204)
29. How is water temperature affected in a water cooler by a faulty start capacitor? (p. 205)
30. What can be wrong if the filter-drier is hot to the touch? (p. 208)

31. What can be wrong if the low-pressure line from the evaporator feels considerably colder than normal? (p. 208)
32. In what position should the suction-line access valve stem be when the system is being charged? (p. 155)
33. In how many locations are access valves usually installed? (p. 155)
34. Where are oil separators installed in a sealed system? (p. 154)
35. How does a solenoid valve operate? (pp. 148 and 149)
36. Where are check valves installed in the sealed system? (p. 144 and fig. 96)
37. How is a low-pressure control checked? (p. 142)
38. Explain how to determine the cut-in and cut-out setting on a low-pressure control. (pp. 141, 142, 140)
39. What is the purpose of the two screws on top of the low-pressure control? (p. 141)
40. How many types of pressure controls are there? (p. 139)
41. How many functions do sight glasses perform? (p. 139)
42. Explain the difference between R-12 and R-22 refrigerants. (p. 137)
43. What refrigerant is used as a flushing agent? (p. 137)
44. What could cause frost on the expansion valve while unit runs on vacuum? (pp. 220, 221)
45. What is the head pressure in a commercial unit using R-22 at an ambient temperature of 90°F? (pp. 133 and 135)
46. What is the recommended refrigerator temperature for a floral display case? (p. 134)
47. What temperature does R-12 produce at 30 psi? (p. 133)
48. What temperature does R-22 produce at 61.5 psi in an evaporator? (p. 133)
49. What does the oil-pressure safety control do when the low-side pressure goes higher than the oil pressure in the compressor? (p. 144)
50. Does the solid-state thermostatic expansion valve operate by pressure in the evaporator? (p. 165)
51. What are the possible causes for the refrigeration unit to run too long? (pp. 219, 220)
52. What are the possible causes for a start capacitor to burn out? (pp. 219, 220)
53. What are the possible causes when a compressor hums but does not start? (p. 219)
54. What are the three rules of thumb for superheat adjustment in different units? (p. 161)
55. When the tubing is kinked, what symptoms manifest themselves in the unit? (p. 208)

# REPAIR TECHNIQUES IN COMMERCIAL AND RESIDENTIAL AIR-CONDITIONING UNITS

This section covers step-by-step procedures in troubleshooting and repair techniques for commercial and residential central air conditioners, wall and window units, heat pumps, rooftop console air conditioners, and split systems.

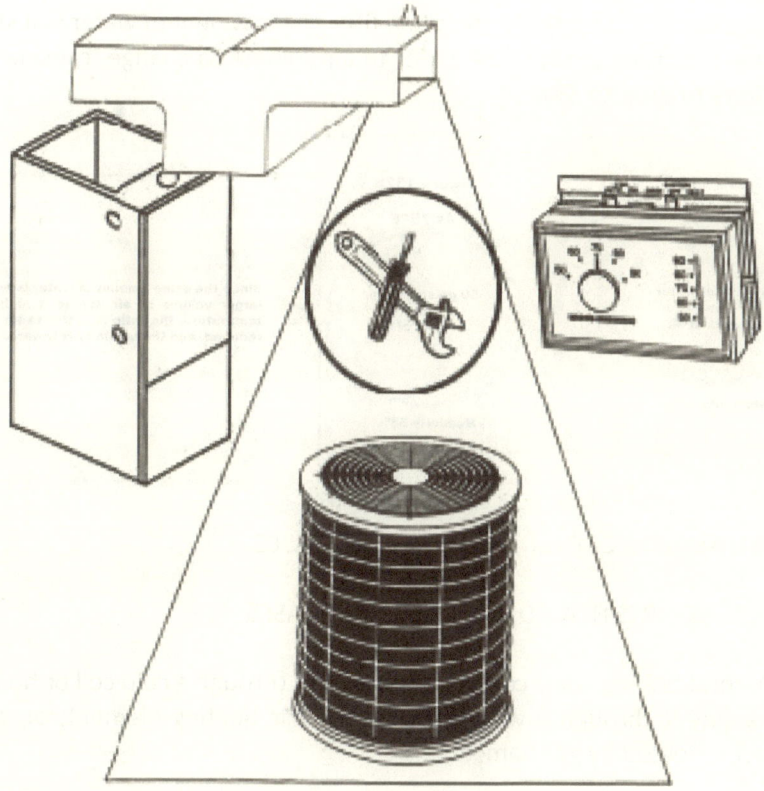

Courtesy of Coleman Heating and Air-Conditioning

Since air has weight, it takes energy to move it, and it is capable of transferring heat. As the temperature of air rises, it expands and occupies more space. When its temperature drops, it contracts, causing it to occupy less space due to the rule of expansion and contraction.

Humidity: The amount of water vapor in the air.

Air consists of several gases, one of which is water vapor. The higher the amount of vapor per cubic foot, the higher the humidity. The average comfortable humidity at 75°F is 50%. As the humidity increases, it creates discomfort, and as it drops, it feels drier. In desert areas, humidity is sometimes under 3%, and that is why air feels so dry and uncomfortable.

Figure 132a shows an imaginary body of fifty grams of air holding fifty grams of water at 50°F (constituting a 50% humidity). As you see in figure 132b, if the temperature of the fifty grams of air is raised to 75°F, it will expand (a larger volume of air holding the same fifty grams weight of water in it stays constant) causing the ratio of water to air volume to change, causing the humidity to drop to 33%.

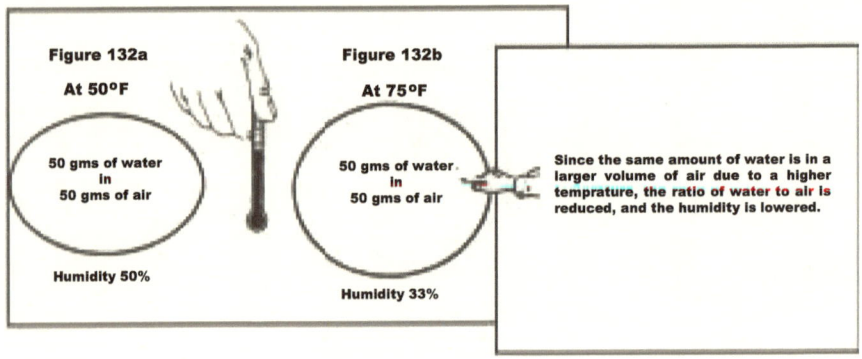

RULE I: AS AIR IS COOLED, HUMIDITY INCREASES.

RULE II: AS AIR IS HEATED, HUMIDITY DECREASES.

In air-conditioning, air is cooled as it is forced through a cold coil or heated by passing it through a warm coil, an electric heating element, or, over surfaces, heated by gas flames.

The capacity of an air conditioner is measured by the British thermal unit (Btu). A Btu is the amount of heat required to raise the temperature of 1 lb of water 1°F.

When an air-conditioning unit is rated at 10,000 Btu, it means that it is capable of removing 10,000 Btu every hour. The larger the Btu capacity, the bigger the area the unit is capable of heating or cooling. Usually, a unit is rated by its tonnage (12,000 Btu are equal to 1 ton). The tonnage can be calculated by dividing the Btu rating on the nameplate by 12,000. If the nameplate indicates that the unit is rated at 24,000 Btu, dividing the 24,000 by 12,000 will determine the tonnage corresponding to the indicated Btu which will equal to 2 tons. A 4000 Btu unit will be equal to four-twelfths of a ton, or one-third ton.

A frequently used term is *energy efficiency ratio* or EER. The ratio of Btu to the wattage rating of the unit is referred to as the *energy efficiency ratio*. This indicates the cooling or heating capability of the unit as compared with the electricity it uses.

### Btu divided by watts equals EER

The higher the ratio, the more efficient the unit. Today, air-conditioning units have an energy efficiency ratios of up to twenty-seven (these are the more sophisticated ones with variable speed fans and more energy-efficient motors).

## LATENT HEAT AND THE THEORY OF CHANGE OF STATE

To vaporize water, it must be brought to a boil, which is 212°F at sea level. At this point, the water is still in a liquid state.

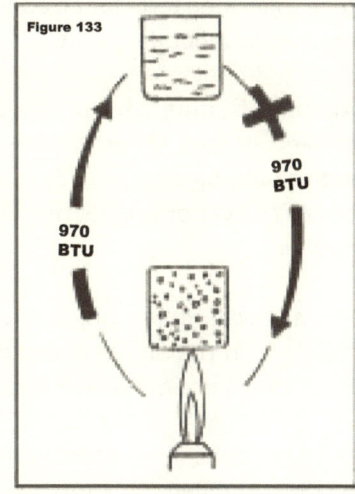

Figure 133

As the water is further heated, the temperature remains at 212°F until an additional 970 Btu are applied. At that time the water begins to vaporize even though the temperature does not increase. In other words, water boils at 212°F and vaporizes at 212°F plus 970 Btu (see fig. 133). The additional 970 Btu applied to the heated water to change its state from liquid to vapor is referred to as *latent heat*.

Latent heat, even though not measurable by a thermometer, makes the change from one state to another possible. Conversely, should the vapor give up 970 Btu, it would change to a liquid state without changing its temperature.

The process whereby latent heat is absorbed by the evaporator and radiated through the condenser is the principle of modern refrigeration and air-conditioning. To change liquid refrigerant to its vapor state when it is forced into the low-pressure environment in the evaporator, it must absorb the needed latent heat from the environment around the evaporator. When the environment around the evaporator loses its heat to the vaporizing refrigerant through the walls of the evaporator coil, it becomes cooler.

When the vapor refrigerant gives up its latent heat going through the condenser, it changes back to its liquid state. This latent heat is then radiated into the surrounding atmosphere.

In an air conditioner, inside air is drawn in and circulated through a cold or warm coil (or an electrically heated element or a gas-type heat exchanger) and forced back into the conditioned area.

In the cold seasons, since the heating of air causes the humidity to drop (figs. 132 a and b), some water is vaporized and mixed with the return air to rehumidify the conditioned area. In summertime, the air conditioner draws the inside air through the evaporator coil and returns it to the conditioned area. As the air cools, its humidity is increased (figs. 132a and 132b). To eliminate the excess humidity, a second cold coil is used to collect condensation as the cooled air is drawn through the unit by a blower fan(s). This moisture, extracted from the chilled air, flows to the outside through a drainpipe. This coil is referred to as a dehumidifier, which is nothing more than a glorified term for an additional evaporator coil installed in the system or as an independent unit.

Air-conditioning consists of six major functions: (1) circulating air by blowers or fans, (2) ventilating air, (3) heating air by bringing it into contact with a heating element, (4) cooling air by bringing it into contact with cold coil of an evaporator, (5) humidifying, and (6) dehumidifying, as explained earlier.

In small wall-mount units, as air comes in contact with the cold coil, the moisture in the air condensates and drips into a pan underneath, then it flows into a drainpipe and to the outside by gravity.

In larger central air-conditioning units, since there is much more air to be dehumidified, an extra cold coil is installed to collect and condensate moisture from the air.

This moisture is collected in a pan and conveyed to the outside by a pump or by gravity. (See fig. 134)

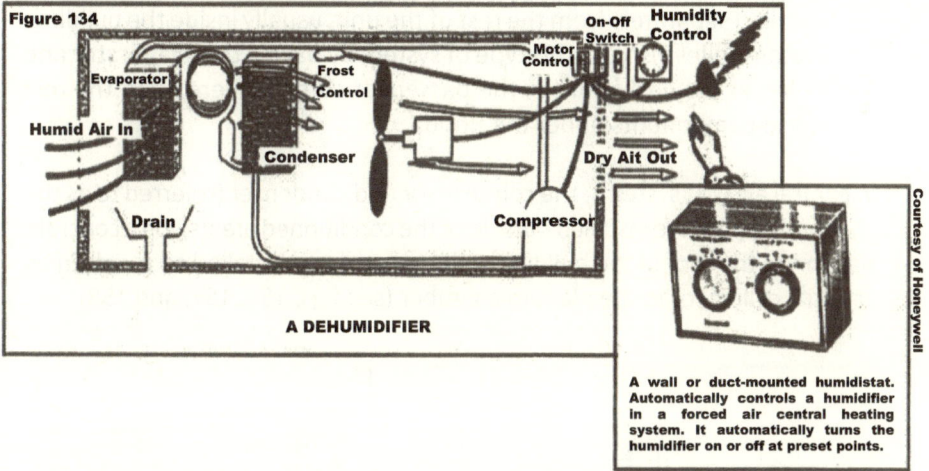

During cold seasons, heated air becomes drier. Some central air conditioners provide added moisture by evaporating water in a pan kept replenished by a float mechanism installed in the pan, which operates a switch in the electrical circuit of a solenoid water valve. This water is vaporized by an electric heating element installed in the pan. The vaporized water is then circulated in the conditioned area by a fan or blower. The operation of the electric heater and the blower is controlled by a humidistat. Figure 135 shows a basic humidifier in which the rotation of a belt made of a porous material and a heater cause a rapid vaporization of the water, which is blown into the conditioned area.

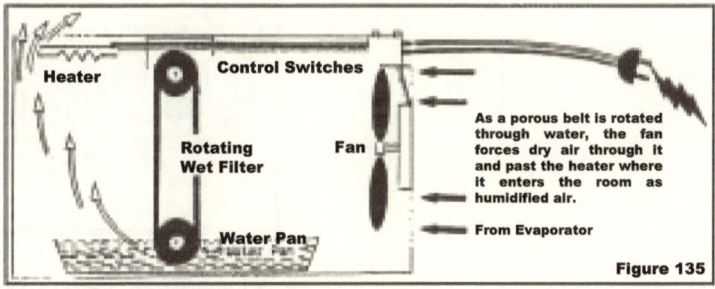

There are basically two different systems employed in air-conditioning: self-contained air conditioners and split systems.

1. *Self-contained air conditioners* are the units in which the compressor, evaporator, condenser, and the blower fans are all contained in a single

housing. A good example of such a unit is a window-mounted-type air conditioner. (Also see fig. 148).
2. *Split systems*. In this system, the evaporator and its blower fan(s) are installed separately from the rest of the unit, usually inside the building. (Automobiles also use this type of system by having the evaporator and its fan in the dashboard in the passenger compartment, with the rest of the components under the hood).

In central air split systems, the compressor and condenser (referred to as the condensing unit) are installed outside of the conditioned area(s), on a concrete slab, or on the roof of the building. The evaporator is installed on a wall, in an artificial ceiling, or in the plenum chamber (see figs. 156, 157, and 158).

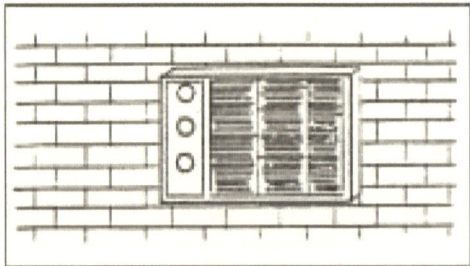

**Figure 136b**

**A window- or wall-mount type air conditioner mounted on the wall**

**Figure 136**

**A typical window-type air conditioner installation**

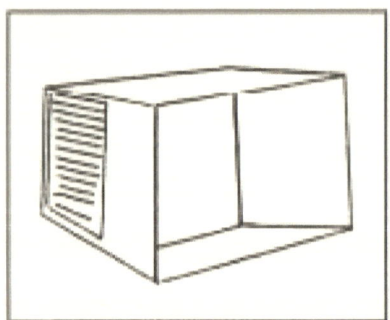

**Figure 136a**

**The outside shell for the window- or wall-mount unit should be well supported**

## WINDOW OR WALL AIR-CONDITIONING UNITS

The type of air conditioner depicted in figures 136 and 137 is usually equipped with a split-phase capacitor run (SPC) motor (see fig. 45). The evaporator in these units is installed in the front and the condenser at the rear outside the conditioned area. A fan motor, usually capacitor run, is installed between the evaporator and the condenser. Two long shafts extend from each end of the motor with a fan blade mounted at the end of each shaft. The fan blade on the condenser side draws air in through the air vents from the outside, forces it through the condenser, and then returns it to the outside. The blade on the evaporator side draws air in through the evaporator from the air-conditioned area and then returns it to the air-conditioned area.

As warm air comes in contact with the cold evaporator coil, the moisture in the air condensates on the coil and drips into a pan in the bottom of the unit where the compressor, evaporator, and condenser sit. Most units are equipped with a slinger fan blade on the condenser side. As the fan turns, it picks up water from the drip pan and slings it through the hot condenser coil to create more efficient cooling. (See fig. 138)

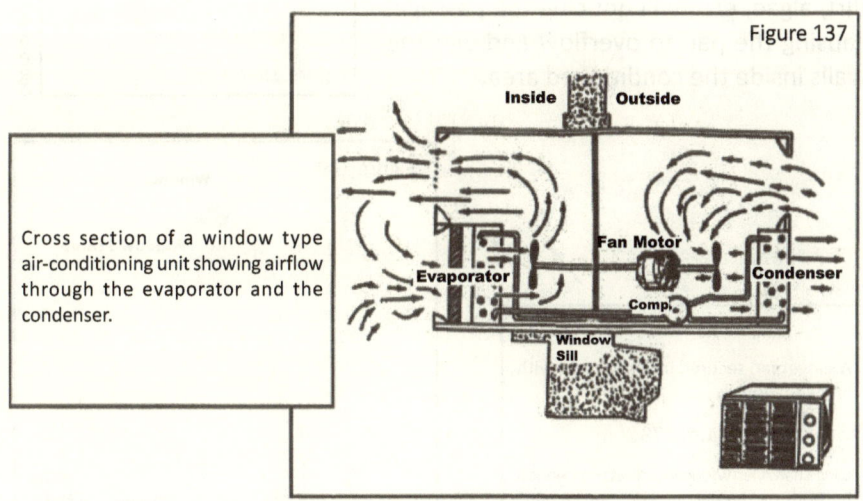

Figure 137

Cross section of a window type air-conditioning unit showing airflow through the evaporator and the condenser.

Thus, a slinger fan performs two functions: (1) it draws air from the outside to cool the condenser coil and (2) picks up the water from the drip pan and throws it through the hot condenser coil, which helps with cooling and evaporation of the accumulated water, reducing the amount of water drip to the ground.

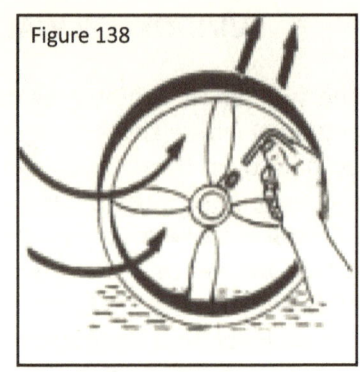

Figure 138

To help facilitate this action, the unit must be tilted one-fourth inch down toward the outside to allow the condensate water to flow toward the rear of the unit and drain out by gravity on the ground through a hole in the side of the pan. (See fig. 139.)

Some units have a hose attached to the drain hole to better control displacement of condensate water. The drain hole in the side of the drip pan and the drain tube must be cleaned regularly. In this way, dirt, algae, etc., will not clog the passage, causing the pan to overflow and wet the walls inside the conditioned area.

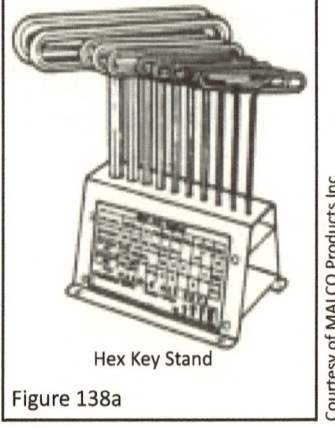

Hex Key Stand

Figure 138a

Courtesy of MALCO Products Inc.

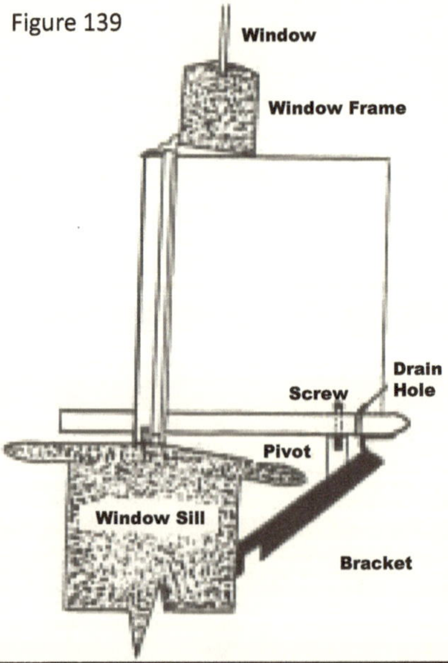

A slinger fan secured by a set screw with an Allen wrench

Figure 138a

Long-shaft Allen wrenches are recommended for both residential and commercial units because of their convenient length

Figure 139

A window unit mounted on an uneven window sill. The rear of the unit should be one-fourth inch lower than the front for good water drainage.

Figure 139

Window

Window Frame

Screw

Drain Hole

Pivot

Window Sill

Bracket

Air conditioner fan blades are always secured to their shaft by an Allen screw or a regular machine screw. A set of long-shanked Allen wrenches is a necessity for routine repair work. Occasionally, the fan hub develops rust between it and the shaft, in which case, using a rust-dissolving fluid such as Screw-Loose becomes necessary. A similar fluid can be made by mixing equal parts of gasoline and kerosene, but it is not as easy to use as the commercially made product in a spray can.

There are two basic types of fans used in air conditioners: the axial (propeller) flow and the squirrel cage (radial) flow. Fan motors are very sensitive to line voltage. A drop in voltage can cause a drop in the fan speed. Pay particular attention to the fan blade pitch and shape when replacing fans. Don't attempt to straighten a bent fan blade as good balance and vibration-free running are nearly impossible to achieve. Always replace a defective blade.

Fans move air by creating above atmospheric pressures (positive pressures) on one side of the blade and below atmospheric pressure (negative pressures) on the other side to force air out or draw air in. This of course, is accomplished by the pitch (the twist in the blade) and the direction of rotation of the shaft.

Occasionally, a blade develops a small crack, especially in units with cast aluminum blades. The defect is easily recognized by the high-pitched noises the fan makes while working.

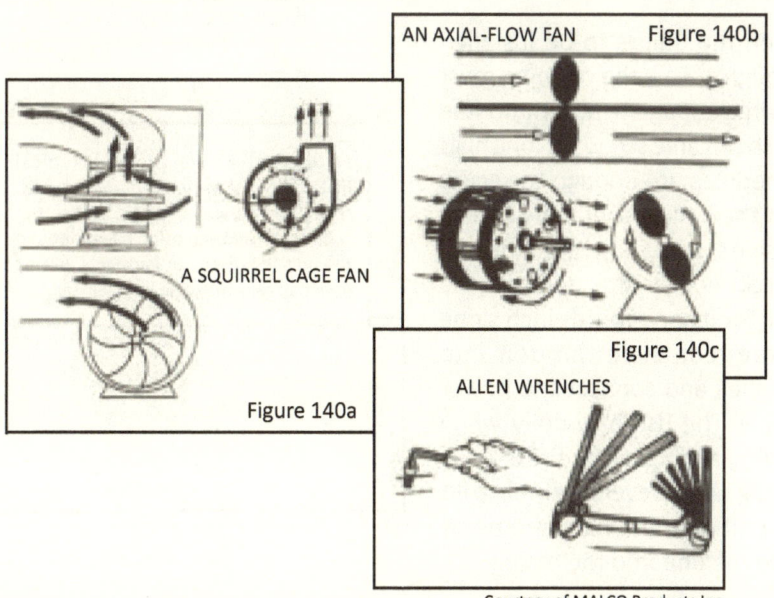

A SQUIRREL CAGE FAN

Figure 140a

AN AXIAL-FLOW FAN   Figure 140b

Figure 140c

ALLEN WRENCHES

Courtesy of MALCO Products Inc.

## INSTALLING WINDOW UNITS

All window units must be installed in a way that the rear is seated about one-fourth inch lower than the front to allow condensate water to drain to the outside. A small carpenter's level will help to achieve this. Since most walls are not thick enough for good support, outside wall brackets are necessary to prevent the unit from falling (see fig. 139).

Once the unit is in place, rubber gaskets, metal plates, sealing compounds, adjustable shutters, filler boards, and sponge strips can be used to make a weather-tight installation.

Wall and window units are mounted inside a shell or casing that come with them. (See fig. 141) The manufacturers attach adjustable shutters on the sides near the front to fill in the spaces to the edge of the window. These shutters are extended to the inside edges of the window and secured to the window. Then a filler board and/or foam or rubber weather seal is used on the top where the window frame comes in contact with the shell.

When the unit is to be installed through a wall, the opening must be large enough to accommodate the shell. Large screws or long nails sometimes are enough to secure it in the wall without the use of outside support brackets if the wall is thick enough. After adjusting the shell for its one-fourth-inch slope to the rear, slide the unit into the shell and connect the power supply. The use of a dolly when moving and installing these units is a big help. Never attempt to lift the unit by the tubing and never force the unit into the casing.

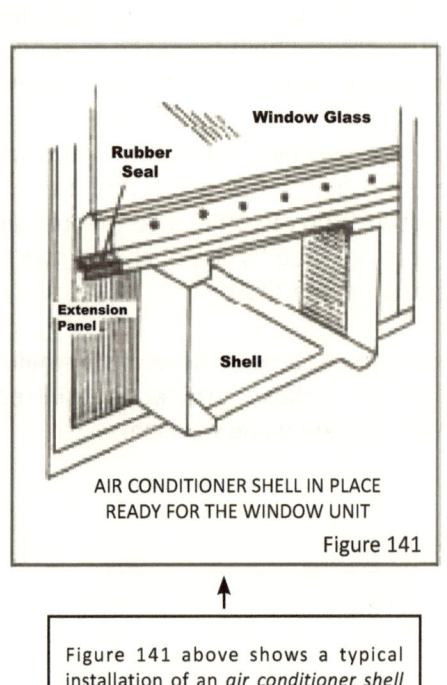

AIR CONDITIONER SHELL IN PLACE
READY FOR THE WINDOW UNIT

Figure 141

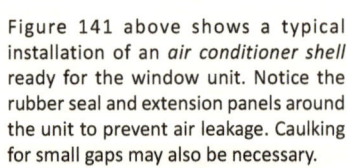

Figure 141 above shows a typical installation of an *air conditioner shell* ready for the window unit. Notice the rubber seal and extension panels around the unit to prevent air leakage. Caulking for small gaps may also be necessary.

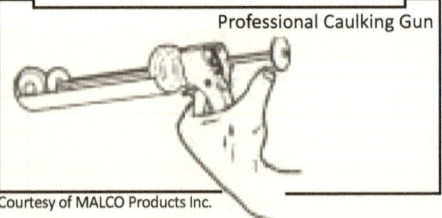

Professional Caulking Gun

Courtesy of MALCO Products Inc.

A polarized circuit (one with a ground wire) must be used for air conditioners. Apartments use mostly wall-mounted units.

The thermostat setting on most units is adjustable between 55°F and 60°F (12°C and 16°C). The thermostat bulb (a tubing) is fastened to the evaporator. Avoid bending the bulb as it may kink or crack and impede its function. The thermostat bulb is covered with a plastic guard. The function of the guard is to protect the bulb from extremes in temperature and to prevent short-cycling by frequent turning on and off of the unit.

The capacitor(s) and relays (for heat pumps) for these units are installed behind the control knobs on the front panel (see fig. 142). This plastic front panel is usually held on by a few small screws or clips. Behind the plastic panel covering the entire front of the evaporator, a dust filter is held in place. Because wall and window units use permanent-split-capacitor-run motors, they do not require a starting relay (see fig. 45). They also use capillary tubes.

NOTE: There is a strainer located two inches below the capillary tube and the condenser coil joint. That strainer is often the cause of a restricted system rather than the capillary tube itself. It can be removed for cleaning or replacement by cutting the tube, leaving the condenser coil two inches below the capillary joint. A replacement screen is not necessary as replacement filter-driers come with a screen.

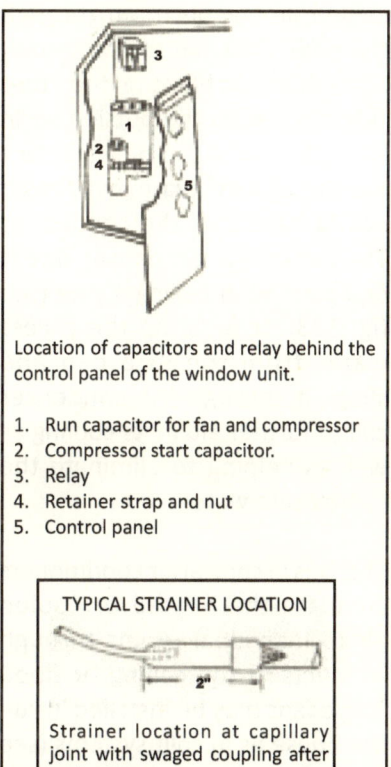

Location of capacitors and relay behind the control panel of the window unit.

1. Run capacitor for fan and compressor
2. Compressor start capacitor.
3. Relay
4. Retainer strap and nut
5. Control panel

TYPICAL STRAINER LOCATION

Strainer location at capillary joint with swaged coupling after replacement

Figure 142

Fig. 45e gives more information about the different sizes and lengths of capillary tubing required for the various horsepower ratings of compressors.

The principle involved in the operation of these air conditioners and the techniques used in their repair is the same for all small refrigeration units.

# FANS FOR AIR CIRCULATION

Fans circulate the air in the environment to be conditioned as well as circulating it through the evaporator and condenser coils. In a simple wall-mounted or window air-conditioning unit, the fan draws the room air in through the cold evaporator coil and returns the chilled air to the room through the grille. This fan motor usually has a shaft and blade on the other side that draws the outside air in through a grille and blows it back out through the warm condenser coil to help cool the condenser. The fan on the condenser side is equipped with a slinger ring (see fig. 138). It picks up the excess water from the drip pan and slings it through the condenser coil to further aid in its cooling as well as helping to eliminate the condensate water.

The fans in central air conditioners circulate the heated or cooled air to individual rooms through air ducts in the ceiling or floor. These fans may be installed inside the house as in split systems (see figs. 156, 157, and 158) or in a self-contained unit where they draw the air in from one side and expel conditioned air from the other. (See figs. 137, 148, and 156a.)

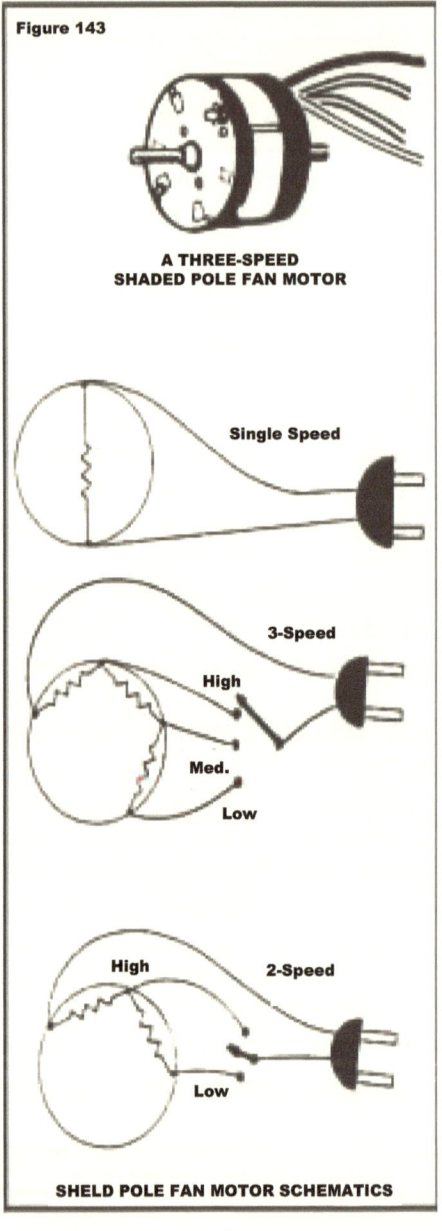

Figure 143

A THREE-SPEED SHADED POLE FAN MOTOR

Single Speed

3-Speed — High, Med., Low

2-Speed — High, Low

SHELD POLE FAN MOTOR SCHEMATICS

Many of the motors use a capacitor to run. They should be lubricated once a year with an oil designed for them. Fan motor oil, obtainable from refrigeration supply dealers, is rather thin as compared to ordinary motor oil.

Sometimes fan motors begin to fail due to an internal short, worn shafts, age, etc. They operate slower than they should and do not provide enough air circulation through the evaporator coil. This becomes evident by the accumulation of frost on the evaporator fins and a sweaty or frosted suction line.

The two most popular types of fan motors are the permanent-split capacitor (PSC) motor and the shaded pole motor.

a. The *permanent-split capacitor motor* costs a little more, but it is much more energy efficient. This type of motor is usually used in the better-quality air conditioners. It comes with a start and a run winding and uses a run capacitor wired in series with its start winding (see fig. 144). Normally, if the capacitor shorts out and the house circuit breaker does not trip as soon as the fan comes on, the fan motor will become extremely hot after a short period of running time, and it will eventually stop operating.

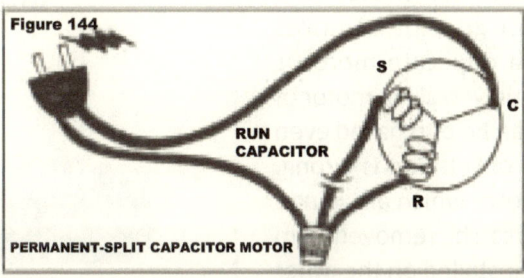

b. The *shaded pole fan motor* does not come with a capacitor since power gets to the motor through a selector switch (a low, medium, or high speed selection). See figures 143 and 145.

Power must go through a high resistance path for the low speed setting, a medium resistance path to operate at medium speed, and a low resistance path to run at high speed.

This type of fan motor is less expensive and not as efficient as the PSC motor. To test a shaded pole motor, simply connect the common wire and any other wire to a proper power source.

Air conditioners have two basic functions: air circulation and cooling or heating. Air circulation is made possible by the use of fans and blowers. A shaded pole fan motor may have a number of speeds. The number of wires coming out of a motor determine the number of speeds. Two-speed motors have three wires, three-speed motors have four wires, etc., (see fig. 143).

To test the PSC motor, fabricate a test cord as shown in figure 146. The illustration is self-explanatory.

As mentioned earlier, there are basically two types of fans:

1. The *axial flow* type.

2. The *radial flow* type fan or squirrel cage (see fig. 140a).

1. The *axial flow* type is known as a "regular"-type fan as air is moved along the axis of the fan motor shaft. Whether air moves into or away from the motor depends upon the rotation of the shaft and the pitch of the blades.

If the evaporator fan motor in a residential refrigerator requires changing and a duplicate motor is not available, a universal fan motor of the same size can be purchased even if the direction of rotation is wrong. The shaft and rotor which are joined in those motors can be removed from one side and reinstalled on the other side of the stator. By doing this, the rotation is reversed. The fan blade also must be reversed on the shaft to keep the airflow in the same direction as it was originally.

In larger air conditioners and commercial fans, the direction of rotation can be reversed simply by reversing the motor polarity. This can be done by connecting the wires of the new motor in accordance with its instructions.

2. *Radial flow* fans are mostly referred to as "squirrel cages" or blowers. In this type of fan, the

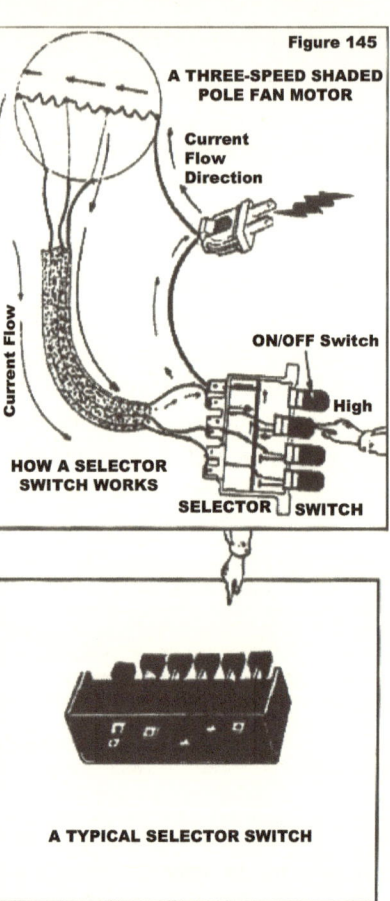

Figure 145

A THREE-SPEED SHADED POLE FAN MOTOR

A TYPICAL SELECTOR SWITCH

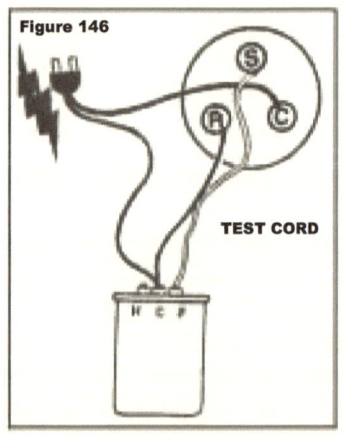

Figure 146

TEST CORD

airflow is perpendicular (at a right angle) to the motor shaft. (See fig. 140a). They are widely used in central air-conditioning systems where the air is required air to be directed vertically (see fig. 147).

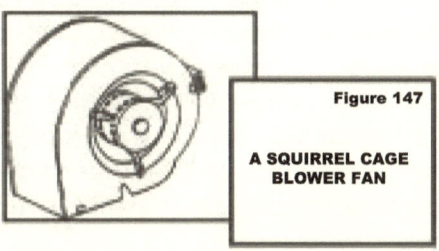

**Figure 147**

**A SQUIRREL CAGE BLOWER FAN**

Sometimes, this type of blower fan is driven by a belt and pulley connected to a separate motor (see fig. 148). Most central air conditioners, swamp coolers, commercial kitchen ventilators, etc., use this type of fan.

**A belt-driven blower fan housing used in central air-condationing units.**

***Adove:*** **A typical radial blower fan draws air from the center and forces it out through the side.**
***Below:*** **Check the manufacture's specifications for lubrication points on the blower motor. Some blower motors have sealed and permanently lubricated bearings and do not require periodic maintenance.**

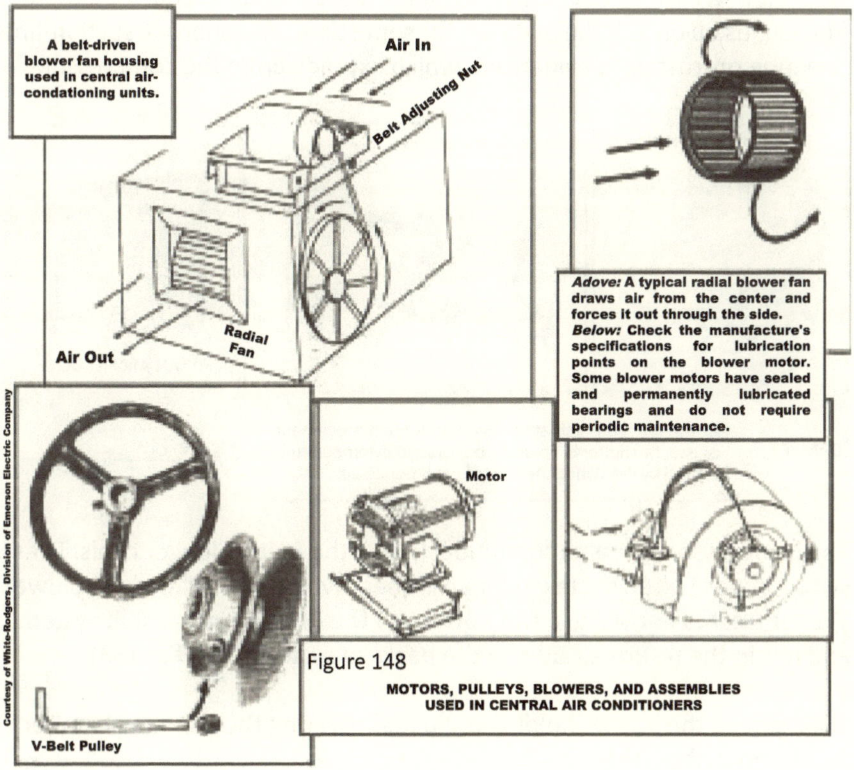

Figure 148

**MOTORS, PULLEYS, BLOWERS, AND ASSEMBLIES USED IN CENTRAL AIR CONDITIONERS**

## 2. COOLING AND HEATING BY AIR CONDITIONERS

In commercial and residential refrigeration, controls can be adjusted to have the evaporator produce and sustain subzero temperatures. But in air-conditioning units, evaporator temperatures do not go that low. Since the most comfortable room temperatures range between 74°F and 80°F at 50% humidity, the evaporator temperature is normally set between 40°F and 50°F. And since that temperature is well above the freezing point, defrosting of their cooling coil is not necessary. As air is moved through the coil, it becomes cooler (or warmer for heating) until the conditioned space reaches the desired temperature at which time the thermostat shuts off the unit.

Since air is moved through the evaporator coil, filters should always be installed in the path of the airflow to prevent the coil from becoming obstructed (see fig. 149). These filters should be cleaned regularly or replaced if cleaning is not practical. Dust and dirt particles collect on the filter or on the evaporator coil if the filter is removed, causing the unit to lose its cooling efficiency. When air circulation through the evaporator coil is restricted, frost builds up on it. If the unit runs for some time, an abnormal suction-line sweating or frosting is evidenced, which extends up to the compressor.

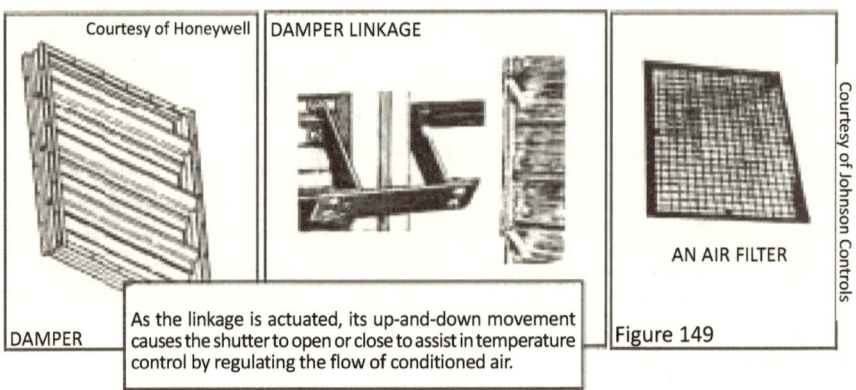

Courtesy of Honeywell | DAMPER LINKAGE

DAMPER | As the linkage is actuated, its up-and-down movement causes the shutter to open or close to assist in temperature control by regulating the flow of conditioned air. | AN AIR FILTER — Courtesy of Johnson Controls | Figure 149

The compressor operates continuously, and the unit no longer cools. These symptoms are often mistaken for an inoperative or slow evaporator blower. The air filter is normally in the bottom of the furnace in the split systems, and it is in the return air duct in the package systems (see fig. 158).

1. Remove the screws holding the panel covering the plenum chamber.
2. Remove the panel.
3. Pull out the evaporator and clean under the coil with a wire brush (you will find a blanket of lint covering the inside of the coil, see fig. 165).

Central air conditioners generally use capacitor-start-capacitor-run compressor motors (see fig. 44), unless a three-phase compressor motor is used (see fig. 178).

## TRANSFORMERS

A transformer is used when there is a need to increase or decrease voltage in an electrical circuit. There are two sets of windings (see figs. 150 and 150a) in a transformer, which are referred to as the primary and the secondary windings.

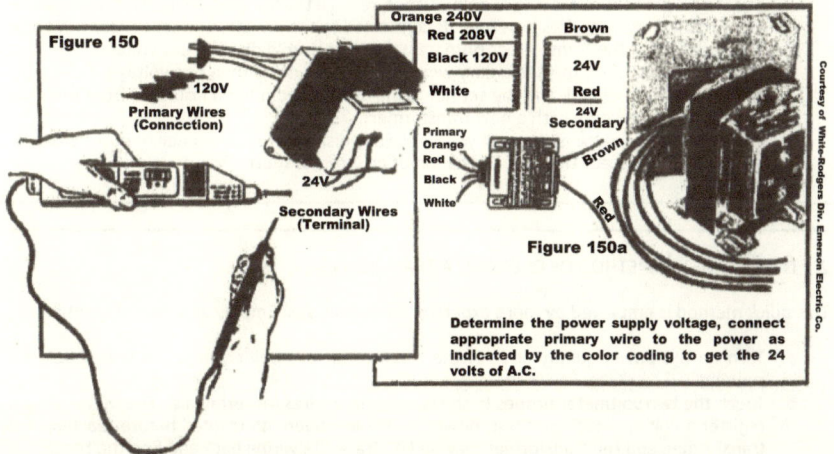

Figure 150

Figure 150a

Determine the power supply voltage, connect appropriate primary wire to the power as indicated by the color coding to get the 24 volts of A.C.

When the primary winding is connected to the available power supply with a particular voltage, the secondary winding generates current with a higher or lower voltage, depending on the number of times the wire is wound as well as the size of the wire used in the winding. If the number of windings of the primary winding is larger than that of the secondary's, the secondary winding produces a lower voltage and it is referred to as a step-down transformer. (For example, converting 110 VAC to 24 VAC.)

If the number of windings in the secondary winding is greater than the number in the primary winding, a higher voltage is produced from the secondary winding, and it is referred to as a step-up transformer. (For example, converting 110 VAC to 120 VAC.)

Since most relays and contactors (coming up in future pages) operate on 24, or 40 VAC in the central air conditioners, step-down transformers are installed in their electrical circuits to convert the 110 VAC, 220 VAC, etc., to that 24, or 40 VAC. The primary winding of the transformer is hooked up to the power line, and the secondary winding terminals are always connected to the load controls, such as wall thermostats, electronic timer defrost boards, contactors, relays, etc.

---

1. THE OHMMETER METHOD OF CHECKING A TRANSFORMER

(Disconnect power from unit)

   a. Disconnect the power.
   b. Disconnect the transformer from the unit.
   c. Set the ohmmeter on its lowest scale.
   d. Touch the probes to the two primary wires (terminals). The meter should register a continue reading, otherwise replace the transformer.
   e. Next, touch the probes to the two wires coming out of the secondary winding (or terminals). You should get an ohm reading as above, if not, replace the transformer. NOTE: There may be more than one set of wires or terminals connected to the primary and secondary windings. This means that the transformer primary winding is usable on different voltages, i.e., 110 VAC or 220 VAC; also, the secondary winding produces more than one voltage. Each set is individually color coded, and the set not being used is tied off and insulated from the rest. (The instructions come with a new transformer).
   f. Visually inspect the transformer. If you notice any discoloration or a smell of burnt Bakelite, it should be replaced. This is an indication of an internal short.

---

2. THE VOLTMETER METHOD OF CHECKING A TRANSFORMER

This quick method is employed by more experienced technicians. Employ all safety precautions.

   a. Turn on the power supply and set the thermostat to a point where the unit should be running.
   b. Touch the two voltmeter probes to the two primary wires (or terminals). The meter should register a voltage reading. If not, power must have been interrupted before reaching the transformer, and the transformer may be OK. Trace the wiring back and find the break.
   c. If there is power at the transformer, touch the two voltmeter probes to the two secondary winding wires (or terminals). If you get no reading, the transformer is bad and requires replacement. Otherwise, it is OK.

---

Since voltage going out of the transformer is proportional to the voltage going in, sometimes the secondary voltage may not be high enough to activate the contactor coil. Check the voltage at the primary terminal, if the voltage reads lower than the required primary voltage (usually) printed on the transformer, there is nothing wrong with the transformer. Have the customer call the power company or an electrician to provide adequate voltage. (See the "Basic Electricity" section for more information about reading wiring diagrams where the transformer voltage requirement can be easily read.)

There are many types, styles, and sizes of transformers available. Your only concern when buying a replacement is the specified primary input and secondary voltage output and rating.

## RELAYS

A relay is an enclosed electromagnetic switch and comes in several types and styles (see fig. 151).

Relays operate on 24 VAC and 40 VAC but seldom on 12 VAC or line voltage. The low voltage operating the relay is created by a transformer installed in the electrical circuit (see fig. 153).

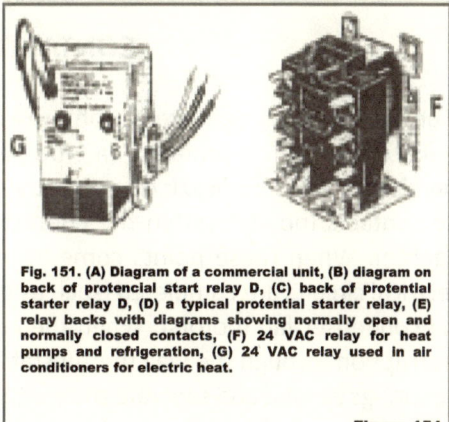

Fig. 151. (A) Diagram of a commercial unit, (B) diagram on back of protencial start relay D, (C) back of protential starter relay D, (D) a typical protential starter relay, (E) relay backs with diagrams showing normally open and normally closed contacts, (F) 24 VAC relay for heat pumps and refrigeration, (G) 24 VAC relay used in air conditioners for electric heat.

Figure 151

Three sets of wires connect to relay terminals: one set carries line voltage or low voltage to energize the relay (depending on the type of relay), another set connects the relay terminals to a different controlled load circuit(s) and a third set of wires connects the relay contacts to the power supply. When the coil is energized, all the normally open contacts close energizing circuits that are interrupted by the relay, and at the same time all the normally closed contacts open to de-energize the circuits that are active.

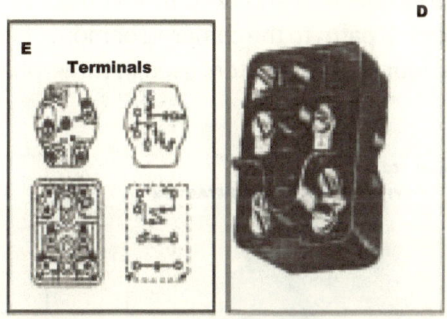

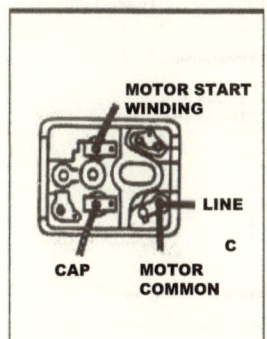
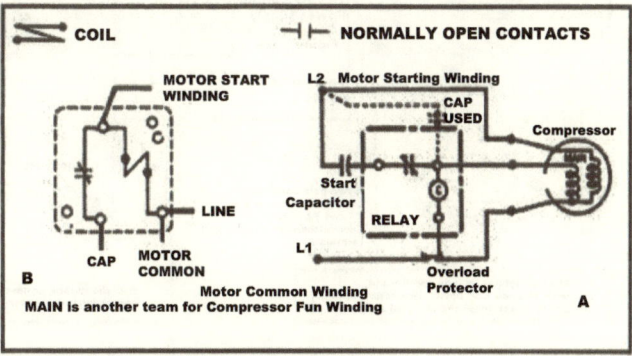

Relays have several contact points which energize or de-energize different components in the unit. The line voltage on each contact point can be 120 VAC, 220 VAC or more. Without the relay, it would be very difficult to activate or de-energize several switches simultaneously. Relays usually operate on the same voltage as the wall thermostat.

## CONTACTORS

In an air-conditioning system, compressor and condenser fan motors are energized through a contactor (a type of relay). When the contactor coil is energized, a magnetic field is created to overcome the spring tension separating the contacts (points) within the contactor assembly, pulling the contacts together. When these points come into contact with each other, certain electrical circuits are completed (such as the ones energizing the compressor and condenser fan motors). When the contactor coil is de-energized by an interruption through the wall thermostat, the magnetic field is removed and the springs expand and separate the contacts. (See figs. 152, 152c and 153). In figures 152a and 152c, note that the terminals in those contactors are normally open unless the thermostat closes the 24 VAC circuit to contactor coil, in which case terminals T1 and T2 come in contact with terminals L1 and L2 and complete electrical paths to the compressor motor, fan motor(s), etc. There are three-pole and four-pole contactors used in three-phase circuits (see fig. 152b).

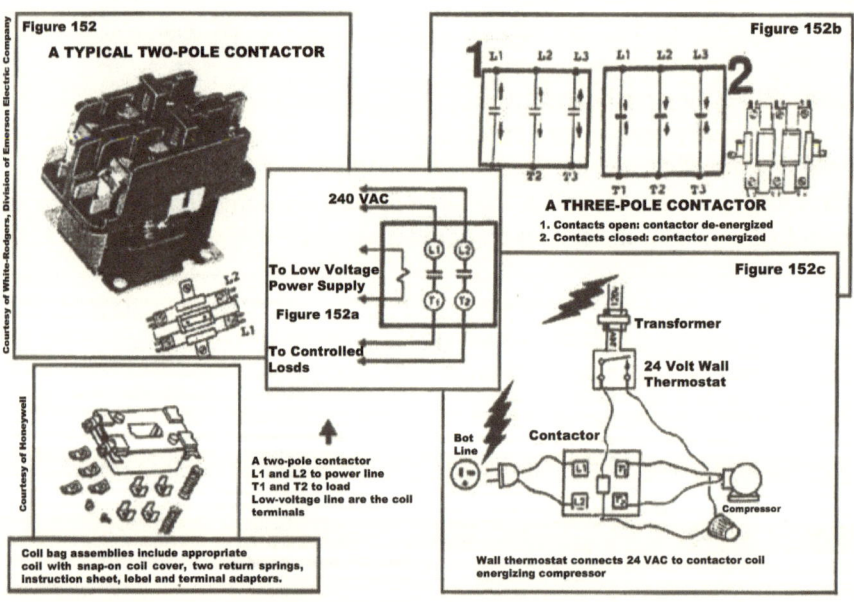

The low voltage comes to the contactor coil through the wall thermostat. When the wall thermostat is turned off or satisfied, the low-voltage circuit activating the contactor coil is interrupted, causing the compressor to shut off.

## HOW TO CHECK A CONTACTOR

Contactors are magnetic switches that are used in applications where simultaneous openings and closings of different line circuits are required. They can become inoperative by a burned point(s) or by a burned coil.

1. During normal operation, the contactor points may become pitted and/or charred by an electrical arcing. When this happens, the metal points become incapable of completing a particular circuit and the component(s) in that circuit will not function. When a contactor point becomes charred, only the particular component(s) which is/are activated by that contact becomes inoperative, while the rest of the unit works.
2. Unlike solid-state relays, the operation of a contactor depends on a magnetic field created by a coil when it becomes energized. When this magnetic field is created, the points are drawn together closing circuits to energize certain components.

To determine if the contactor is the problem when a unit becomes inoperative (which in many cases it is), find the condensing unit (where the compressor is located) and remove its side panel. There is a junction box where the main power lines come into the condensing unit. The contactor is usually installed inside that box (figs. 152c and 152d). With the wall thermostat inside the building turned on, watch the contactor to see if the points close. If not, touch the two probes of a voltmeter to the two wires connected to the coil terminals to check for proper voltage. If power reaches there, but the points are not pulled together; or if it pulls for a short while and then releases, the contactor must be replaced. If the meter registers no reading at the coil terminals, the wall thermostat and the transformer will have to be checked as either one could cause an interruption of power to the coil terminals.

If the contactor hums but does not pull, it may mean either insufficient voltage is reaching the coil terminals or the coil is bad. Check the voltage as outlined above. If the meter registers voltage below the requirement of the contactor coil, tell the customer to call the local power company to remedy the low power input to the property. The voltage produced by the secondary winding is directly proportional to the input voltage to the primary winding of the transformer. If the primary voltage is too low, the secondary voltage will be too low. Determine the required voltage that energizes the contactor (per instructions beginning on page 218) or by checking the back of the contactor. If the contactor pulls but the unit will not start, the problem can be due to an interruption in the compressor and condenser fan power supply line.

---

**A QUICK WAY TO CHECK FOR A BAD FUSE OR CIRCUIT BREAKER**

1. Turn the unit on and use your digital multimeter on the volt setting (fig. 152f).
2. Touch the two probes on 1 and 2 and 3 and 4. If you get a voltage reading on either one of them, replace the blown fuse. (A good fuse should read zero on the voltmeter).
3. Touch 5 and 6; the meter should register a reading of 220/240 VAC. And touch 6 and 7 or 5 and 7; you should get a 110/130 VAC reading. Otherwise, call a licensed electrician to remedy the problem. Sometimes the main circuit breaker trips, causing the disruption of power to the unit. Find the circuit breaker box and return the tripped switch to the "on" position.

---

Pay particular attention to all safety precautions while working with contactors.

## TESTING A CONTACTOR WITH AN OHMMETER

1. Disconnect the power supply.
2. Turn the wall thermostat off.
3. Remove the screws in back of the contactor that hold it in place. There could be two, three, or four screws.
4. Remove the contactor and disconnect the two wires to the coil.
5. Set the ohmmeter on its lowest scale and touch the probes to the two coil terminals (or the wires coming from it). The meter should register a continuity reading. If not, replace the contactor or the coil. (Usually, contactors are relatively inexpensive, and it would not be worth the time to just replace the coil. The cost to the customer would be about the same—the points could go bad during the warranty period for the coil. Save a callback and the trouble of explaining to the customer the difference between a coil and the contacts.)

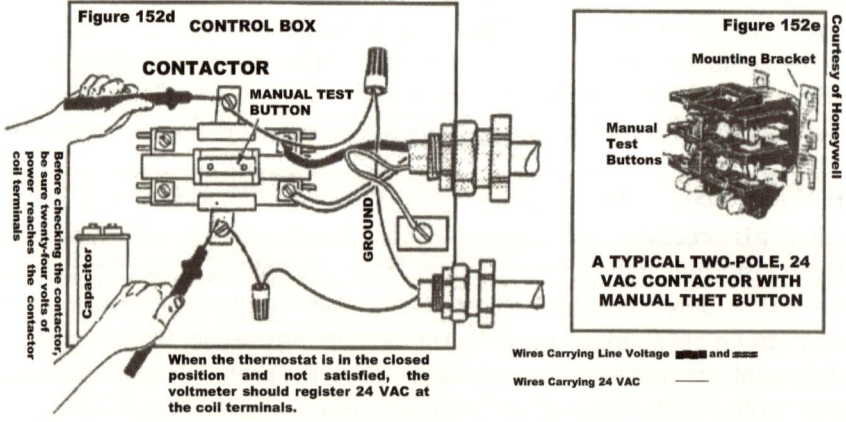

6. Check the contacts carefully. If there is any pitting or discoloration on or around the points, replace the contactor even if it is working. These are signs that the contactor will not go much longer without becoming defective.

Figure 152h

Older circuits may use the plug type fuse with a screw base like a light bulb. If you see any discoloration or a broken filament, the fuse must be replaced.

*Below*: USING A VOLTMETER TO CHECK A FUSE BOX

Touching the probes at the points indicated should get the following results:
Between
5 and 6 = 220 volts AC
5 and 7 = 110 volts AC
6 and 7 = 110 volts AC
(If not, call the local power company).
1 and 2 = 0 volts AC
3 and 4 = 0 volts AC
(If a voltage registers, replace burnt fuse).

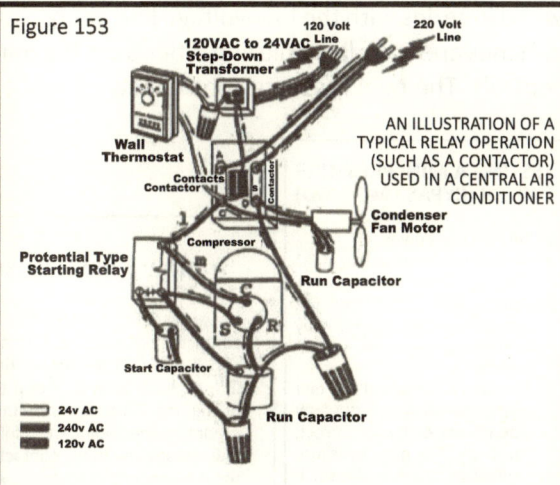

Figure 153

AN ILLUSTRATION OF A TYPICAL RELAY OPERATION (SUCH AS A CONTACTOR) USED IN A CENTRAL AIR CONDITIONER

Note the direction of current flow in the line voltage and low-voltage circuits. A step-down transformer converts 110 VAC to 24 VAC. The coil becomes energized when the thermostat closes the 24 VAC circuit, creating a magnetic field in the contactor (or relay) coil causing movable contacts to connect with the stationary ones in the 220 VAC circuit. At this time, the compressor and fan motor will start operating. (The line voltage plugs are only symbols for power sources.)

Time-delayed relays come with thermal (current) protection devices to prevent abuse of the compressor start winding from frequent start-ups. Every time the coil is energized, the temperature of its heating element rises and disconnects the primary power until the coil cools. As a result, the air-conditioning system pressures will have time to equalize to prevent excessive pressures on the compressor during the next start-up on this White-Rodgers's time-delayed contactor. Make the connections this way:

COIL CONNECTIONS

Figure 152g

To achieve time delay, connect field wiring to terminals designated A and B.

To eliminate time delay, connect field wiring to terminals B and C.

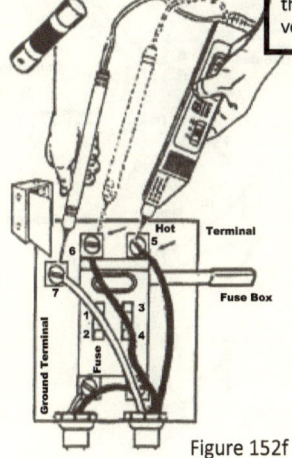

Figure 152f

*Courtesy of White-Rodgers, Division of Emerson Electric Company*

# MINI PRESSURE CONTROL (SWITCH)

A mini low-pressure control is used in a central air-conditioning unit to prevent extreme low pressures in the sealed system. It is mounted on an access valve in the suction line (mini high-pressure controls are mounted on an access valve on the discharge line to shut down the system when head pressure reaches a dangerous level). If a mini pressure switch is used, it is wired in series with the low-voltage line energizing the contactor coil. These switches are not adjustable and differ from the manually adjustable pressure controls. The mini pressure control acts only as a safety device.

## TESTING A MINI PRESSURE SWITCH (BYPASS METHOD)

1. Turn off the power.
2. Set the wall thermostat to AUTO and the coldest position.
3. Remove wires e and f from the minicontrol and bypass it by connecting the wires e and f together.
4. Turn on the power. If the unit starts operating and the low-side pressure checks correct, the defective mini pressure control must be replaced. If the low-side pressure checks abnormally low, find the cause, remedy the problem, and place the control back in the circuit, as it most probably is good. The same checking procedure is applied to a high-pressure mini control with the difference that the high-side pressure must be checked.

OBSERVE ALL THE PERSONAL SAFETY MEASURES

## HOW A MINI PRESSURE CONTROL WORKS

In figures 154 and 155, the 24-volt current magnetizes the contactor coil causing it to make contact between A and C, and B and D. The current flows in the circuit through A to C, l, m, b, D, back to B and then it flows to the neutral side of the line. If the pressure in the discharge line falls below a predetermined point, the mini pressure control disconnects the low-voltage line to the contactor, stopping the compressor and the condenser fan motor(s). A defective mini pressure switch can shut down the system.

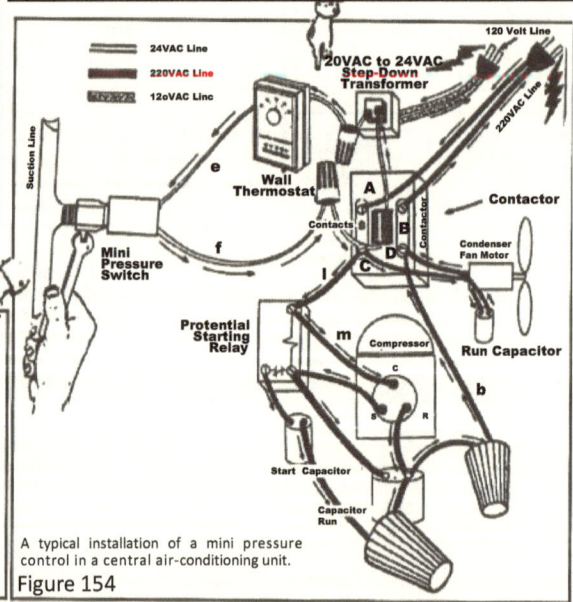

Figure 155

A Mini Pressure Switch

Courtesy of Johnson Controls

A typical installation of a mini pressure control in a central air-conditioning unit.
Figure 154

# CENTRAL AIR CONDITIONERS SPLIT (REMOTE) SYSTEMS

1. Condensing unit
2. Service valves
3. Suction line
4. Liquid line
5. Filter-drier
6. Duct work
7. A-type evaporator coil
8. Expansion valve
9. Condensate drain
10. Gas heating element
11. Blower
12. Air handler

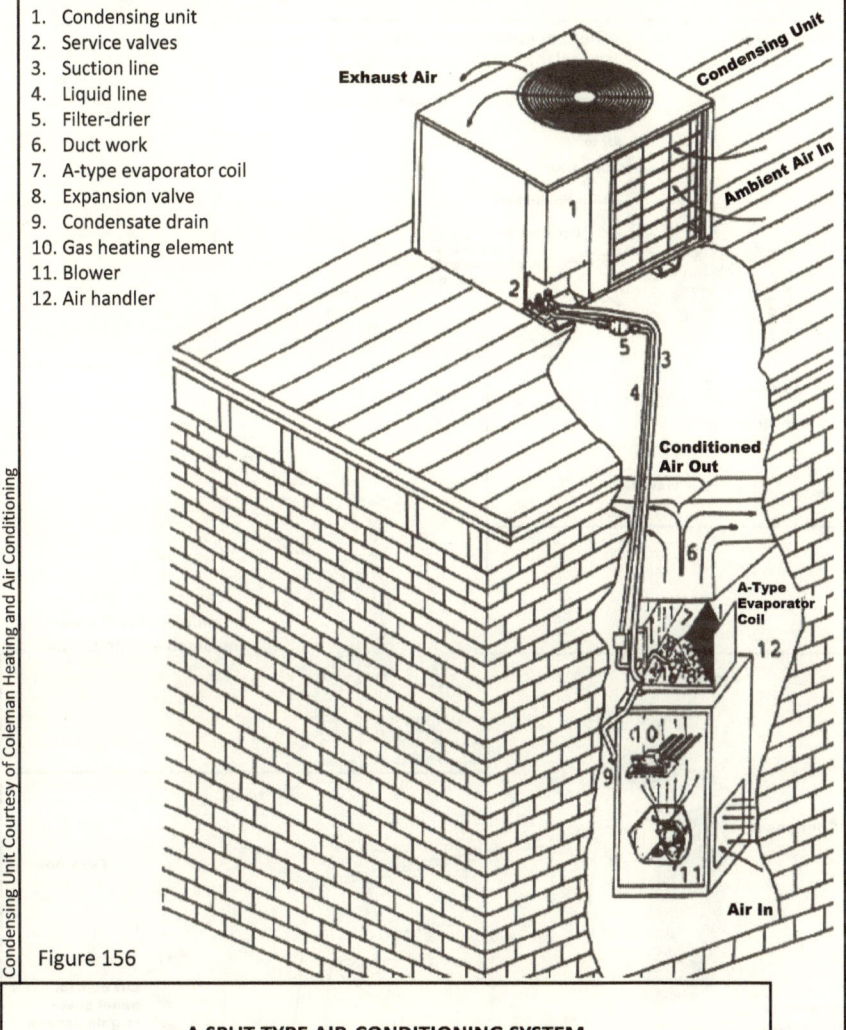

Figure 156

## A SPLIT-TYPE AIR-CONDITIONING SYSTEM

In a typical split system, the compressor and condenser are installed together (called the condensing unit), outside the structure with the evaporator in the inside, as opposed to the *console system* where all components are housed in a single unit. The condensing unit is usually installed outside on a concrete slab or on the roof, with the evaporator on the wall, in the air handler, in an attic, or in an artificial ceiling with a fan circulating the air through the cold evaporator fins inside the conditioned area. (See figs. 156, 157, and 158.)

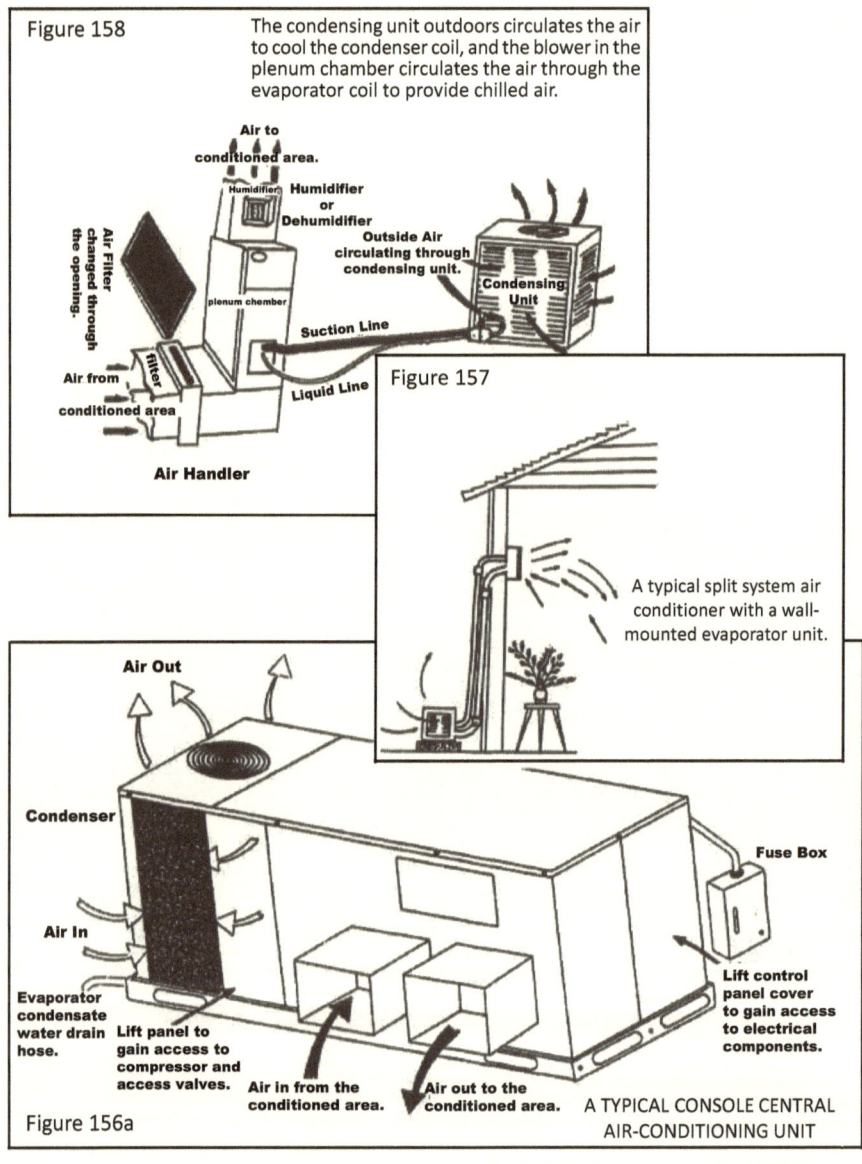

There are three types of evaporator coil used in central air conditioners.

1. Regular, finned evaporator as in window units, consoles, or split systems (see figs. 137 and 157).
2. Slanted-style evaporator coil used in split or console units (see fig. 159).
3. A-coil-style evaporator which is widely used in split systems with plenum chambers. (See figs. 163 and 164.)

The *plenum chamber* (Fig. 165) is the center of airflow inside the air-conditioned area, and it is usually installed in a closet. The suction and the liquid lines are connected to the evaporator coil in the plenum chamber (see fig. 158). A blower fan draws the air from the conditioned area into the plenum chamber, passes it through the evaporator coil, and expels it back into the conditioned area through insulated metal or flexible ducts. (See figs. 161, 162, and 165.)

An A-coil or slant type evaporator is used inside the plenum chamber. Condensate water from either of these coils is collected in a drip pan underneath and evacuated by a drainpipe. (See fig. 164)

Central air-conditioning systems use either capillary tubes or expansion valves. On the outside, the condenser fan forces air through the condenser to cool it. Inside the furnace chamber (see fig. 158) an air filter (see fig. 149) is installed in the path of the return air where

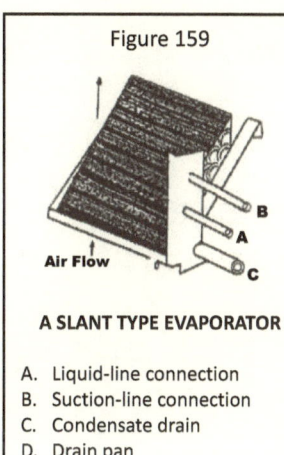

Figure 159

A SLANT TYPE EVAPORATOR

A. Liquid-line connection
B. Suction-line connection
C. Condensate drain
D. Drain pan

A WALL-MOUNTED EVAPORATOR USED IN A SPLIT SYSTEM

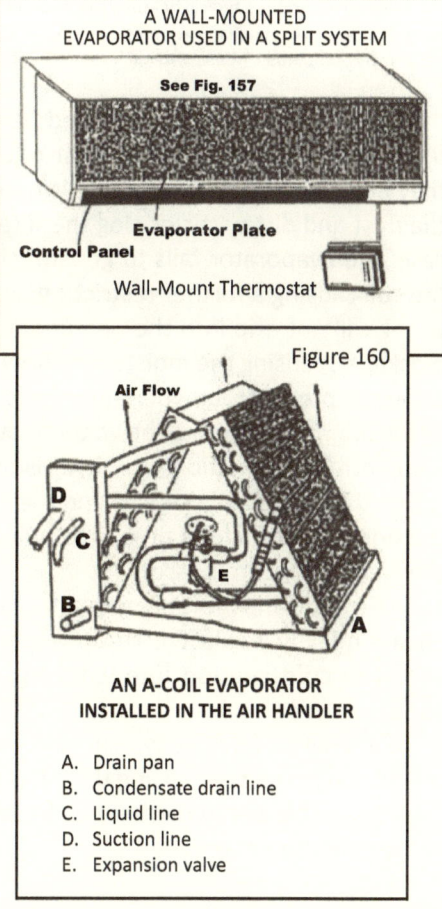

Figure 160

AN A-COIL EVAPORATOR INSTALLED IN THE AIR HANDLER

A. Drain pan
B. Condensate drain line
C. Liquid line
D. Suction line
E. Expansion valve

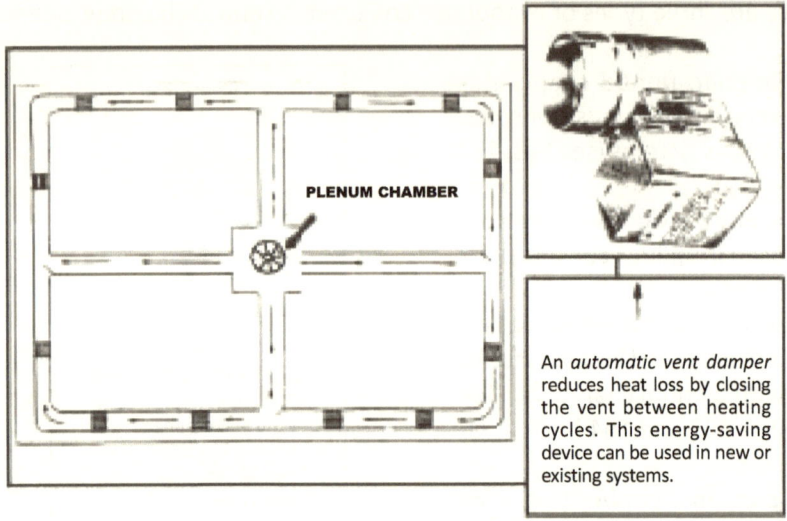

An *automatic vent damper* reduces heat loss by closing the vent between heating cycles. This energy-saving device can be used in new or existing systems.

This drawing illustrates a typical heating/air-conditioning duct work mounted in the ceiling or floor of a home or office building. Also known as a perimeter loop system, it carries the conditioned air from the furnace and evaporator housing to each room or office through this duct work.

dust and dirt particles are trapped before reaching the evaporator. These filters should be replaced at least once a month or cleaned that often if they are of the permanent type. When this is not done on a regular basis, the dust and dirt particles clog the filter, restricting the air passage. In this case, the evaporator fails to get sufficient air circulation. Ice builds up on the coil causing a further restriction, and the unit fails to cool. Condensate water will not drip into the pan under the evaporator coil (because of ice blockage), causing the moisture to flow outside and run onto the floor. (The same symptoms occur when sludge and dirt accumulate inside the drain pan or the drainpipe, causing the water to overflow in the pan and wet the floor.) In a central air-conditioning unit, a visible sign of a linted evaporator or filter is that the suction line sweats and the unit becomes incapable of cooling. If the unit runs for a long time, frost will cover the suction line extending all the way to the compressor.

The suction line entering the furnace and evaporator housing is always heavily insulated. This prevents warm air from coming in contact with the cold suction line, causing a water leak on the floor due to condensation.

Generally, in a split system, the contactor, drier, and the compressor are placed in the condensing unit. The transformer is installed somewhere close to the blower in the furnace housing. The wall thermostat is installed in the conditioned area.

The blower fan and the transformer in the furnace compartment are connected to a separate power source independently. The fan and transformer in residential units require 115 VAC, and the condensing unit is normally hooked up to 220/240 VAC. The transformer reduces 120 volts to 24 or 40 volts of alternating current (rarely, 12 VAC) to energize the contactor through a wall thermostat to control the operation of the compressor and the condenser fan. Figure 162 illustrates another configuration of a typical duct system.

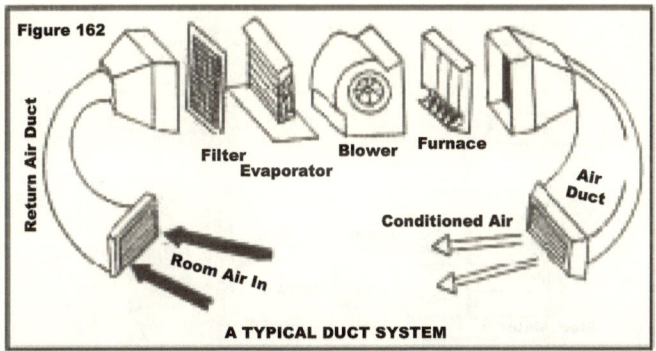

A TYPICAL DUCT SYSTEM

In the cold seasons, when the thermostat is in the *on* and *heat* position, it closes the 24/40 volt circuit to a heating relay, causing a gas valve in the furnace compartment to become energized (to open the flow of natural gas), heating the conditioned area until the temperature sensed by the wall thermostat rises above a preset point (see fig. 156 and 162). In the warm seasons, when the wall thermostat is set in the On and Cool position, it closes the 24/40 volt circuit to the contactor coil, energizing the compressor and the condenser fan, causing the unit to cool. Once the conditioned area reaches a predetermined temperature, the thermostat opens the 24/40 volt circuit and the contactor coil becomes de-energized, and consequently, power to the compressor and the condenser fan motors is interrupted.

Each wire in the wall thermostat is color coded and fastened to its designated terminal. There is more detail about this on the pages about wall thermostats.

Inside the furnace compartment (air handler), a fan relay is placed in series with the fan circuit to start the evaporator blower fan when the thermostat closes the 24-volt circuit.

A condensing unit is hooked up to a 220 VAC circuit breaker installed next to it. This makes it possible to disconnect power to the unit if needed.

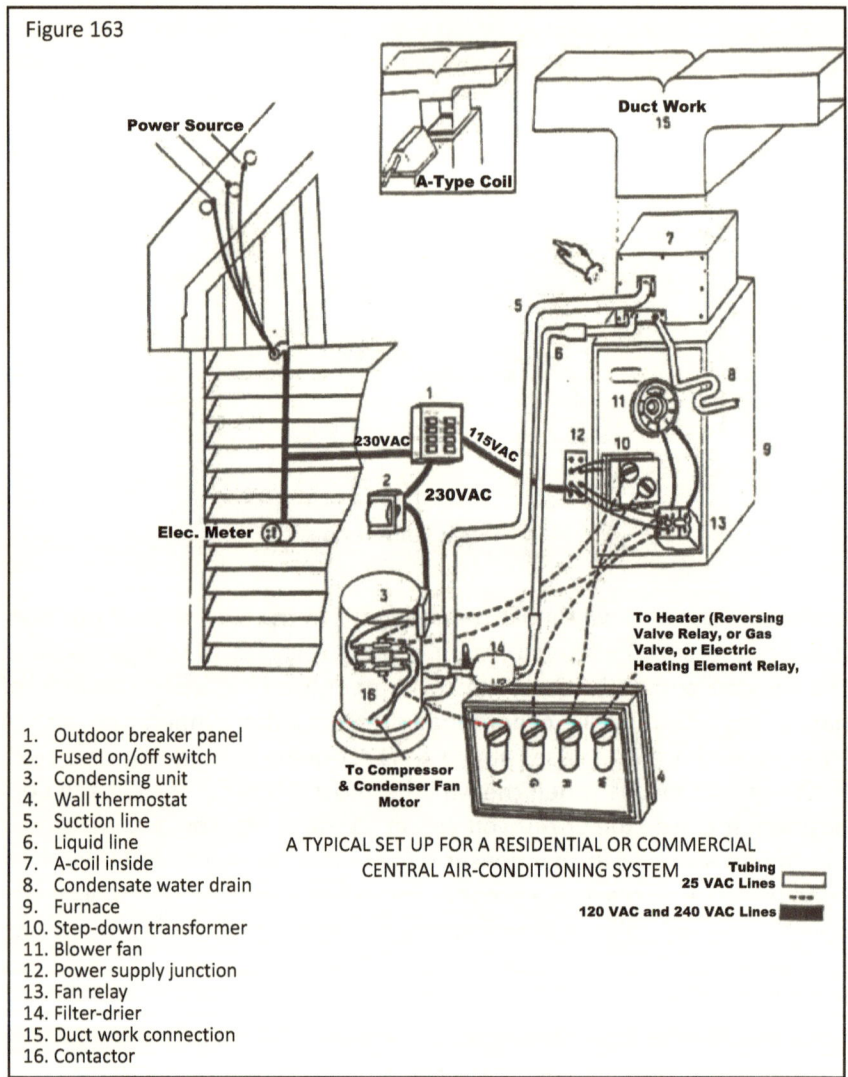

Figure 163

1. Outdoor breaker panel
2. Fused on/off switch
3. Condensing unit
4. Wall thermostat
5. Suction line
6. Liquid line
7. A-coil inside
8. Condensate water drain
9. Furnace
10. Step-down transformer
11. Blower fan
12. Power supply junction
13. Fan relay
14. Filter-drier
15. Duct work connection
16. Contactor

A TYPICAL SET UP FOR A RESIDENTIAL OR COMMERCIAL CENTRAL AIR-CONDITIONING SYSTEM

## A COMPLETE WIRING METHOD FOR A CENTRAL AIR-CONDITIONING SYSTEM
### (Unit in Cooling Mode)

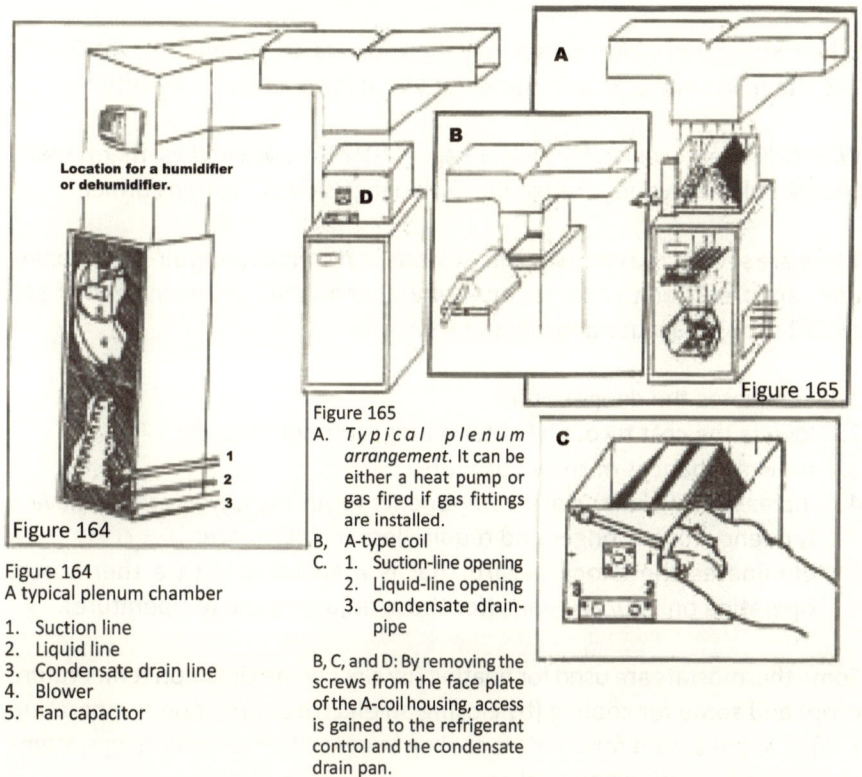

Figure 164
A typical plenum chamber

1. Suction line
2. Liquid line
3. Condensate drain line
4. Blower
5. Fan capacitor

Figure 165
A. *Typical plenum arrangement.* It can be either a heat pump or gas fired if gas fittings are installed.
B, A-type coil
C. 1. Suction-line opening
2. Liquid-line opening
3. Condensate drain-pipe

B, C, and D: By removing the screws from the face plate of the A-coil housing, access is gained to the refrigerant control and the condensate drain pan.

## AIR CONDITIONER WALL THERMOSTATS

The thermostat is the temperature-sensing part of an air-conditioning unit. It is mounted on a wall in the air-conditioned area, and it is sensitive to the changes in the air temperature. It turns on or off the relay(s) and the contactor(s) responsible for energizing the heating or cooling section of the unit, depending on how it is adjusted.

In the heating cycle of the unit, heat is produced by a gas flame, a warm coil, or an electric heating element. The thermostat activates a relay, which in turn connects power to an igniter and a gas jet or to the electric element, or energizes the reversing valve and the compressor in heat pumps (covered on pp. 265 through 271).

That is the primary difference between the thermostat used in a central air-conditioning unit and one used in refrigeration which has a sensing bulb attached to the evaporator to detect temperature changes and operates on line voltage instead of 12, 24, or 40 volts.

When wiring these low-voltage wall thermostats, use number 16 AWG wire. Use the next smaller size (number 18 AWG) for lengths under fifty feet.

WARNING: Never connect the line voltage (110/220) to low-voltage thermostats! They should always be wired to the low-voltage side of the transformer.

There are several reasons why thermostats and contactors in air-conditioning units are preferred to operate on a 24 VAC or 40 VAC system instead of 110 or 220 simply because a low-voltage system:

1. eliminates the danger of fire;
2. lowers the cost through lighter wiring, components, etc.;
3. reduces chances of relay burnouts;
4. increases reliability since components requiring lower voltage have a tendency to last longer and require less maintenance;
5. eliminates the shock hazard that is associated with a thermostat operating on 110/220 volts (for those adjusting the temperature).

Some thermostats are used for heating (by closing the circuit on temperature drop) and some for cooling (by closing an electrical circuit on temperature rise). The ones used for comfort cooling in central air-conditioning systems control both heating and cooling.

Basically, wall thermostats are divided into two groups:

1. Thermostats operating on line voltage (110 VAC, 220 VAC, etc.).
2. Thermostats operating on low voltage (12 VAC, 24 VAC, or 40 VAC) generated by a step-down transformer.

Some thermostats used in more sophisticated air-conditioning units close circuits to operate shutters and dampers in the duct work to regulate airflow to the conditioned areas. In the more elaborate systems, solid-state thermostats provide a high degree of efficiency in monitoring the ambient temperature without mechanical movement. They operate on the principle of change of resistance in a thermister with the change in temperature. Lately transistors, triads, and amplifiers are used in the more sophisticated solid-state thermostats. Whereas, the operation of conventional ones depends on the physical qualities of metals and gases by their expansion and contraction.

## AIR CONDITIONER WALL THERMOSTATS

A thermostat could be referred to as a "command center" for any refrigeration or air-conditioning unit. When used in air-conditioning, the temperature in the air-conditioned space is automatically controlled and maintained by the thermostat setting. Thermostats are designed to control heating, cooling, or both. They come in a variety of styles and sizes.

The wire terminals in wall thermostats are color coded. Each colored wire is connected to its appropriate terminal, which is marked with a corresponding color. Normally, R stands for the red wire representing the common hot wire. G stands for the green wire, which energizes the plenum fan motor relay; Y stands for the yellow wire energizing the compressor contactor in the cooling cycle; W stands for the white wire energizing the appropriate contactor or relay(s) for heating; O for the orange wire, which energizes the reversing valve relay to take the heat pumps into the cooling cycle (if the reversing valve is energized on a call for cooling). The wire connected to terminal B is energized to take the heat pump into the heating mode by energizing the reversing valve (if the valve is energized on a call for heat). (See figure 173 and figure 4 on page 352.) However, if the wall thermostat is not used in a heat-pump system, terminals O and B can be used for such applications as electronic air cleaners, zone damper controls, or humidifiers (see fig. 1 on page 353). RH and RC shown on the thermostat sub-base are connected to the red wire! RC represents the thermostat common (hot) terminal for the cooling cycle. RH represents the thermostat common (hot) terminal for the heating cycle. (See figs. 166 and 167.)

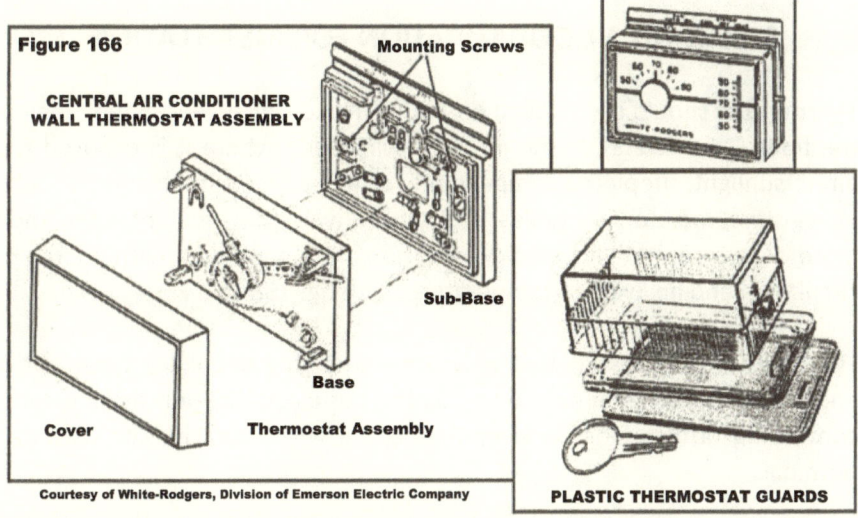

Figure 166 — CENTRAL AIR CONDITIONER WALL THERMOSTAT ASSEMBLY

Mounting Screws, Sub-Base, Base, Cover, Thermostat Assembly

Courtesy of White-Rodgers, Division of Emerson Electric Company

PLASTIC THERMOSTAT GUARDS

As illustrated in figure 166, a wall thermostat consists of three parts:

1. A back plate, or sub-base, is fastened to the wall where the wires for the thermostat protrude (see fig. 167).
2. A base is installed on the back plate with the wires running through a hole in the center and connected to the thermostat terminals on this base. (In some models, the sub-base and base are combined into a single unit.)
3. A cover is snapped onto the base. It is very important to install the base absolutely level; particularly in mercury-bulb types as it will not function accurately if it is canted to one side or the other.

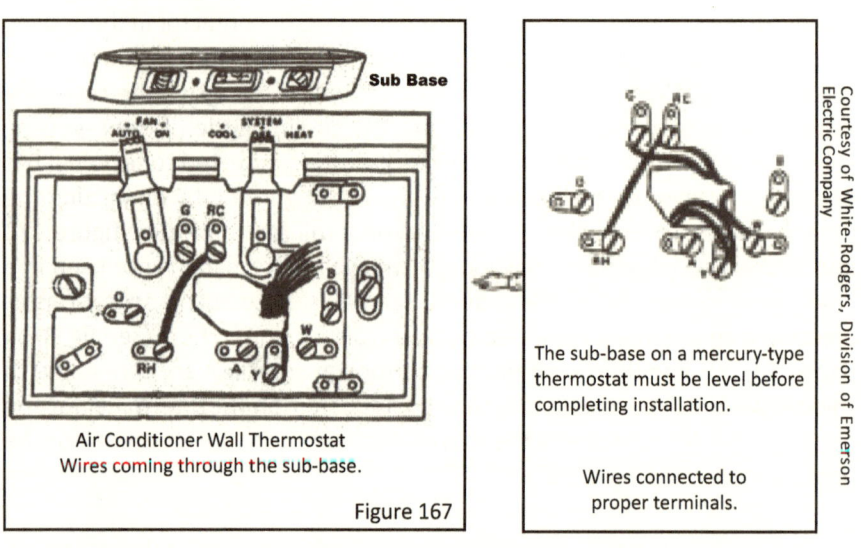

Air Conditioner Wall Thermostat
Wires coming through the sub-base.
Figure 167

The sub-base on a mercury-type thermostat must be level before completing installation.

Wires connected to proper terminals.

Courtesy of White-Rodgers, Division of Emerson Electric Company

## SELECTING A GOOD LOCATION FOR INSTALLATION

Thermostats should be installed on a solid inside wall at least 5 feet above the floor (1.5 meters). It should not be installed where it is exposed to direct sunlight, fireplaces, lamps, draft, or any heat-emitting source such as registers, radiators, or grilles. The chosen wall should be unheated and more or less centrally located within the conditioned area with good air circulation and an average temperature. (See fig. 168.)

The types of thermostats that only control heating or cooling come with only two or three terminals; whereas the basic combination thermostats controlling both heating and cooling come with four, five, or six wire terminals.

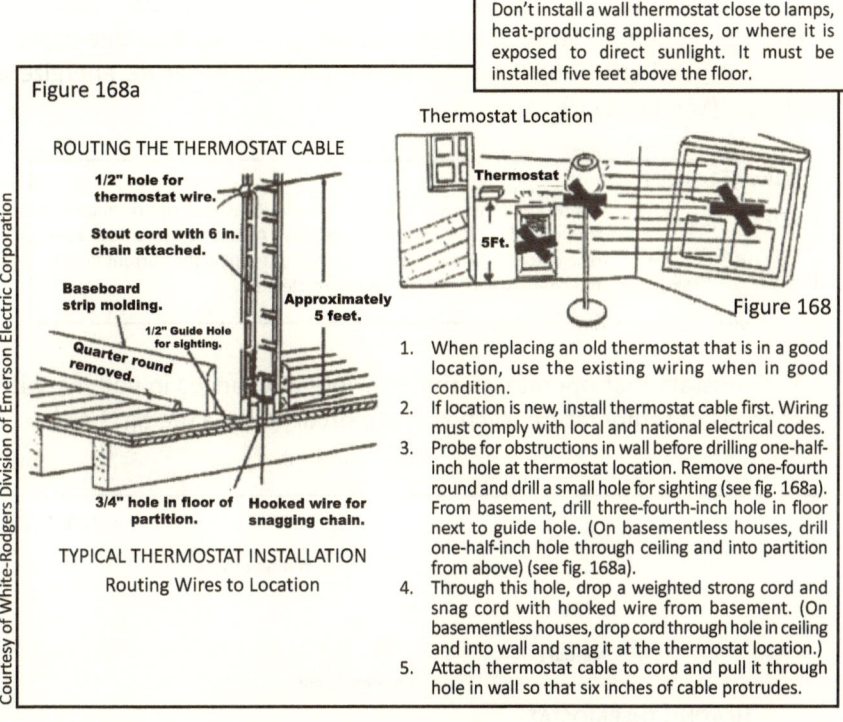

Figure 168a

ROUTING THE THERMOSTAT CABLE

TYPICAL THERMOSTAT INSTALLATION
Routing Wires to Location

Thermostat Location

Don't install a wall thermostat close to lamps, heat-producing appliances, or where it is exposed to direct sunlight. It must be installed five feet above the floor.

Figure 168

1. When replacing an old thermostat that is in a good location, use the existing wiring when in good condition.
2. If location is new, install thermostat cable first. Wiring must comply with local and national electrical codes.
3. Probe for obstructions in wall before drilling one-half-inch hole at thermostat location. Remove one-fourth round and drill a small hole for sighting (see fig. 168a). From basement, drill three-fourth-inch hole in floor next to guide hole. (On basementless houses, drill one-half-inch hole through ceiling and into partition from above) (see fig. 168a).
4. Through this hole, drop a weighted strong cord and snag cord with hooked wire from basement. (On basementless houses, drop cord through hole in ceiling and into wall and snag it at the thermostat location.)
5. Attach thermostat cable to cord and pull it through hole in wall so that six inches of cable protrudes.

## HOW THERMOSTATS WORK

There are four basic types of thermostats in common use:

1. Thermostats that operate on the principle of the different expansion rates of different metals (like the in-line thermostat mentioned in a previous chapter). (See fig. 169.)

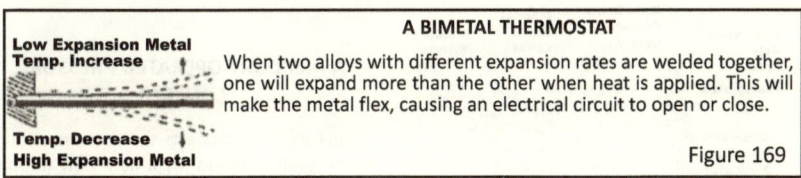

A BIMETAL THERMOSTAT

When two alloys with different expansion rates are welded together, one will expand more than the other when heat is applied. This will make the metal flex, causing an electrical circuit to open or close.

Figure 169

2. Thermostats that operate on the principle of gas expansion by heating. (See fig. 170.)

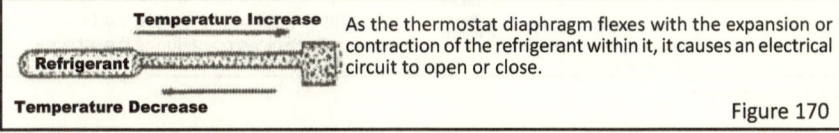

As the thermostat diaphragm flexes with the expansion or contraction of the refrigerant within it, it causes an electrical circuit to open or close.

Figure 170

3. Thermostats that operate by using a thermister. Thermister resistance changes as temperature changes, causing an increase or decrease in current passing through the thermister to energize or de-energize a circuit. (See fig. 171.)

---

Figure 171

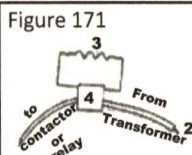

1 and 2 are the same low-voltage wire lines that run through the thermostat. 3. Thermister. 4. Amplifier. As temperature decreases, resistance in the thermister increases, stopping the current flow to the control relay and the cooling cycle ends.

---

4. Thermostats that operate on the principle of changes in volume with changes of temperature, such as the mercury type. (See fig. 172.)

---

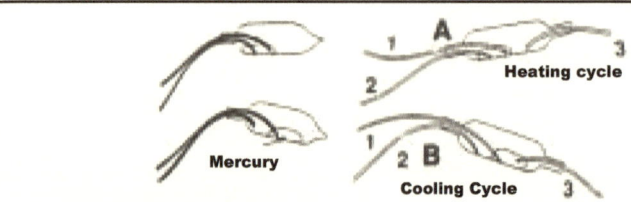

Figure 172

**HEATING THERMOSTAT**

When room temperature rises, mercury expands causing a change in the balance of the bulb. As the bulb tilts, the mercury rolls away from the wire, opening the circuit.

Figure 172 shows a small pool of mercury in a glass bulb. As the room temperature goes up, the volume of mercury increases, causing the balance to change and the bulb tilt, connecting or disconnecting a circuit.

Because mercury is a liquid metal, it is an excellent conductor of electricity.

Mercury-operated combination Heating and cooling thermostat

A. Heating cycle: lines 1 and 2 connected
B. Cooling cycle: lines 1 and 3 connected

---

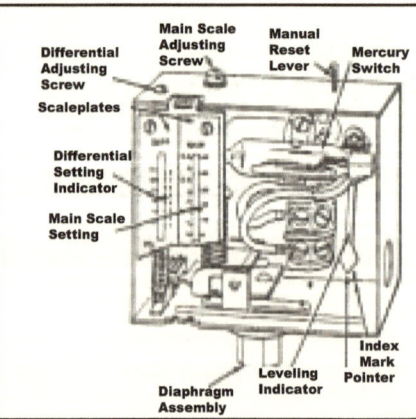

**A MERCURY-OPERATED PRESSURE CONTROL**

A capillary tube connects the pressure control to the sealed system (by a flare nut) through an access valve. The two manually operated adjusting screws on top regulate the pressures in the system. One is the cut-in, the other is the differential adjustment. When the pressure in the system reaches a predetermined point, the mercury switch tilts, causing the electrical circuit to the compressor to shut off until the pressure rises to a predetermined point during the off cycle.

Courtesy of Honeywell

It has been mentioned that the mercury-type wall thermostat is very sensitive to proper leveling and should be accurately installed to operate correctly.

Since homeowners are primarily concerned about the high cost of heating and cooling, the solid-state programmable thermostat is becoming more popular. The heating and cooling cycles can be separately programmed to economically regulate temperatures for daytime and evening operation, from day to day during a whole week, summer, or winter. They come with detailed instruction manuals. They are not sensitive to level adjustment.

## INSTALLING A NEW THERMOSTAT

When installing a new thermostat, determine whether it is to be used for heating or cooling (or both) and find the most suitable location as suggested on page 259. Always follow the instructions that are included with each new unit.

## INSTALLING A REPLACEMENT THERMOSTAT

1. Disconnect the power. Where a split system is used, be sure to disconnect power to the furnace (air handler) as the transformer is energized there.
2. Remove the snap-off thermostat cover.
3. Remove the wires from the base.
4. Remove the sub-base and the base.
5. Run the wires from the wall through the hole in the new sub-base and fasten it to the wall. If the wall is thin, use behind-the-wall expanding nuts (moly-bolts) being careful not to tighten them too much. Very often regular wood screws work as well.
6. Run the wires through the base. Fasten the base loosely (just to hold it in place while connecting the wires to their designated and marked terminals). Be sure that each wire is confined within its proper area so that the cover can be mounted properly.
7. Adjust the base with a small carpenter's level and tighten the screws. Check the level again after tightening the last one. If it is not perfectly level, loosen the screws and try again. (Step 7, of course, only applies to the mercury bulb thermostats). Skip step 7 when installing solid-state wall thermostats because they are not sensitive to accurate leveling.
8. Finally, snap on the cover, reconnect the unit to the power supply and start the unit.

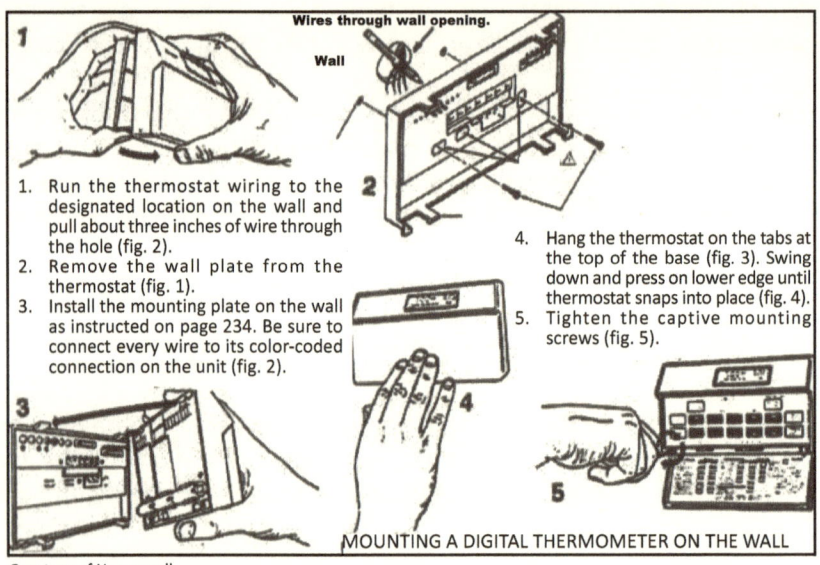

1. Run the thermostat wiring to the designated location on the wall and pull about three inches of wire through the hole (fig. 2).
2. Remove the wall plate from the thermostat (fig. 1).
3. Install the mounting plate on the wall as instructed on page 234. Be sure to connect every wire to its color-coded connection on the unit (fig. 2).
4. Hang the thermostat on the tabs at the top of the base (fig. 3). Swing down and press on lower edge until thermostat snaps into place (fig. 4).
5. Tighten the captive mounting screws (fig. 5).

MOUNTING A DIGITAL THERMOMETER ON THE WALL

Courtesy of Honeywell

## RECALIBRATING THE THERMOSTAT THERMOMETER

If it is suspected that the thermometer of the thermostat is not giving an accurate reading (it may seem too warm or too cool for the indicated temperature), check it first with a thermometer known to be accurate. If there is a significant temperature difference, the thermometer can be recalibrated.

1. Remove the cover. The thermometer adjusting screw is in the back of the cover. (See fig. 173.)

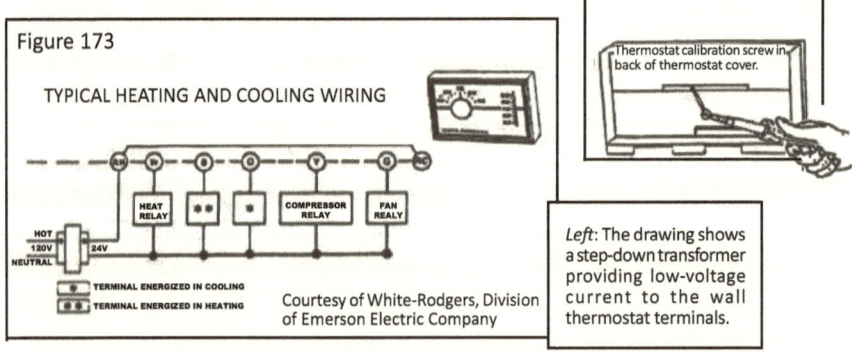

2. Place the thermostat thermometer close to the accurate one. Wait for about three minutes for the temperatures to stabilize.

3. With a small screwdriver, recalibrate the thermostat to match the temperature reading of the accurate thermometer.

---

**TESTING A WALL THERMOSTAT**

When (a certain component in) the unit fails to start and the wall thermostat is suspected to be the problem, disconnect the unit from the power supply (in split systems disconnect the furnace from the 120 VAC power).

1. Remove the thermostat cover and the base from the sub-base.
2. Remove the wires that energize the cycle that has failed and connect them directly together, then reconnect the power. The thermostat is now bypassed, and if the unit starts, that proves the thermostat is bad and will have to be replaced. Otherwise, the thermostat is OK, and the answer to the problem is to be sought elsewhere. Reconnect the wires to their appropriate terminals and put back the thermostat base and cover. Before running these tests, bear in mind that only the red wire is hot. Other wires will only become energized when they are connected to the red wire. Disconnect power to the unit, connect the appropriate wires together, and then reconnect power to the unit to check for results.

   ★ Connect red and yellow wires to energize the contactor coil which will energize the compressor and the condenser fan.
   ★ Connect red and green wires to energize the plenum blower relay causing the plenum blower to operate.
   ★ Connect the red, yellow, and green wires together to energize the entire cooling system (see fig. 1 on page 352).
   ★ Connect red and white wires together to energize the gas valve or the electric heater relay, causing the gas valve, the compressor (for heat pumps), or the electric heater in the furnace housing to operate.
   ★ Connect the red, white, and green wire together to energize the entire heating system (see fig. 3 on page 352).

To test the wall thermostat for a heat pump reversing valve connections, refer to page 352 for the proper connections.

---

## TIME-DELAYED RELAYS

Wall units usually use capillary tubes and central air conditioners may use capillary tubes, thermostatic, or automatic expansion valves. Since most air conditioners run on refrigerant R-22, the pressures in both the high and low sides are considerably higher than in the regular refrigeration units. That is why the diameters of the tubing used in air conditioners are larger. Because of the high pressures in the sealed system, larger units use relays that are time-delayed. That means when the unit is turned on, the time-delayed relay takes a little longer to energize the compressor, giving the high pressures in the system a chance to equalize. This prevents the exertion of those high pressures against the compressor during start-ups. Due to the higher load on the compressors in these units, the circuit breakers used in their electrical circuits must be no less than 20 amperes (20A) time-delayed.

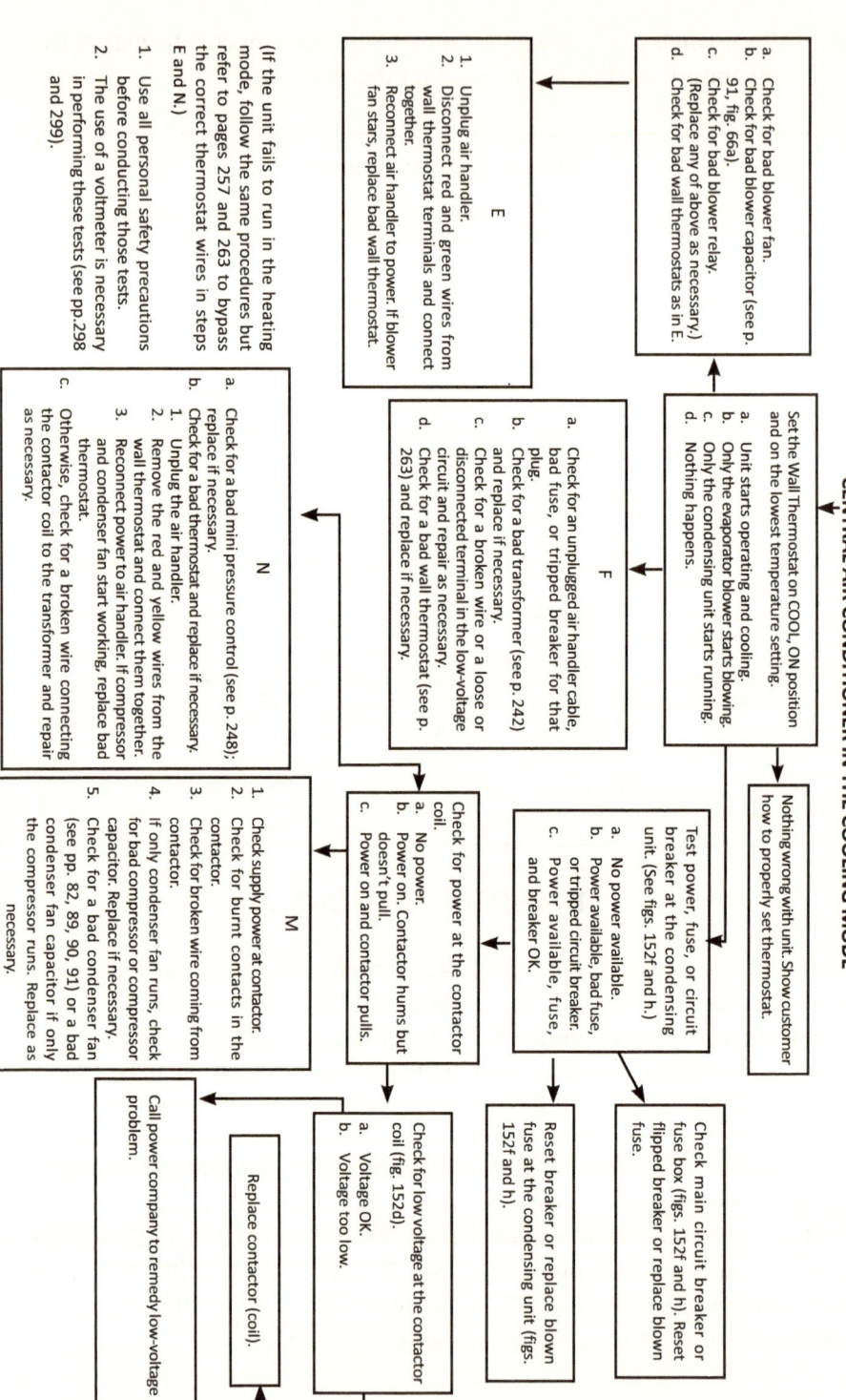

## HEAT PUMPS

In the evaporator coil of a regular air-conditioning unit, the vaporizing refrigerant absorbs heat from the conditioned area. This heat is then transferred to the condenser where it is radiated to the outside environment. In other words, the heat picked up from the inside is transferred to the outside.

Heat pumps are basically like regular air conditioners with the difference that they can reverse the action of the evaporator and condenser simultaneously by reversing the direction of the refrigerant flow. When this happens, the evaporator becomes the condenser and the condenser coil becomes the evaporator. This is made possible by using a reversing valve.

When the reversing valve is de-energized, it connects the outside coil to the suction line of the compressor (changing it to an evaporator) and the inside coil to the discharge line of the compressor (changing it to a condenser). With the evaporator now on the outside, the unit picks up the available heat from the ambient air and transfers it to the inside coil (now the condenser) and gives up that heat to the conditioned area.

A heat pump unit has a few extra components as compared with those found in ordinary air conditioners.

1. A reversing valve, which is also known as a four-way valve. See figures 103, 103a, and 103b. When the heating switch on the control panel of the unit (or on the wall thermostat) is turned on, a relay automatically deactivates the reversing valve. Additionally, other primary controls (relays, contactors, etc.) energize the compressor, the condenser, and evaporator fans.
2. As illustrated in figures 174 if expansion valves are used, two will be needed in a heat pump. Only one expansion valve becomes active at a time (whether the unit is in the heating or cooling mode). One is installed before the inside coil, and one is installed before the outside coil. The pressure-limiting type of expansion valves are best suited for heat pumps.
3. Since the direction of refrigerant flow changes when a heat pump is used, the installation of a bidirectional filter-drier is necessary.
4. The function of check valves is to ensure that the refrigerant flows in the proper direction during the heating or cooling cycles. One check valve controls the refrigerant flow in the heating cycle, and the other one controls the flow in the cooling cycle. When using a capillary tube in a heat pump, a strainer must be installed at each end. Because the flow of refrigerant can be reversed in heat pumps, capillary tubes become an excellent control. (See fig. 174b.)

# HEAT PUMP IN THE HEATING CYCLE

When the reversing valve is de-energized, it connects the inside coil to the compressor discharge line (acting as a condenser) and the outside coil to the compressor suction line, causing the unit to heat. (See figs. 103 and 104.)

As illustrated in figure 174, heat pumps using expansion valves are required to have check valves installed.

In locations with severe weather, the outside coil is placed underground because the temperature there is higher in the winter and lower in the summer. They are referred to as ground coils. Some units use circulating water as a heat-absorbing agent to cool the outside coil when the unit is in the cooling mode. When the unit is in the heating mode, circulating water is a good method to give up its stored heat to the outside coil. (See figs. 104 and 105.)

## HEAT PUMP IN THE COOLING CYCLE

When the reversing valve is energized, a magnetic field is created inside the reversing valve coil, causing the armature to lift or slide, opening one port and closing the other one. This connects the inside coil to the compressor suction line (changing it to an evaporator) and the outside coil to the compressor discharge line, enabling the unit to cool. Thus, the unit becomes a regular air-conditioning unit.

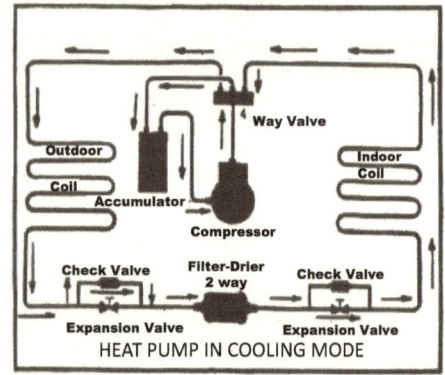

HEAT PUMP IN COOLING MODE

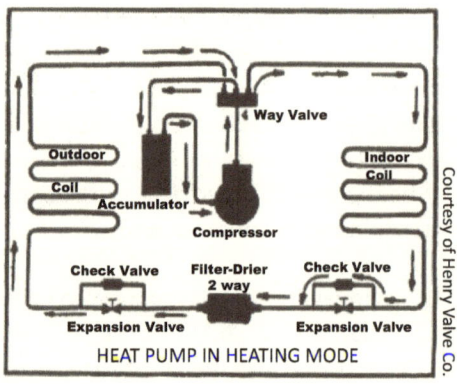

HEAT PUMP IN HEATING MODE

Courtesy of Henry Valve Co.

Figure 174

Note the direction of flow of refrigerant in cooling and heating modes. Only two-directional driers can be used in heat pumps as refrigerant flow changes in each mode. Notice that the coil inside the house acts as an evaporator in the summer and a condenser in the winter; the opposite being true in the case of the outside coil. Note also that the reversing valve (four-way valve) does not affect the flow of Freon into or out of the compressor; it only creates a change in the direction of flow for each coil causing it to act as a condenser or evaporator as required to heat or cool.

NOTE: Whether the unit is in the heating or in the cooling mode, the flow of refrigerant to and from the compressor does not change. Only the direction of refrigerant flow to the coils changes.

When the unit is in the heating mode, the evaporator temperature should be set 20°F below the average ambient temperature for the most efficient operation. This temperature difference is heat which is absorbed by the outside coil.

With the outside temperature at 50°F and the evaporator temperature set at 30°F, the refrigerant condenses in the condenser at 110°F, which is enough to warm the conditioned area.

Despite extremely cold weather, when the unit is in the heating mode, heat can always be extracted as long as the evaporator temperature is set below the outside ambient temperature.

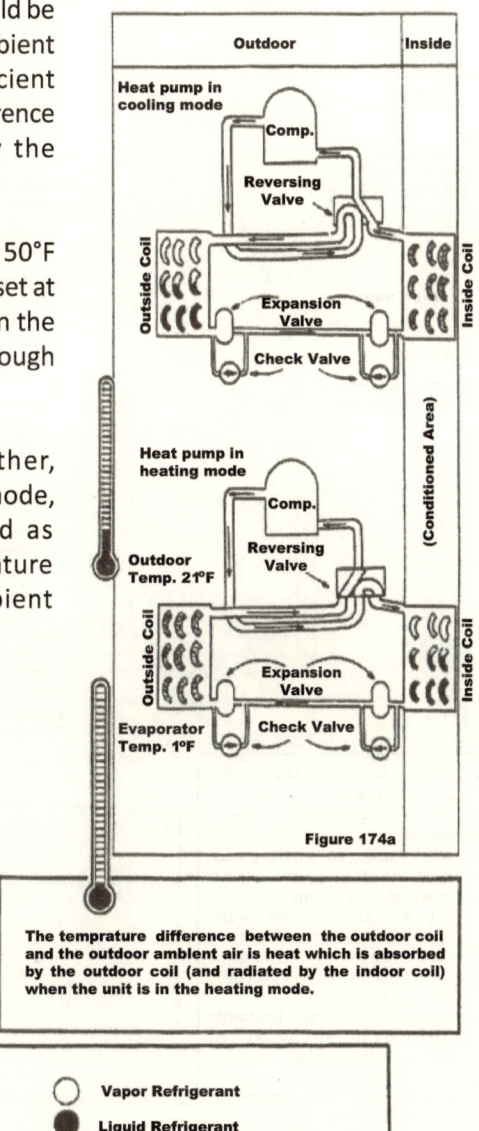

Figure 174a

The temprature difference between the outdoor coil and the outdoor ambient air is heat which is absorbed by the outdoor coil (and radiated by the indoor coil) when the unit is in the heating mode.

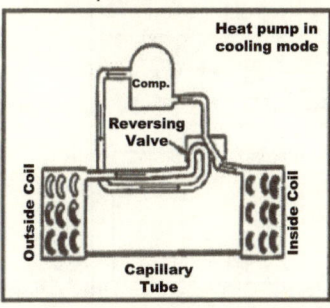

Figure 174b

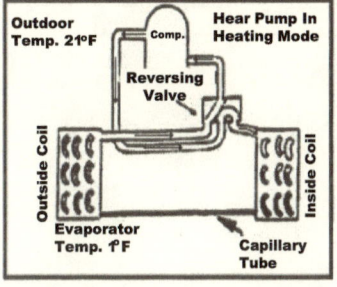

○ Vapor Refrigerant
● Liquid Refrigerant
◉ Liquid Vapor Refrigerant

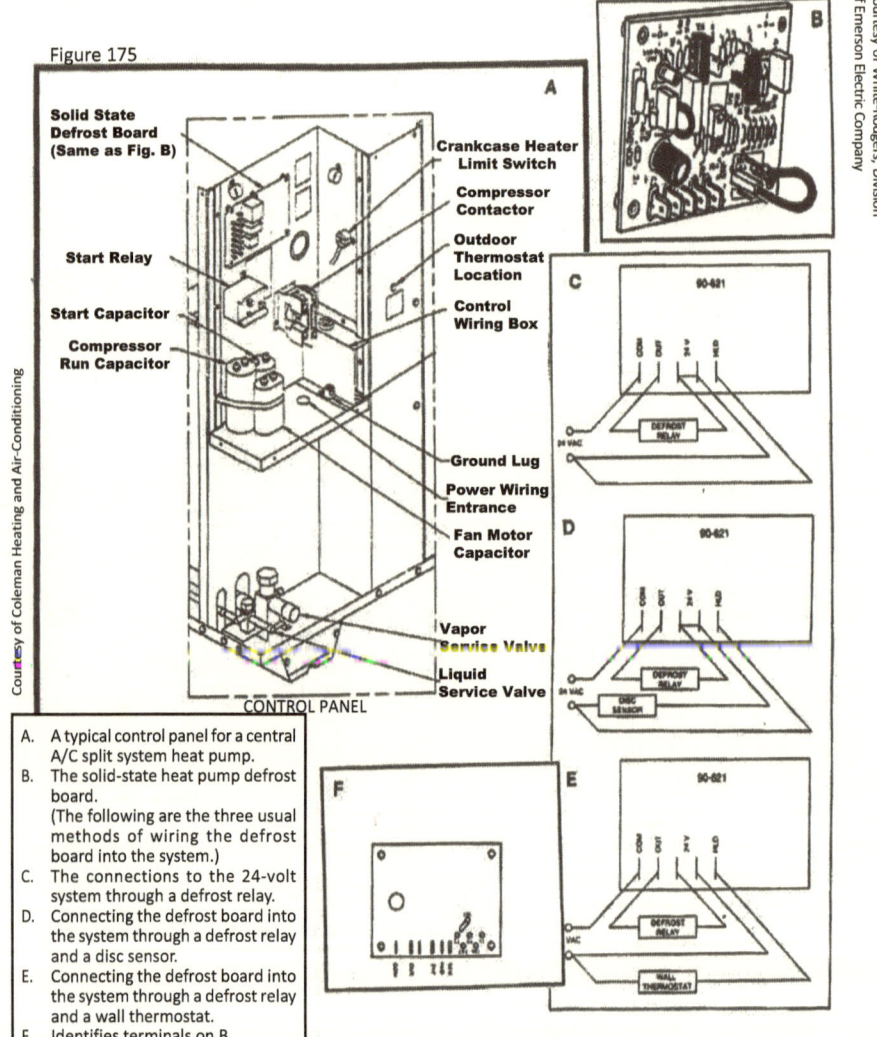

Figure 175

A. A typical control panel for a central A/C split system heat pump.
B. The solid-state heat pump defrost board.
(The following are the three usual methods of wiring the defrost board into the system.)
C. The connections to the 24-volt system through a defrost relay.
D. Connecting the defrost board into the system through a defrost relay and a disc sensor.
E. Connecting the defrost board into the system through a defrost relay and a wall thermostat.
F. Identifies terminals on B.

## THE DEFROSTING OF THE HEAT PUMPS

Heat pumps are used in the cold season, and the outside coil of these units acts as an evaporator when the unit is in the heating mode. Due to the low ambient and coil temperatures, ice tends to build up on the outside coil, and it should be defrosted often enough to maintain the optimum performance of the unit.

When the unit is taken to the defrost mode, the reversing valve is de-energized, and the direction of the refrigerant flow is reversed with the compressor running. During this time, electrical heating elements are energized to supplement the heat to the air-conditioned area.

Every time the unit is shifted to the defrost cycle, a surge of liquid refrigerant is forced back to the compressor; this is called "flood back." An accumulator is installed on the suction line to compensate for this flood back. In fact, the accumulator acts as a receiver to store excess refrigerant so that it gets a chance to vaporize before going into the compressor.

Solid-state timers (boards) are widely used to take heat pumps to the defrost mode (see fig. 176). In these defrost timers, a thermistor senses the difference in air temperature entering the evaporator coil and leaving it. When the thermistor senses a temperature difference of about 25°F, the solid-state board takes the unit into the defrost cycle. Unlike mechanical timers that put a unit into the defrost mode every so many hours, solid-state defrost systems put the unit into the defrost mode only when needed (but as often as necessary).

The defrost boards are equipped with testing posts through which the board can be easily tested for proper operation. The testing procedure comes with most new boards.

Here is the testing procedure of the defrost system in a Coleman Heat Pump:

Operate the heat pump in the heating mode for about five minutes.

Short across the two SEN JMP posts and the two SPEED-UP posts to put unit into defrost (fig. 176). Maintain the jumper on the test pins until the defrost cycle begins. As

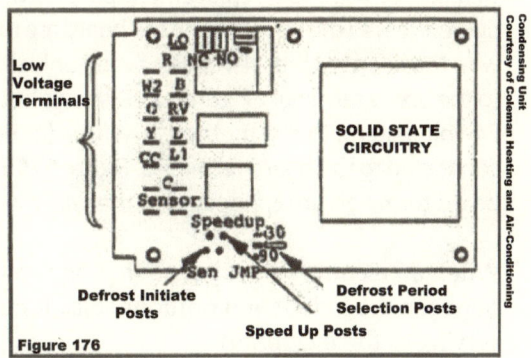

Figure 176

the unit goes into the defrost cycle, remove the short from the SPEED-UP and SEN JMP posts.

Allow the unit to terminate the defrost cycle automatically. This will occur

1. immediately upon removal of the SEN JMP short, (only if coil temperature is warm enough to signal a complete defrost);
2. when the defrost control receives the temperature terminate signal from the coil thermistor; and
3. when the onboard override timer signals for a timed termination (about ten minutes).

In humid conditions with temperatures near 32°F, during normal operation, with the heat pump in the heating mode, the out door coil will gather frost causing the temperature of the coil to drop. This in turn causes the defrost timer to be activated in response to the control thermistor (located in the condensing unit).

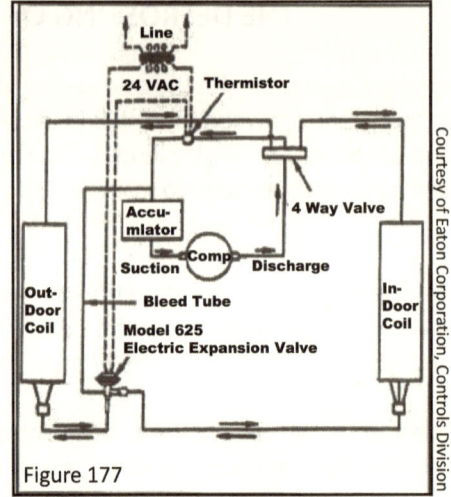

Figure 177

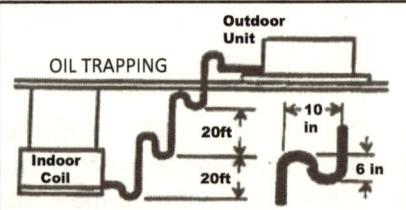

When installing a heat pump (or regular air conditioner) with the outdoor unit above the indoor coil, oil trapping is necessary. An oil trap should be provided for each twenty feet of rise. Be sure to seal the holes in the structure made for the condensate drain and refrigerant lines.

There are three different methods of connecting a solid-state defrost timer (board) to a heat pump system. Figure 175 B is a White-Rodgers defrost board that can be used as a replacement part for most air conditioners. As shown in figure 175 F, all the terminals on the board are marked. Figure 175 C shows how to wire the board to the 24-volt line when only the defrost relay is to be connected to the board terminals. Figure 175 D shows how to wire the defrost board to the 24-volt system. In this method, a sensor (a thermistor) and a defrost relay are connected to the board terminals. Figure 175 E shows the method by which the board terminals are connected to the wall thermostat and the defrost relay.

When an electric valve is used in a heat pump, only one valve is installed between the indoor and outdoor coils. It maintains 0°F in whichever coil is serving as the evaporator.

A bleed tube should be installed between the valve and the suction line to prevent liquid migration and chilling the bimetal chamber within the valve. (See figs. 177 and 177a.)

The electric valve on heat pumps can be controlled by two methods:

1. A thermistor is installed in the common suction line between the reversing valve and the compressor. (See fig. 177.)
2. Two thermistors, wired in series with the electric valve, are installed at the outlet of each coil. (See fig. 177a.)

The valve is controlled by either one of the thermistors. In the cooling mode, thermistor number 1 controls the electric valve, and thermistor number 2 has no effect on valve operation because it senses high temperature gas and is self-heated to a negligible resistance. In the heating mode, thermistor number 2 controls the valve and thermistor number 1 is self heated to a negligible resistance. No check valves are required when electric valves are used.

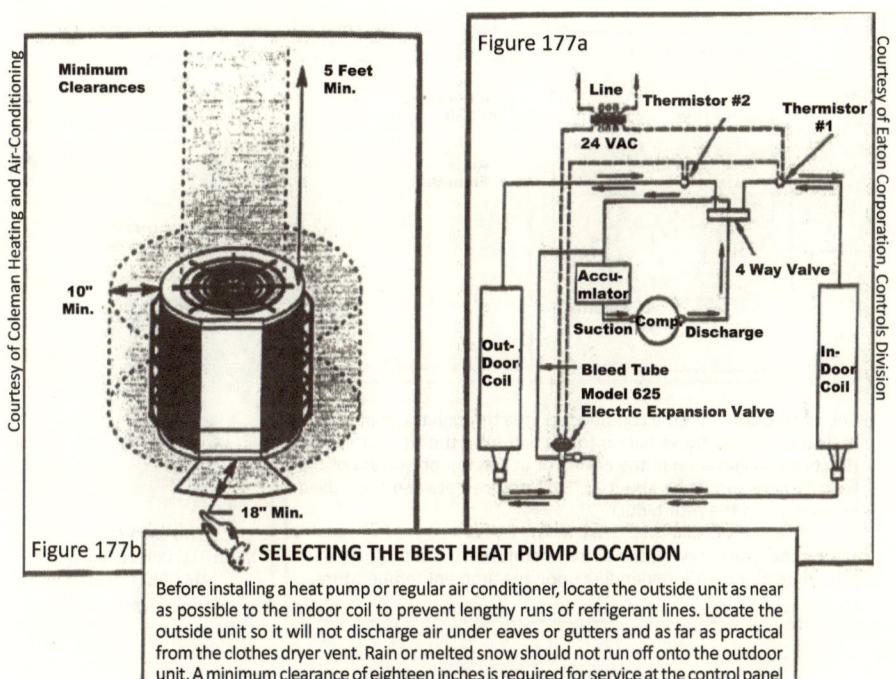

Figure 177a

Figure 177b

**SELECTING THE BEST HEAT PUMP LOCATION**

Before installing a heat pump or regular air conditioner, locate the outside unit as near as possible to the indoor coil to prevent lengthy runs of refrigerant lines. Locate the outside unit so it will not discharge air under eaves or gutters and as far as practical from the clothes dryer vent. Rain or melted snow should not run off onto the outdoor unit. A minimum clearance of eighteen inches is required for service at the control panel and compressor compartment access. A ten-inch clearance is required for the air inlet to the outdoor coil around the perimeter of the unit. A minimum of five inches should be maintained between the top of the unit and overhead obstructions.

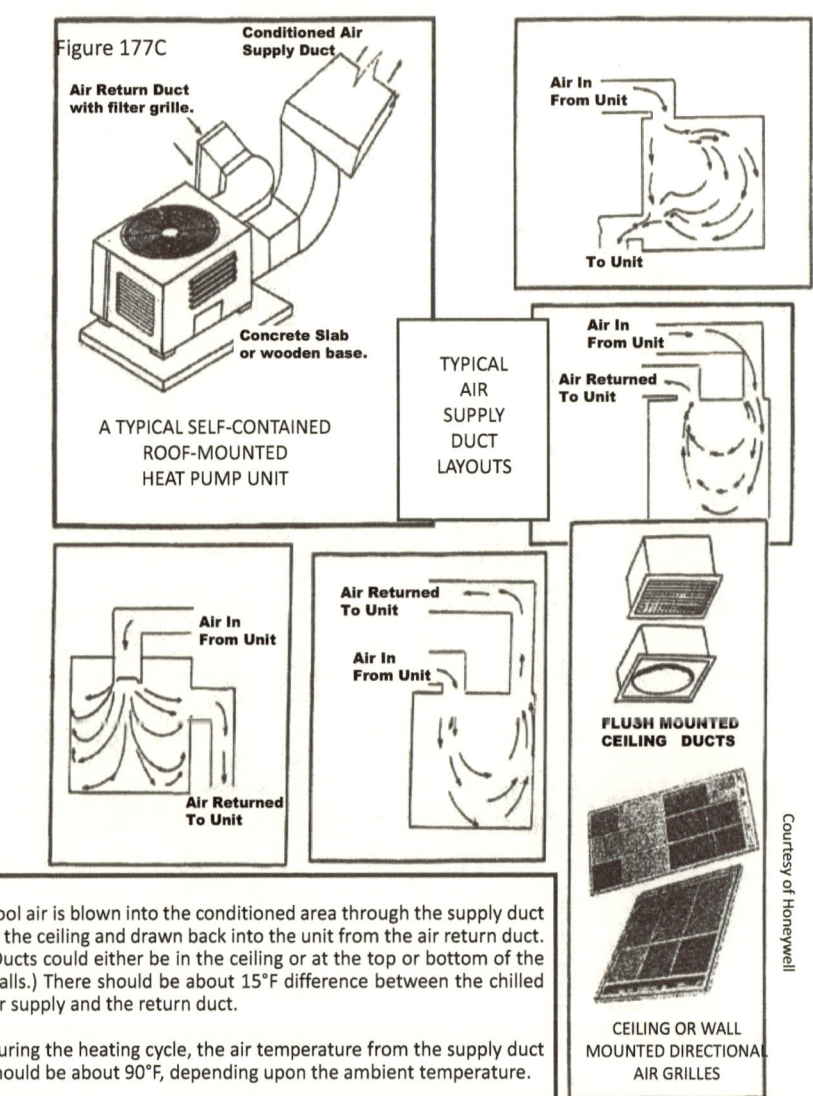

Figure 177C

Cool air is blown into the conditioned area through the supply duct in the ceiling and drawn back into the unit from the air return duct. (Ducts could either be in the ceiling or at the top or bottom of the walls.) There should be about 15°F difference between the chilled air supply and the return duct.

During the heating cycle, the air temperature from the supply duct should be about 90°F, depending upon the ambient temperature.

## INSTALLING A THREE-PHASE COMPRESSOR

The installation of a three-phase compressor is a rather simple job since three "hot" wires make a three-phase power supply. The power supply lines should be hooked up directly to the three compressor terminals.

Since the compressor consumes the most power during the start-up, the power supply line having the highest voltage must be hooked up to the compressor terminal marked S. (See figs. 38 and 39 again if the compressor terminals are unmarked.) This line can be determined by using a voltmeter to measure the voltage in the three power supply wires two by two until the "highest leg" is found. Normally, in a three-phase circuit, voltage in one of the power supply lines is up to 3% higher than the other two. Figure 152b shows a three-terminal contactor. It connects directly to the three terminals of a three-phase compressor. A typical connection is illustrated in figure 178 below.

When a new three-phase compressor is installed, check the voltage between the three terminals. If the voltage varies more than 3%, check the terminals for a loose connection. If the connections are OK, check the power at the main fuse box. If the problem is there, call the power company to correct it before turning on the unit.

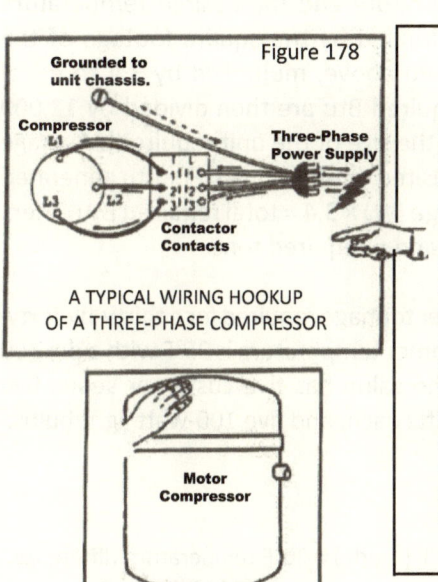

Figure 178

A TYPICAL WIRING HOOKUP OF A THREE-PHASE COMPRESSOR

Three hot line wires are connected to the compressor terminals through the contactor. Power is connected to contactor terminals L1, L2, and L3. When contactor coil is energized, those terminals pass the power to the compressor through terminals T1, T2, and T3.

If power reaches the compressor and it short-cycles, if no unusual pressures are observed to indicate a restriction and the condenser is clean, then the compressor has an internal problem, and it must be replaced. If the unit is equipped with an external overload protector, check that too. However, if low voltage is suspected, check the circuit breaker (or fuse box). If the compressor feels cool to the touch and does not operate, check the circuit breaker for interrupted power, the mini pressure switch, the contactor, and the transformer.

When the compressor short-cycles, it feels very hot to the touch.

# DETERMINING THE TONNAGE REQUIRED FOR A NEW AIR CONDITIONER

The geographical location and construction of the building in which a new air-conditioning unit is to be installed are important factors in determining the cooling requirements. The better insulated the building, the less heat penetration in summer and penetration of cold air in winter; hence, the lower the tonnage requirement for an air conditioner. A building in San Francisco requires less energy to cool than one in San Antonio, Texas, because the climate in San Francisco is mild. The number of people that occupy the conditioned area and their activity are other factors that affect the heating and cooling requirements for determining the size of the unit. Also take into consideration the number of light bulbs and heat-producing appliances involved. In the case of heat-producing appliances, the total load can be determined by multiplying the wattage of all the appliances by 3.4. (EXAMPLE: Ten light bulbs of 100 watts each would be 1000 watts × 3.4 = 3,400 Btu).

Consider 500 to 1000 Btu for each person who occupies the building, depending on his or her activity. Naturally, if a person just sits or sleeps, the body gives off less Btu than if engaged in a physical exercise. A rule of thumb for determining the required tonnage for a new unit is to find the difference between the average outside temperature and the desired temperature inside the air-conditioned area multiplied by the square footage of the building plus all the extras mentioned above, multiplied by 3.4 to equal the total required Btu. The total required Btu are then divided by 12,000 to determine the required tonnage (the size of the unit required). Average outdoor temperature (°F) − indoor desired temperature (°F) + Btu generated by the occupants + equipment wattage (W) × 3.4 = total required Btu. Then, total required Btu ÷ by twelve thousand = required tons.

EXAMPLE: Determine the air conditioner tonnage required for a thirty-by-forty-feet hair salon where the average summer temperature is 95°F with a desired conditioned temperature of 75°F. The salon has five customer seats, five technicians, five hair driers at 700 watts each, and five 100-watt light bulbs.

SOLUTION:

95°F (Average outside temperature) - 75°F (inside) = 20°F temperature difference.

| | | |
|---|---|---|
| 30' × 40' = 1200 ft². 1200 ft² × 20°F = | 24000 | Btu |
| 5 hair driers × 700 watts = 3500 watts × 3.4 = | 11900 | Btu |
| 5 × 100 watt light bulbs = 500 watts × 3.4 = | 1700 | Btu |
| 10 people × 500 Btu = | 5000 | Btu |
| TOTAL: | 42,600 | Btu |

Total 42,600 Btu divided by 12,000 = 3.55 tons. Thus, a 3½- to 4-ton unit should do the job quite well. Choose the smaller unit if the shop is well insulated from the outside (as in a mall) or use the next larger unit if the area to be air-conditioned is in a free-standing building.

The chart on the next page is used to quickly determine the general tonnage requirement of an air conditioner for two small rooms or a small house. When deciding on a unit, keep in mind that the percentage difference is important and not the number of tons. For example, if a thirty-ton unit is determined to be suitable for a building and the closest unit available is rated at thirty tons, the result will be satisfactory because the difference between thirty and thirty-two tons is less than 10%; whereas, if a two-ton unit is substituted for one-ton equipment, the difference will be 100%. In which case, the conditioned area will feel uncomfortable because of the excessive humidity. The air conditioner will cool the area too fast and the thermostat will become satisfied too soon, shutting off the compressor too quickly. This will cause the unit to stay off most of the time. During the off cycle, the evaporator is no longer cold enough to condense the excess humidity. That is why the conditioned area will remain too humid. Thus, a two-ton unit will cool the area too fast, preventing the humid air from condensing on the cold evaporator coil. (See p. 226 to review the relationship between air temperature and humidity.)

If an undersized air conditioner were used, it would have to run longer than normally required, possibly continuously, possibly it would never bring the temperature to a comfortable level, and most assuredly, it would increase the energy consumption and raise the utility bill.

So when the required tonnage is determined, find a unit with a rating closest to it. Normally, a small percentage difference is unavoidable with no adverse result on the effectiveness of the unit. This advice applies to units using a heat pump, gas-fired heaters, or electric heating elements too.

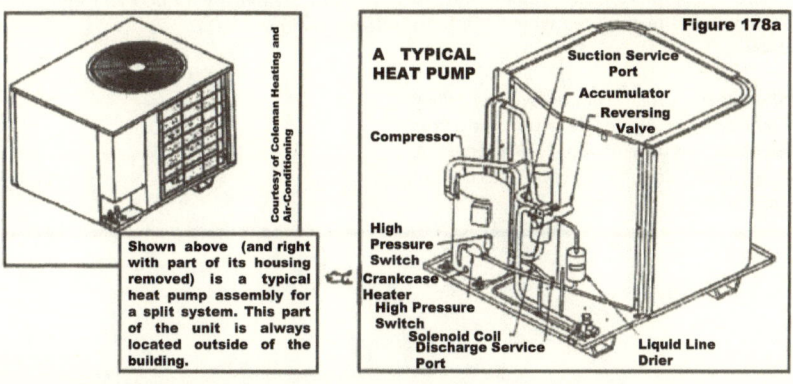

Figure 178a

Shown above (and right with part of its housing removed) is a typical heat pump assembly for a split system. This part of the unit is always located outside of the building.

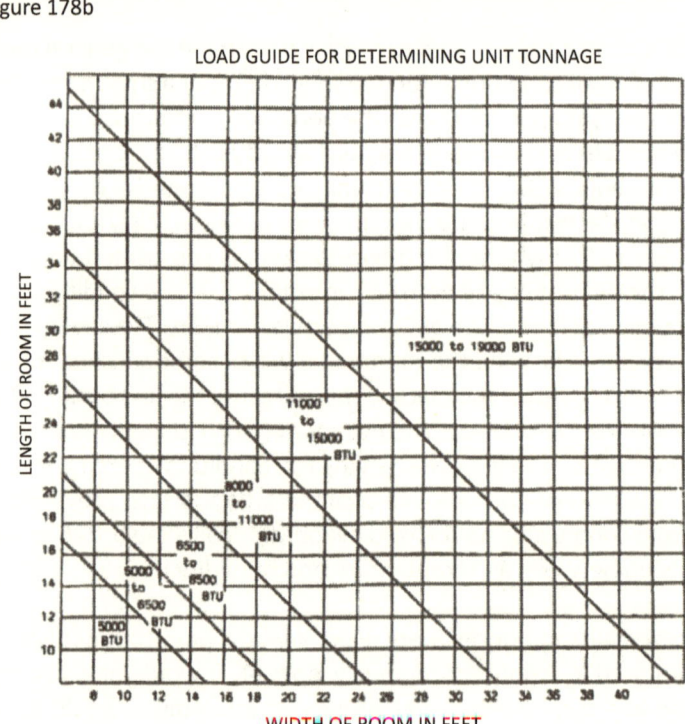

Figure 178b

## DIRECTIONS

Measure the width and length of the room to be conditioned.

Find the corresponding length and width on the chart.

See the minimum and maximum Btu recommended in the area where the lines intersect.

**EXAMPLE:** For a room with an average ceiling height of eight or nine feet that measures seventeen feet wide and thirty-one feet long, an air-conditioning unit of 11,000 to 15,000 Btu would be required.

(NOTE: Units with less than the minimum Btu recommended will cool less efficiently and run longer. Units that exceed the maximum Btu recommended will make the conditioned area too damp.

The Btu divided by 12,000 will equal the tonnage required:

$$\frac{15,000}{12,000} = 1.25 \text{ tons}$$

# TROUBLESHOOTING CENTRAL AIR CONDITIONERS

| Problem and Possible Cause | Remedy |
|---|---|
| **Compressor Motor and Fan Do Not Start** | |
| 1. Circuit breaker off or fuse bad. | 1. Reset breaker or check fuses. |
| 2. Thermostat set too high. | 2. Reset thermostat to lower temperature. |
| 3. Bad thermostat switch or wiring. | 3. Short between Y, R, and G; if unit starts, the thermostat must be replaced. |
| 4. Loose connections. | 4. Check wire terminals and tighten. |
| 5. Bad transformer. | 5. Check primary voltage (110/220), if power is there but no voltage on secondary wires (24 VAC), replace transformer. |
| 6. Bad compressor and/or fan. | 6. Check and repair/replace as necessary. |
| 7. Bad relay in control panel. | 7. Check or replace as necessary. |
| 8. High-pressure switch open. | 8. Check for dirty, linted, or obstructed condenser, bad condenser fan relay or motor, or bad high-pressure switch. |
| 9. Low-pressure switch open. | 9. Check for poor airflow through evaporator, bad low-pressure switch, low Freon or pressure, dirty filter. |
| 10. Contactor winding shorted. | 10. Check continuity in winding; replace contactor if necessary. |
| 11. Bad wall thermostat | 11. Check or replace as necessary. |
| **Condenser Fan Runs, But Compressor Does Not Start** | |
| 1. Compressor motor bad. | 1. Check or replace as necessary. |
| 2. Overload protector or capacitor defective. | 2. Check or replace defective part. |
| 3. Loose connection. | 3. Check and tighten as necessary, particularly compressor terminals. |
| 4. Defective contactor. | 4. Check or replace contactor. |
| **Compressor Runs, But Evaporator Fan Motor Does Not Start** | |
| 1. Bad fan motor or capacitor. | 1. Replace defective part. |
| 2. Loose connection. | 2. Check and tighten. |
| 3. Bad fan relay. | 3. Check and replace as necessary. |
| 4. Fan blade or blower obstructed. | 4. Adjust motor mounting to clear fan blade or blower wheel. |
| **Compressor Stops Before Thermostat Is Satisfied** | |
| 1. Condenser dirty. | 1. Clean condenser coil and fins. |
| 2. Low voltage. | 2. Check for required voltage. |
| 3. Condenser fan speed too slow. | 3. Check for loose blower wheel. |
| 4. Dirty filter(s). | 4. Clean or replace. |
| 5. Defective run capacitor. | 5. Check and replace if necessary. |

| Problem and Possible Cause | Remedy |
|---|---|
| 6. Defective compressor motor. | 6. Check for proper voltage. Allow enough time for overload to reset. If condenser pressure is normal, but compressor draws more than rated amperage, compressor is defective. |
| 7. High- or low-pressure control switch inoperative. | 7. Check control switches, refer to nos. 7 and 8 on page 277. |

Unit Does Not Cool Sufficiently

| | |
|---|---|
| 1. Thermostat set too high. | 1. Adjust to desired temperature. |
| 2. Thermostat improperly located. | 2. Relocate thermostat away from drafts, out of direct sunlight, etc. |
| 3. Compressor and/or condenser fan not running. | 3. Check cause. |
| 4. Dirty condenser and/or evaporator. | 4. Clean condenser and/or evaporator. |
| 5. Dirty filter(s). | 5. Clean or replace. |
| 6. Blower wheel slips on shaft. | 6. Check and tighten Allen screw. |
| 7. Refrigerant low as shown by low amperage, evaporator not cold, or large portion of condenser cool. | 7. Recharge after checking for restriction in capillary tube, strainers, TEV and filter-drier. Amperage should not be lower than FLA shown on condensing unit nameplate. |
| 8. Lack of insulation on ducts. | 8. Replace loose or missing insulation. |
| 9. Air leaks in ducts. | 9. Check and repair. |
| 10. Insufficient air from evaporator. | 10. Make sure duct dampers are open, duct runs are not too long or too small. Adjust blower speed. |

Compressor Does Not Shut Off

| | |
|---|---|
| 1. Dirty condenser. | 1. Inspect and clean. |
| 2. Unit too small for structure. | 2. Check for required cooling capacity. |
| 3. Low charge of refrigerant. | 3. Add refrigerant after checking for leaks and refrigerant restriction. |
| 4. Control does not shut off. | 4. Check thermostat and contactor. |

Unit Is Excessively Noisy

| | |
|---|---|
| 1. Air noise in duct work. | 1. Add stiffener to duct metal. |
| 2. Mechanical noise in unit. | 2. Check compressor shock mounts, loose parts, blower fan mounting, etc. |

Unit Does Not Run

| | |
|---|---|
| 1. No power to unit. | 1. Check plug and cord, circuit breaker, or fuse. Fuse should be right capacity and proper time delay. |
| 2. Low voltage. | 2. Be sure voltage is no more than 10% below that specified on nameplate. |

| Problem and Possible Cause | Remedy |
|---|---|
| 3. Broken wire or bad component. | 3. Check wiring and connections: check compressor, capacitor, switches, relays, overload protector, and thermostat. |

**Compressor Runs But Fan Does Not**

| | |
|---|---|
| 1. Fan motor burned out. | 1. Check continuity in fan motor. |
| 2. Fan blade or blower wheel restricted. | 2. Check mounting or adjust blade. |
| 3. Broken wire or bad component. | 3. Check wiring and connections, fan capacitor, defrost control, and all switches. |

**Fan Runs But Compressor Does Not**

| | |
|---|---|
| 1. Power supply faulty. | 1. Check for proper voltage to unit. |
| 2. Bad compressor or component. | 2. Check compressor, start and run capacitors, overload protector, relays, switches, and wiring. |

**Unit Runs But Does Not Cool**

| | |
|---|---|
| 1. Compressor not pumping. | 1. Check for restriction, loss of Freon, or lowered capacity of compressor. |
| 2. Restricted airflow. | 2. Check for obstruction in air passage and dirty filter. |

**Unit Runs But Does Not Cool Enough**

| | |
|---|---|
| 1. Restricted airflow. | 1. Check for dirty filter, restricted air passage, open exhaust door, dirty fans, condenser, and/or evaporator. |
| 2. Fan motor running slowly. | 2. Lubricate fan motor. Check for proper fan blade(s) or blower wheel; be sure blade or wheel is not binding. Check for a short in the fan motor. |
| 3. Voltage too low. | 3. Voltage should not be lower than—10% of that shown on nameplate. |
| 4. Poor seals. | 4. Seals missing or improperly located. Doors or windows open. |
| 5. Compressor not operating at full capacity. | 5. Check for restriction in sealed system loss of refrigerant or low voltage. |
| 6. Conditioned area too large. | 6. Check dimensions of area to be cooled and Btu rating of unit. Advise customer of proper unit for area. |

**Unit Short-Cycles**

| | |
|---|---|
| 1. Low supply voltage. | 1. Check for proper voltage. |
| 2. Temperature set too high. | 2. Reset thermostat, instruct customer. |

| Problem and Possible Cause | Remedy |
|---|---|
| 3. Bad thermostat(s). | 3. Check defrost thermostat (in heat pumps) and wall thermostat. |
| 4. Unit restarted too soon. | 4. Advise customer to allow two to three minutes for pressure to equalize. |
| 5. Faulty fan motor. | 5. Check for wrong or binding fan or blower wheel or wrong fan motor. |
| 6. Sensing element improperly located. | 6. Check location of sensing element or "comfort guard" (a plastic sleeve) missing from sensing element. |
| 7. Defective compressor or attached components. | 7. Check compressor for short or ground, weak connections, broken wires, run capacitor, overload protector, etc. (Check start capacitor and start relay, if unit has these.) |

Evaporator Has Excessive Frost Buildup

| | |
|---|---|
| 1. Outside temperature below 70°F. | 1. Advise customer unit not designed to operate at low ambient temperature. |
| 2. Sensing element improperly located. | 2. See number 6 above. |
| 3. Faulty thermostat. | 3. Check thermostat; see that temperature is not set too low. |
| 4. Insufficient airflow. | 4. Check for dirty filter, condenser, or blower wheel. Loose fan or wheel, or exhaust door open. |
| 5. Faulty unit. | 5. Check for loss of refrigerant or restriction in sealed system |
| 6. Fan running too slow. | 6. Advise customer to run fan at higher speed; check for low voltage. Check for binding blower wheel or shaft, wrong motor, or blower wheel. |

Unit noisy

| | |
|---|---|
| 1. Loose parts or mounting. | 1. Check for loose parts, tubing vibrating against sides or components, loose fan, blower, or mountings. Check for worn fan shaft bearing(s) or loose parts in the unit. |
| 2. Faulty compressor. | 2. Could have internal parts worn, low on oil or low voltage. Correct as necessary. |

Circuit Breaker Tripping or Fuses Blowing

| | |
|---|---|
| 1. Faulty wiring. | 2. Check condition of wiring and connections, look for short-circuiting, and repair as necessary. |
| 2. Restarting too soon. | 2. Advise customer to wait two to three minutes before attempting restart. |
| 3. Wrong fuses or circuit breaker. | 3. Check fuse for proper type and time delay; check circuit breaker size for adequate amperage. |
| 4. Incorrect voltage. | 4. Check power source for voltage no more than 10% ± of that on nameplate. |

| Problem and Possible Cause | Remedy |
|---|---|
| 5. Faulty component. | 5. Check capacitors, thermostats, overload protector, relays, switches, fan, and compressor motor, etc. |

### Moisture Drips Inside of Room

| Problem and Possible Cause | Remedy |
|---|---|
| 1. Unit improperly leveled. | 1. Unit should be one-fourth inch lower at rear to allow moisture to run outside. |
| 2. Drain hole(s) clogged. | 2. Clean drain hole(s) of debris. |
| 3. High humidity. | 3. Reduce door openings, repair or improve sealing around unit, and advise customer of any abnormal operating conditions. |

### Unit Does Not Work

| Problem and Possible Cause | Remedy |
|---|---|
| 1. No power. | 1. Check power supply at wall receptacle, check plug and service cord, circuit breaker, or fuse; make sure circuit is not overloaded with appliances. |
| 2. Faulty wiring. | 2. Be sure unit is wired according to its wiring diagram; check electrical components for correct wiring and operation. If all is in proper working order and unit still does not run, check for faulty compressor. |
| 3. Bad start relay. | 3. Check start relay on compressor as instructed in section on relays. |
| 4. Humidity control. (When used; also called a humidistat.) | 4. Unplug unit, set control on *dry* and check for continuity across the plug prongs. If there is no continuity, the control is bad and needs to be replaced. Next, turn control toward *wet* until you hear a click; then check again for continuity. If there is continuity, replace the control. (Relative humidity in room must be between 20% and 80%—the extremes in average control knob settings.) |
| 5. Defective defrost bimetal. (If equipped.) | 5. Check function of bimetal switch as you would on any refrigerator. |
| 6. Defective pressure control switch. (If equipped.) | 6. Unplug unit, bypass pressure control switch with jumper wire, connect power, and if unit starts, replace switch. |

### Unit Short-Cycles

| Problem and Possible Cause | Remedy |
|---|---|
| 1. Humidity control. (If equipped.) | 1. Check as in number 4 above. |
| 2. Restriction in sealed system. | 2. Check head and suction pressures; clear restriction as previously instructed in the refrigeration section. |
| 3. Low voltage. | 3. Check power source with voltmeter. |
| 4. Faulty compressor. | 4. Check for short or ground. |
| 5. Faulty start relay. | 5. Check by bypassing relay. |
| 6. Weak overload protector. | 6. Check by bypassing protector. |

| Problem and Possible Cause | Remedy |
|---|---|
| 7. Dirty or linted condenser. | 7. Clean condenser coil and fins. |
| 8. Extension cord too long or wire gauge too light for load. | 8. Place unit closer to a wall outlet or use a shorter and heavier cord. |

**Compressor Runs But Fan Does Not**

| | |
|---|---|
| 1. Defective fan motor. | 1. Check for current to fan motor, check for continuity in fan motor; repair or replace as necessary. |
| 2. Defective fan relay. | 2. Bypass relay; if fan runs, relay must be replaced. |

**Unit Operates But Doesn't Dehumidify**

| | |
|---|---|
| 1. Defective fan motor. | 1. Check motor for continuity. If there is none, replace fan. (Add a few drops of oil every six months. Check to see that shaft or blade is not binding.) Be sure blade is fastened securely to the shaft. |
| 2. Restriction in sealed system. | 2. Check high—and low-pressures to determine if there is a restriction. Repair if necessary. |
| 3. Poor air circulation. | 3. Check to see that no furniture or other objects are placed closer than six inches from end grilles and that the grilles are clean and unobstructed. |
| 4. Area too large. | 4. Average area to be dehumidified should be no larger than 10 Mft$^3$ (10,000 cubic feet) for a single unit. |
| 5. Humidity too low. | 5. Unit is effective only when temperature is above 65°F and relative humidity is above 60%; otherwise, the air is too cool and dry for efficient operation. |

**Excessive Amount of Frost on Evaporator**

| | |
|---|---|
| 1. Poor air circulation. | 1. See number 3 above. |
| 2. Area too large. | 2. See number 4 above. |
| 3. Humidity to low. | 3. See number 5 above. |
| 4. Defective defrost bimetal. | 4. Check function of bimetal switch as you would on any refrigerator. |

As you can see from the foregoing troubleshooting chart and figure 143, a dehumidifier is nearly the same as any refrigerating unit but with a different function. The troubleshooting procedure is the same. If the unit runs with abnormal noise, check for loose parts or mounting or an out-of-alignment fan blade, which will have to be secured or replaced.

# USING A CHARGING CHART TO CHARGE ROOF TOP AIR CONDITIONERS

The operating amperage is increased as more refrigerant is put into the system. Therefore, overcharging the system drastically reduces the life of the compressor motor.

The correct amount of refrigerant in central air conditioners is directly proportional to the amount of piping used between the evaporator and the condenser. As the length of the tubing increases, the amount of refrigerant charge increases. (A/C manufacturers usually attach a step-by-step charging chart inside an access panel of the condensing unit, such as the following method of charging, which is practiced most by the service personnel who work on Carrier A/C units):

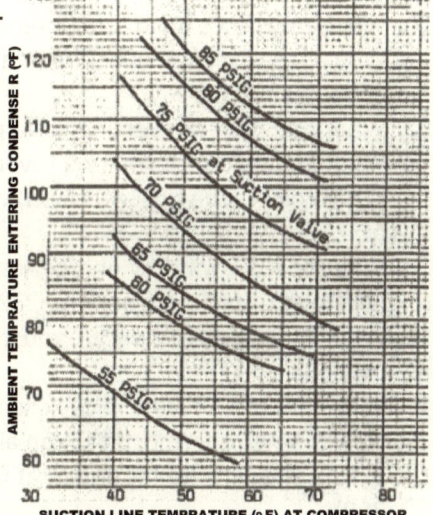

Figure 178c

1. Run unit about fifteen minutes before beginning this charging method.
2. Connect your gauge manifold to the suction-line access valve and get a pressure (or vacuum) reading.
3. Attach a thermometer close to the suction-line access valve and insulate it to get an accurate reading.
4. Use another thermometer to determine the ambient air temperature entering the condensing unit.
5. On the chart, figure 178c, find the air temperature entering the condenser and move horizontally until you intersect the curve indicating the suction-line pressure registered in step 2. Then move downward from that point on the curve to determine the corresponding suction-line temperature. If the suction-line temperature reads higher than the indicated figure on the chart, add refrigerant until the temperature falls within the parameters of the chart. Bleed some refrigerant off if the suction line has a temperature lower than that shown on the chart.

If you are using the weight method of charging (the amount of charge is sometimes printed on the nameplate), the system must be thoroughly evacuated. Weigh the refrigerant tank carefully and subtract the amount of charge from the total weight of the refrigerant tank. Leave the tank on the scale and charge the unit until the scale indicates that the proper amount of refrigerant is added to the system. For more accuracy in charging a system by weight, a Dial-A-Charge cylinder can be purchased from refrigeration supply firms. Of course, if the unit is equipped with a sight glass, charge the unit until the bubbles in the sight glass disappear.

## A TEST OF KNOWLEDGE
## On Air Conditioners

1. What are the five reasons for a dehumidifier to run, but not dehumidify? (p. 282)
2. What are the possible causes of a circuit breaker tripping as soon as a window air conditioner is turned on? (p. 280)
3. What happens when there is a restriction in the airflow in a central air conditioner? (p. 279)
4. What four causes would keep a compressor from shutting off in a central air-conditioning system? (pp. 277, 278)
5. In a central air conditioner, what are the possible causes for the fan to run when the compressor does not? (p. 277)
6. Why is humidity increased when air is cooled? (p. 226)
7. How is Btu converted to tonnage? (p. 227)
8. Why are there two expansion valves installed in a heat pump? (p. 262)
9. What is the function of check valves in a heat pump? (p. 265)
10. What are the six main functions of an air conditioner? (p. 265, 228)
11. What is the function of a slinger fan blade in a window or wall unit? (pp. 231, 232)
12. What role does a reversing valve play in a heat pump? (p. 265)
13. What is the other term for a reversing valve? (p. 265)
14. What color is the hot (common) wire connected to a wall thermostat? (p. 257)
15. Why should a window unit be installed with the rear edge one-fourth inch lower than the front? (p. 232)
16. What two types of fans are most commonly used in air conditioners? (p. 233)
17. Why are time-delayed relays used in air conditioners? (p. 263)
18. How is a wall thermostat for an air conditioner tested? (p. 263)
19. Where should a wall thermostat for an air conditioner be installed? (p. 259)
20. What kind of a compressor motor is used in a wall air conditioner? (p. 235)
21. Where is the thermostat bulb fastened in a wall air conditioner? (p. 235)
22. What color is the evaporator fan wire in a wall thermostat? (p. 257)
23. In a wall thermostat, what does the white wire represent? (p. 257)
24. How is the required tonnage determined for a new air conditioner? (p. 274, 275, and 276)
25. What are the two methods for checking a transformer? (p. 242)
26. Where are contactors installed? (p. 245)
27. What is an A coil? (p. 274)

28. Where is an A type coil or slant type coil installed in a central air-conditioning unit? (p. 274)
29. What are the two major causes for a contactor becoming inoperative? (p. 245)
30. What are the two basic differences in wall thermostats? (p. 256)

## CORRECT PIPING METHOD FOR REFRIGERATION AND AIR-CONDITIONING

*Liquid-line size.* It is important to use the proper size tubing. A liquid line that is too small causes excessive pressure drop reducing the volume of liquid refrigerant that flows into the evaporator, resulting in reduced unit cooling capacity. It is a good practice to use about 20% oversized liquid-line tubing. Every added tee, valve, and 90° elbow adds to the resistance exerted against the liquid refrigerant flow.

Figure 178d shows the increased length of tubing equivalent to the resistance offered by a valve, tee, or 90° elbow. Figure 178e shows the amount of liquid line charge in pounds per foot of tubing that has to be calculated when designing a sealed system. As stated earlier, a vertical rise offers some increased resistance and consequently a pressure drop in the liquid line. As a rule of thumb, one should consider 1/2 psi pressure drop for every foot of rise. (This is due to the weight of the refrigerant that reduces the velocity of the up flow.) Figure 178f shows the amount of pressure drop according to the unit size and the liquid-line tubing size in a system using R-12.

---

As an example, suppose you were to calculate the pressure drop in a two-ton system using R-12 that has one hundred feet of one-half inch tubing in the liquid line, and has a ten feet vertical rise, a valve, four 90° elbows, and two tees. Figure 178d shows that a valve on one-half inch tubing has an equivalent pressure drop to two feet of added tubing, four 90° elbows have the equivalent of four feet of tubing, and two tees, also equal to four feet. The ten-feet rise adds the equivalent of another five feet of tubing. Add them all together to determine the liquid-line pressure drop in this particular system.

| | |
|---|---|
| one-hundred feet liquid line | 100.0 |
| valve | = 2.0 |
| four elbows | = 4.0 |
| two tees | = 4.0 |
| a ten feet rise | = 5.0 |
| Total: | 115.0 |

According to figure 178f, 115 feet of 1/2 inch liquid line has a pressure drop of 1.84 lbs if the system uses R-12.

---

To give you a better understanding, according to figure 178f the pressure drop in a three-eighth-inch liquid line of a 48,000 Btu/h (four ton) unit circulating

R-12 is 0.450 psi/ft. If 100 feet of liquid line were used, a 45 psi (100 × 0.450) pressure drop would be applied at the TEV. If the unit were to normally operate with a 130 psi head pressure, the head pressure in the liquid line at the TEV would be reduced to 85 psi (130-45 = 85). According to the chart on page 124, at 84.2 psi (the closest to 85 psi), R-12 boils at 80°F. If the unit operated under this condition in a humid ambient of about 90°, liquid-line sweating would become evident (a sign of too much pressure drop). Pressure drop can also be caused by tees, elbows, rises, etc. (That has to be taken into consideration when designing a refrigeration or air-conditioning system.)

Other ways to compensate for the effect of liquid-line pressure drop is to add to the amount of refrigerant in the system, reduce the number of tees and elbows, or increase the diameter of the tubing.

Oil traps must be installed at the bottom of a suction line vertical rise to preclude the return of a large quantity of oil to the compressor at the moment of starting.

Generally, when multiple evaporators are installed in series, the size of the suction and liquid lines must be increased for each evaporator added. The cross-sectional areas of every additional line feeding multiple evaporators is added to the liquid and suction lines.

Figure 178d

| | \multicolumn{13}{c|}{OUTSIDE DIAMETER OF TUBING (INCHES)} |
|---|---|---|---|---|---|---|---|---|---|---|---|---|---|
| | 1/4 | 3/8 | 1/2 | 5/8 | 3/4 | 7/8 | 1-1/8 | 1-3/8 | 1-5/8 | 2-1/8 | 2-5/8 | 3-1/8 | 3-5/8 | 4-1 |
| | \multicolumn{13}{c|}{EQUIVALENT RESISTANCE IN FEET OF TUBING PER FITTING} |
| 90° Elbow | 0.75 | 0.75 | 1.0 | 1.0 | 1.5 | 1.5 | 2.0 | 2.5 | 3.0 | 4.0 | 5.0 | 5.5 | 6.5 | 7.5 |
| Tee | 1.5 | 1.5 | 2.0 | 2.0 | 2.5 | 3.0 | 4.0 | 5.0 | 6.0 | 7.5 | 9.0 | 11.0 | 13.0 | 15.0 |
| Valve | 1.5 | 1.5 | 2.0 | 2.0 | 2.5 | 3.0 | 4.0 | 5.0 | 6.0 | 7.5 | 9.0 | 11.0 | 13.0 | 15.0 |

Figure 178e

| | \multicolumn{7}{c|}{LIQUID LINE REFRIGERANT CHARGE (IN LBS/FT)} |
|---|---|---|---|---|---|---|---|
| Liquid-line size: | 1/4" | 3/8" | 1/2" | 5/8" | 3/4" | 7/8" | 1-1/8" |
| Refrigerant charge: | 0.015 | 0.043 | 0.086 | 0.134 | 0.202 | 0.269 | 0.458 |

Figure 178f

← LIQUID-LINE PRESSURE DROP (USING REFRIGERANT R-12)

| UNIT Btu | \multicolumn{7}{c}{LIQUID-LINE TUBING SIZE IN INCHES (OD)} |
|---|---|---|---|---|---|---|---|

| UNIT Btu | 1/4 | 3/8 | 1/2 | 5/8 | 3/4 | 7/8 | 1-1/8 |
|---|---|---|---|---|---|---|---|
| 9,000 | 0.250 | 0.021 | | | | | |
| 12,000 | 0.420 | 0.036 | | | | | |
| 18,000 | | 0.75 | 0.010 | | | | |
| 24,000 | | 0.127 | 0.016 | | | | |
| 36,000 | | 0.260 | 0.033 | 0.012 | | | |
| 48,000 | | 0.450 | 0.054 | 0.020 | 0.010 | | |
| 60,000 | | | 0.080 | 0.030 | 0.014 | 0.009 | |
| 84,000 | | | 0.150 | 0.054 | 0.025 | 0.015 | |
| 120,000 | | | 0.280 | 0.100 | 0.049 | 0.028 | 0.009 |
| 240,000 | | | | 0.350 | 0.160 | 0.095 | 0.029 |
| 360,000 | | | | | 0.340 | 0.200 | 0.058 |
| 480,000 | | | | | | 0.340 | 0.100 |

(To determine the total pressure drop in the liquid line, multiply the corresponding figure by the length of the liquid line.)

# BASIC ELECTRICITY

This section provides simplified instruction in basic electricity and illustrates typical circuits and symbols.

You will learn step-by-step, how to read refrigeration and air-conditioning schematic wiring diagrams through typical examples of the normal operation of these units.

It teaches how to use diagnostic tools such as ohmmeters, ammeters, and voltmeters to troubleshoot electrical malfunctions that commonly occur in refrigeration and air-conditioning units.

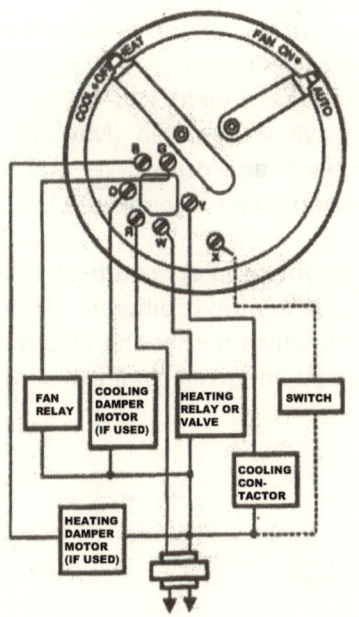
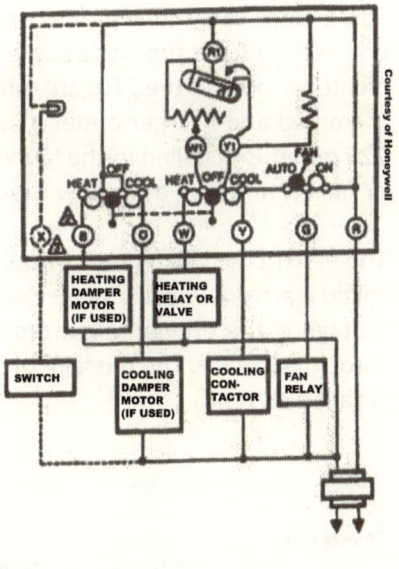

# BASIC ELECTRICITY

The electricity for mass consumption is created by large generators and conducted to final destinations with the help of transformers to maintain its force and energy.

The large electric generators in a city's power and light company move current through heavy cables to transformers, which in turn parcel out the current to homes and businesses. It enters those homes and businesses through their electric meters and then to circuit breaker boxes. (Fuse boxes as prime receivers are almost extinct now.) The main wires go to a "bus bar" (a common conductor in the circuit breaker box) for several circuit breakers. Wires from each circuit breaker supply current to one particular circuit in the home or business. During the construction of the home or business building, the electrical contractor determines which rooms to be wired to a particular circuit breaker and also the number of outlets in each room. Usually one to three rooms in a home are wired to one circuit breaker; the kitchen and laundry room may have two or three circuit breakers due to the increased load. A standard circuit breaker is capable of 110 and 220 VAC circuits.

Three wires carry current into a house: two 110 VAC and a neutral. When one of the *110*s is run into a socket along with a neutral wire, a *110* socket is created. If both of the *110*s are run into an electrical outlet, a 220 VAC socket is created and a larger outlet of a different style will be used. (Normally a 220 outlet is installed in the laundry room where an electric dryer is used, or on a wall where a window-mounted air conditioner is connected.)

The electricity which flows through wires is composed of minute particles called electrons. Electrons are exactly alike even in wires of different metals. Voltage is the energy (electromotive force), which pushes the electrons through a circuit. The passage of electrons past any given point in a circuit is called *current*.

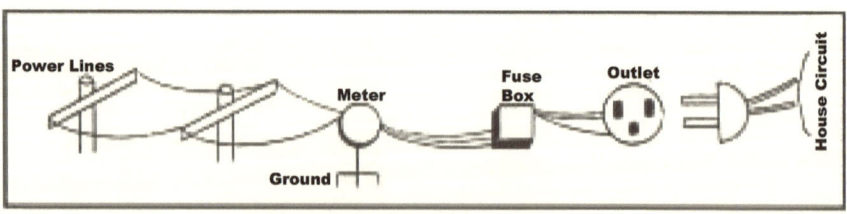

TYPICAL CURRENT FLOW TO HOUSE CIRCUITS

Current is measured by the ampere. One ampere is about the amount of current carried by a medium-size light bulb.

Any source of voltage can move electrons, such as a car battery, a flashlight cell, or even a wire passing through the magnetic field of a small handheld magnet although its current is very tiny. And that example is an oversimplification of how electricity is created. A generator has hundreds of wires in its winding (called a *rotor*) with magnets placed around the inside of the housing (called a *stator*) to create a magnetic field. When the rotor rotates, those hundreds of wires pass through that magnetic field hundreds of times per minute creating an electrical current.

Any material with low resistance to the flow of current is called a conductor. Materials that are poor conductors are referred to as insulators. Wires of various sizes and materials are used as conductors today. There are four basic elements in an electrical circuit:

1. A power source (i.e., 110-440 volts of electromotive force).
2. A set of conductors (the wires).
3. A load (any device in a circuit that converts electrical energy to useful work, i.e., a relay, a timer, a heater, etc.).
4. The means to control the current flow (switches, circuit breakers, thermostats, resistors, pressure controls, fuses, etc.).

Since current is measured in amperes, the measuring device is called an ammeter. (A flux-type meter is widely used.) (See figs. 71 and 72 and p. 305.) The device to measure electromotive force (voltage) is called a voltmeter. (See pages 298 and 299.)

An ohm is the unit of measurement of the amount of resistance to the flow of current in a circuit. Any resistance slows the number of electrons passing any given point. An ohmmeter is used to measure the amount of resistance in a circuit. (There are pocket-size combination meters on the market today that measure volts, ohms, and amperes (see p. 303).

A load can become energized only when it is in the path of a completed electrical circuit. Again, a complete circuit must have the four elements: power (also called potential in some books), uninterrupted conductors (wires), the load, and the means to control the flow of current. Resistors are used in many circuits to control the flow of current. They are used to reduce the amount of energy going to a particular part of a circuit or to a particular component.

There are three basic methods employed in the wiring, and each one has certain advantages and disadvantages: (1) series circuits, (2) parallel circuits, and (3) series-parallel circuits.

1.  *Series circuit* (Components connected in series). In this type of wiring, current must flow through every component (load) in the entire system to complete the circuit.

Since a series circuit provides only one path for current to flow, the advantage of this type of wiring is the on/off controllability of the whole circuit with only one switch. In the illustration below, power flows from L1 through the switch, the fuse, the heater, and the light bulb, then returns to the L2 side of the source, completing the circuit. If the circuit is broken (opened) by the switch, a blown fuse, burned-out heater, or a broken wire anywhere in the circuit, the entire electrical path will become interrupted and nothing in the entire circuit will work because the flow of current will no longer be able to return to L2 to complete the circuit.

The main characteristic of a series-wiring method is that the components with the greatest amount of resistance consume the greatest amount of power and leave too little power to energize the loads with minimal resistance. When a load with a high resistance value is installed in a series circuit, current flows and passes through other components without activating them as the insufficient remaining voltage does not meet their minimum voltage requirements. This characteristic is evidenced in faulty household wiring where a light bulb starts glowing dim or goes out as soon as an iron or a clothes dryer is plugged into that circuit. This principle has led to the development of the wiring method in many commercial ice machines and cycle-defrost refrigerators. The forthcoming pages (315, 320, 321, and 322) will explain how the manipulation of this series-circuit quality has led to the development of the electrical mechanisms of those units.

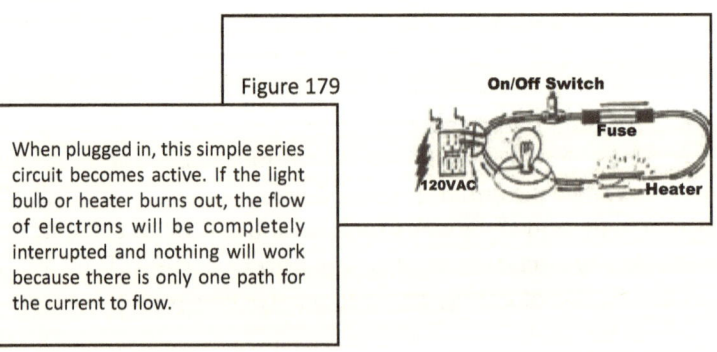

Figure 179

When plugged in, this simple series circuit becomes active. If the light bulb or heater burns out, the flow of electrons will be completely interrupted and nothing will work because there is only one path for the current to flow.

2. *Parallel circuit.* In this circuit, there is more than one path for the current to flow. In figure 180, below, if current flow were interrupted at point a or point b, nothing in the entire circuit would work. Current starting at L1 would not be able to reach L2. If power were interrupted at point c, only the forward heater and the reversing valve would remain active, while the rest of the circuit behind point c would not be energized. If the interruption were at point d, only the compressor would become de-energized while the rest of the system would remain active because the current flow would reach L2 through other branches in the circuit.

This type of wiring is used for applications in which the whole system is not intended to shut down when only one component becomes de-energized or defective. When this happens, current continues to flow through the other paths to complete the circuit. This type of circuit is found in household refrigerators. For example, when the refrigerator lamp becomes de-energized when the door is shut, other components such as the motor compressor, etc., in the system stay energized. Note that in the simple parallel circuit of figure 180, the broken wire near the motor does not interrupt the current flow to the rest of the circuit. The heaters, the fan, and the reversing valve remain active because current can still flow through these components and complete its unbroken paths.

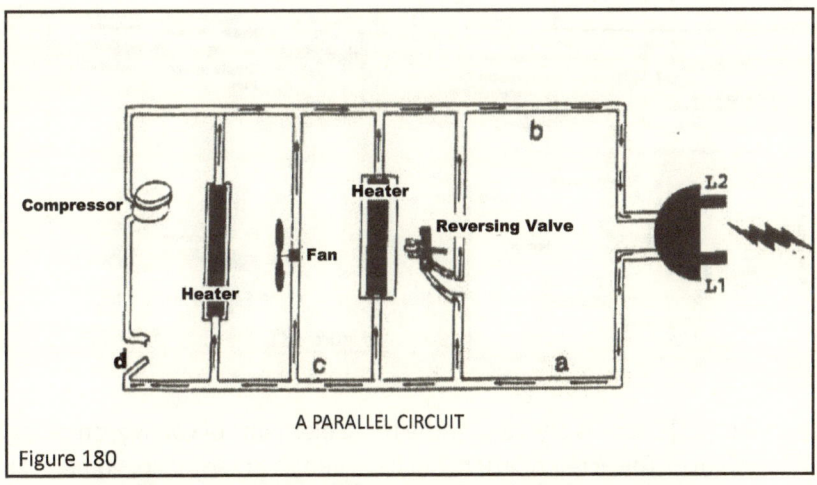

A PARALLEL CIRCUIT

Figure 180

In parallel wiring, if one segment of the circuit is turned off, other components stay unaffected because there are still unbroken paths through which current flows.

3. *Series parallel circuit.* As the name implies, it is a combination of series and parallel circuits. In this type of wiring, any one load (component) or a set of loads can be controlled by different switches, resistors,

solenoids, etc., while the rest of the system remains unaffected. In figure 181 below, circuits A-B, C-D, E-F, and G-H are, individually, series circuits. If the fuse blows, the entire system will become de-energized. If the solenoid burns out, it will cause the fan motor to shut off and nothing else in the other circuits will be affected since current is able to complete its path through the E-G-H-F circuits. This is because all the series circuits are wired in parallel.

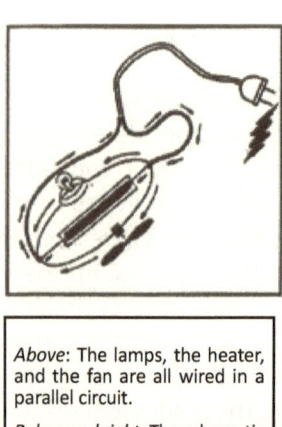

*Above*: The lamps, the heater, and the fan are all wired in a parallel circuit.

*Below and right*: The schematic wiring diagram below is the same as the series-parallel pictorial diagram in figure 181.

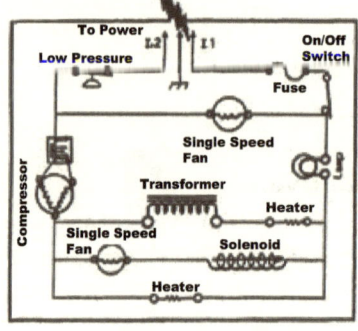

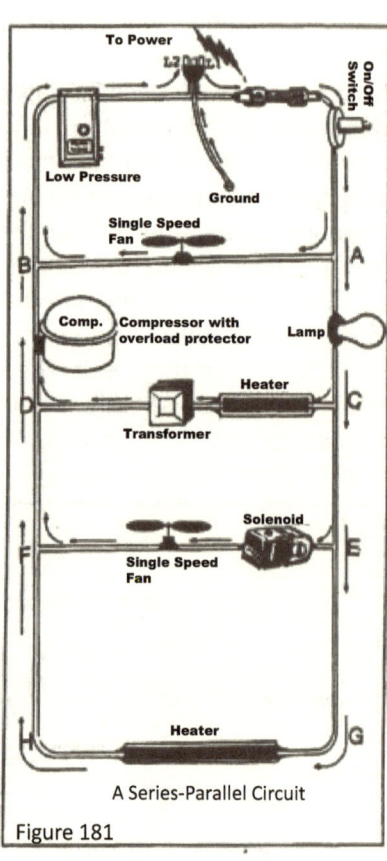

A Series-Parallel Circuit

Figure 181

Every electrical circuit is affected by the three elements of voltage, amperage, and resistance. There is a relationship between the resistance, voltage, and amperage which is shown by the equation $E = I \times R$

Where E is electromotive force (EMF) (voltage),
      I is the intensity of an electrical current (amperage),
and    R is the resistance in a circuit.
This is known as Ohm's law.

By using this equation, one unknown factor can be determined if the other two are known. Graphically, it looks like this:

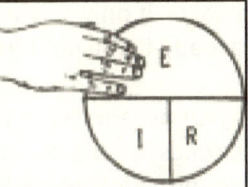

If the amperage and the amount of resistance in a circuit are known, to find the voltage, put a finger over E. This will determine that the amperage (I) must be multiplied by the resistance (R). To determine the resistance, put a finger over R; divide the voltage by the amperage. To determine the amperage, put a finger over I; divide the voltage by the resistance.

For instance, there is a need to determine the amperage expected in a particular circuit. The circuit is supplied with a 220-volt line and the schematic diagram shows 22 Ω listed under a heater as its resistance. Looking at the illustration above, the amperage equals the voltage divided by the resistance (I = E ÷ R), or 220 divided by 22 equals 10 A. If the ammeter reads only 0.5 (1/2) A, it will mean that there is a short in that particular circuit as there should be a reading of about 10 A.

(EXPECT THE INDUCTIVE REACTANCE IN THE WINDINGS OF A LOAD TO CAUSE THE AMMETER READINGS, AND OHM'S LAW MATH, TO BE SLIGHTLY DIFFERENT).

The most important law in electricity: "The current which flows in a circuit is directly proportional to the applied voltage and inversely proportional to the resistance." Simply stated, when voltage is increased, the current increases. If the voltage is constant, the current changes as the resistance changes, but in the opposite direction. The more resistance applied, the lower the amount of current flow.

Ohm's law may be easier to understand if a circuit is thought of as a plumbing system. With the water valves fully open (no resistance), the water (current) is free to flow at whatever pressure (electromotive force) is applied. As one or more valves are partially closed (adding resistance to the system), the amount of water (current) passing through the system (or out the faucet) decreases.

The flow of electrical current always follows the path with the least amount of resistance.
    As electrical current reaches a junction in the circuit, if the difference in the resistances between the paths branching off is substantial, current takes the path(s) with the least resistance and does not flow through the branch with the substantially higher resistance.

An electrical circuit with no resistance (no load) is a short. Think about it. It is just like inserting the two ends of a piece of wire into the two sides of an electrical outlet which will cause a short and blow a fuse, or trip a circuit breaker. If only a light bulb, no matter how small, is hooked up on the wire (a load), it will not blow the fuse.

When an appliance is properly grounded, any electrical short is carried harmlessly to ground since a ground has little or no resistance. A good ground can protect both the equipment and the technician should the current be accidentally short-circuited into the housing. Since electricity follows the path of least resistance, should anyone touch a shorted piece of equipment, current will not flow through a body that has several thousand ohms of resistance; instead, it is bled to ground because a grounded circuit has very little resistance.

> **EQUIPMENT IS GROUNDED BY THE FOLLOWING METHOD**
> 1. All electrical components in the unit are connected to the chassis by a piece of wire (called a ground wire, color-coded green).
> 2. The chassis is grounded in the power receptacle (see fig. 182 in which the opening marked G is the ground connection). The receptacles are connected to ground by a wire leading to an eight-foot copper rod which is buried in the earth outside. If the receptacle being used is not grounded, secure a piece of wire to the chassis and connect the other end to the nearest cold-water faucet. This will do the job as well since cold water pipes run under the ground. It is the amperage, not the voltage, which is the dangerous ingredient of electricity. A very small amount of amperage driven by 120 volts of 60 hertz current can be fatal.

Dry skin may have from 100,000 to 600,000 Ω of resistance. This resistance is reduced to as little as 1,000 Ω when the skin is wet (as from perspiration). Perhaps less, if it is perspiration, because sweat contains salt, and salt makes moisture a more efficient conductor. Using Ohm's law (volts divided by ohms equals amperes), it can be determined what amperage will be dangerous. Experience has shown that 5/10000 to 2/1000 A is just noticeable, 5/100 to 2/10 A causes irregular twitching of the heart muscles with no pumping action, and anything more than that causes paralysis of breathing.

With dry skin, the current flow through the body would be 120 V per 100,000 Ω or 0.0012 (12/10,000) A, barely noticeable. Now if the resistance were lowered to 1000 Ω through perspiration, according to the formula, 120 V per 1000 Ω, or 0.120 (120/1000 A would flow through the body; more than enough for a lethal dose.

However, if the unit is adequately grounded, the current is bled harmlessly to ground with hardly a tingle felt.

Remember, with 120 volts, it only takes as little as 0.025 A (25/1000) to cause death.

Connecting the chassis to a cold water pipe with a length of wire is an excellent ground for the unit. The reason a cold water pipe is used instead of a hot water pipe is that hot water pipes are usually installed inside the structure, whereas the cold water pipe runs directly into the ground where it is connected to the main water line.

## USING THE OHMMETER

CAUTION: The ohmmeter is to be used only on circuits with no current. Be sure the equipment is unplugged before using an ohmmeter as current flow through an ohmmeter will destroy it. Any component to be tested must first be isolated from the rest of the circuit by removing any wiring connected to it.

Several examples of the use of the ohmmeter and voltmeter have already been given in this book. The best diagnostic instrument is a combination voltmeter/ammeter/ohmmeter, which is referred to as a multimeter (see pp. 303 and 305). All of the functions are indispensable to a refrigeration and air-conditioning technician.

Before setting the multimeter to check the ohm rating of a component, disconnect the power to the unit and isolate the component from the rest of the circuit by disconnecting its terminals from all other wiring. The current necessary to operate the meter is supplied by a dry cell battery within the meter.

The ohmmeter is used to check the continuity or the resistance of a circuit, or a load in a circuit. (Continuity means a continuous, uninterrupted circuit.) Today, however, the pocket-size digital multimeters capable of reading volts and ohms (and in some models, amperes) are becoming increasingly popular because of their convenience and accuracy. Rather than a specific scale with an indicator needle, digital meters produce a direct number reading like a calculator. Unlike the conventional meter that registers 0 Ω for a continuous path, the multimeter is extremely sensitive and displays the actual ohm value of the path being measured.

The meter probes must not touch anything other than the part or section being tested to prevent an erroneous reading.

The reason for isolating the component to be tested is that if it is left in the circuit, the reading registered will be erroneous as the resistances of every component in that circuit will be included.

When the needle of a conventional ohmmeter points to *0*, continuity in the circuit being checked is indicated. When it points to a specific ohm reading, the amount of resistance in that circuit is indicated.

When the needle points to the *infinity* symbol, it indicates an interruption in the circuit being checked.

## USING A VOLTMETER

A *voltmeter* registers the potential difference (voltage). In figure 182, if the two voltmeter probes were to touch the H or N terminal of the 110 VAC receptacle, no voltage would be registered on the meter. But if one probe were to touch the H and the other one touch the N terminal, the meter would register a voltage reading because of the existence of a potential difference between these two terminals. In wiring diagrams, the two terminals are represented as L1 (the *hot* terminal) and L2 (the *neutral* terminal in a 110 VAC circuit, and another hot terminal in a 220 VAC receptacle). Knowing this simple principle is essential in using a voltmeter in any circuit.

In figure 179, if L1 and L2 were plugged in, the switch closed (turned on), and the two voltmeter probes touched the two power terminals on the switch, the meter would register 0 volts. It would be just like the two probes were touching only the L1 side of the receptacle. As soon as the switch were opened (turned off), the meter would register a voltage reading because the open switch would separate the L1 and L2 sides of the circuit and a potential difference would be detected. Any open switch or load separates the line. In the case of a *good* fuse, if the voltmeter probes were to touch the two terminals, a zero voltage reading would be registered. But if the fuse were blown (burnt out), the meter would register a voltage reading because of the separation of the L1 and L2 sides of the circuit by the blown fuse creating a potential difference. The heater is a load. A load is a device that converts electrical energy to useful work. Any load separates the L1 and L2 sides of the circuit whether it is operational or nonoperational due to the disconnection of power within itself, such as a burnt and disconnected element in a heater. When the probes touch the two sides of the heater, it registers voltage whether or not the heater is operational.

When the probes touch the two terminals of the light bulb, a voltage reading is registered because the bulb is a load. If there is a short in the lamp socket through which *current is bypassed*, the voltmeter registers no voltage because no potential difference is created.

In figure 181, with the switch turned off and its two terminals contacted by the probes, the meter registers a voltage reading, but when the other components are checked, no voltage reading is registered because no power reaches any other component to create a potential difference. With the switch closed (ON), when the probes touch the two terminals of the lamp, solenoid, heater, fan, or compressor, the meter registers voltage. The meter should register zero voltage when its probes are placed across a blown fuse. With the power ON and the voltmeter probes touching the following points, expect these readings:

1. The two terminals of a closed switch reads 0 volts because no potential difference is created.
2. The two terminals of an open switch reads voltage.
3. The two terminals connecting a load to the power produce a voltage reading.
4. The two sides of a broken wire registers a voltage reading.
5. The two terminals of a blown fuse gives a voltage reading.
6. The two terminals of a good fuse produce a zero reading because no potential difference is created.
7. The two sides of a load, whether or not it is operative, produce a voltage reading.

Make sure to set the conventional-type voltmeter on the proper scale for the expected voltage to prevent damage to the instrument. These, or similar instructions are included with all new meters. As mentioned earlier, there are also pocket-size digital multimeters that are inexpensive. With these handy meters, the necessity of selecting a particular scale is eliminated as they produce a direct reading up to their maximum capacity. These meters are much more convenient and are pictured in pages 303 and 305.

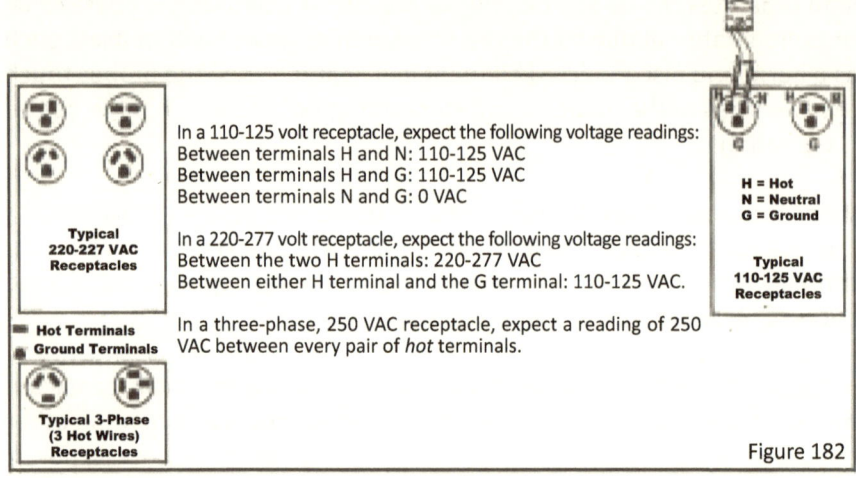

Figure 182

## HOW TO CHECK A RECEPTACLE FOR A CIRCUIT GROUND

In a 110-125 volt receptacle, one of the two top slots are the hot wire terminal (the smaller opening H) and the neutral wire terminal (the larger opening N), and the bottom terminal (G) is connected to ground (G). (See fig. 182.)

1. Place one voltmeter probe in H (the hot side terminal) and the other one in N (the neutral terminal). The meter should register a reading of 110-125 VAC.
2. Place one voltmeter probe in slot H and the other one in slot G. The meter should register a reading of 110-125 VAC; otherwise, the receptacle is not grounded, and a licensed electrician should be called to remedy the problem.
3. Place one voltmeter probe in slot N and the other one in slot G. The meter should register a 0 volt reading; otherwise the polarity of the receptacle is reversed due to improper wiring which will have to be remedied by a licensed electrician. In 220-277 volt receptacles, the two top terminals are hot (H). The voltmeter should register a reading of 220-277 VAC between the two H terminals, and a reading of 110-125 VAC should be registered between either of the H terminals and the ground terminal. Otherwise the receptacle is not grounded.

In the schematic wiring diagrams, the hot and neutral receptacle terminals are shown as L1 and L2 (in a 220-277 VAC system, the two *hot terminals* are shown as L1 and L2), and the three hot terminals in a three-phase receptacle are shown as L1, L2, and L3.

A voltmeter is widely used in diagnosing any interruption in an electrical circuit. It registers a voltage reading only when there is a potential difference between the locations where its probes touch. Potential difference in an electrical circuit is created across any point of disconnection, such as a burnt fuse, an open or defective switch, a broken wire, etc. When the two probes touch across an open switch, the meter registers a voltage reading, but when the switch is closed, the meter registers 0 volt reading. A voltmeter always registers a voltage reading when its two probes touch across the terminals of a load, whether or not the load is functional. When the two voltmeter probes touch the two terminals of a fuse, a voltage reading will mean that the fuse has burnt out (an interruption in the circuit). If the meter reads 0 volts, it will mean there is no disconnection between the two fuse terminals. Therefore, one can conclude that the fuse is good.

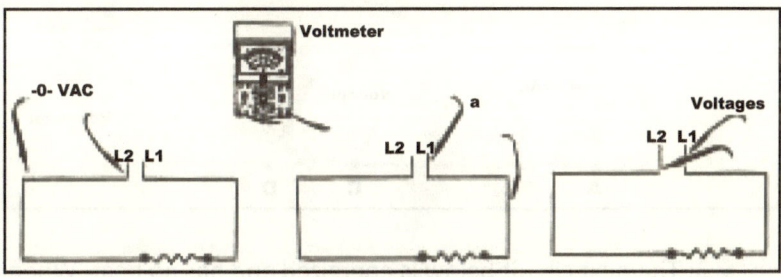

PROBLEM 1: In the bottom illustration, the power is on, and there is no heat from the heater. Placing the probes across the heater (C-D) a 0 volt reading is registered. This will mean power does not reach the heater because if it did, the meter would register a reading even if the heater were defective. (You should always get a voltage reading when placing the voltmeter probes across a load.)

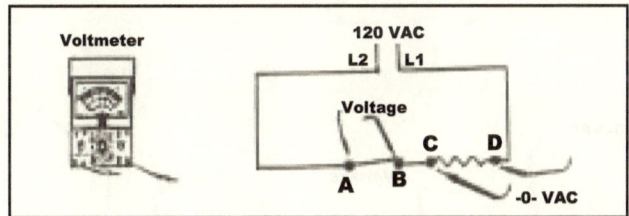

The voltmeter registers 120 volts when placing the probes across the closed switch (A-B), this means there is an interruption between A and B. A good switch that allows current to pass through when it is closed, placing a probe on either side, should register a 0 volt reading because current is not interrupted between its terminals to create a potential difference.

Since there is a voltage reading, this will mean the switch is defective because power reaches one terminal of the switch, but because of a break in the circuit inside the switch, power cannot reach the other terminal even with the switch closed. With the voltmeter probes across a good switch in the ON position the meter should read zero voltage. (No potential difference.)

PROBLEM 2: In the circuit below, power is on and there is no heat. When the circuit is checked with a voltmeter, A-B shows a 0 volt reading. C-D shows a voltage reading.

CONCLUSION: Since power reaches the heater but the heater does not heat, the heater element is bad (shorted and burnt).

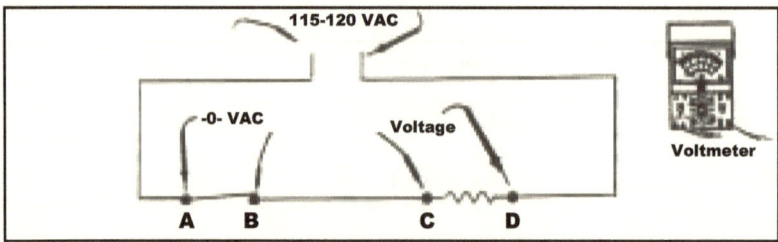

PROBLEM 3: The next circuit below is energized and has no heat. When the circuit is checked with a voltmeter, A-B shows 0 volts. C-D reads voltage. E-F also shows 0 volts.

CONCLUSION: Switch C-D is defective (or open) because voltage is stopped there and cannot pass.

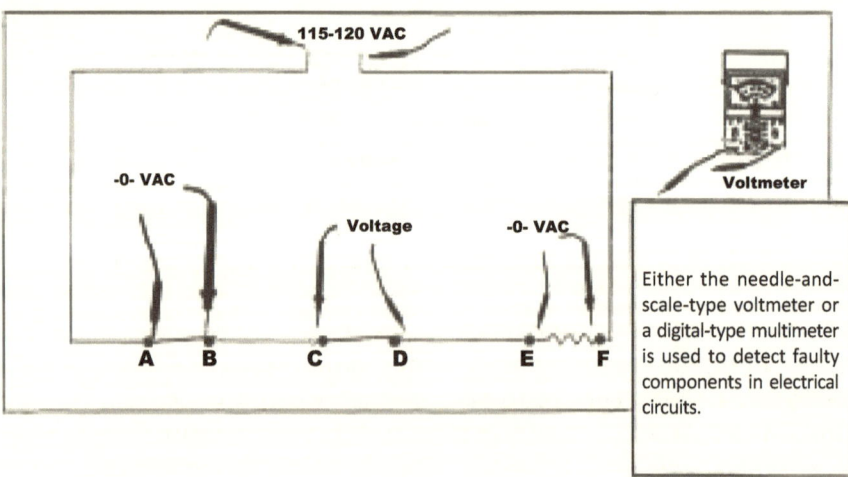

Either the needle-and-scale-type voltmeter or a digital-type multimeter is used to detect faulty components in electrical circuits.

The circuit below shows four positions for voltmeter readings. These positions produce the following readings:

1. It's 115 to 120 VAC when power is on. Zero voltage when power is off.
2. No reading if switch is closed or a voltage reading if switch is open.
3. A voltage reading across the resistor (with the switch closed). No voltage reading across the resistor (with the switch open).
4. A voltage reading across the lamp (with the switch closed). No voltage reading across the lamp (with the switch closed) if lamp is shorted. A voltage reading across the lamp with the switch closed if the lamp is burnt out or inoperative.

The total voltage across numbers 2, 3, and 4 should be equal to the reading across L1 and L2 if the switch at number 2 is closed and power is on.

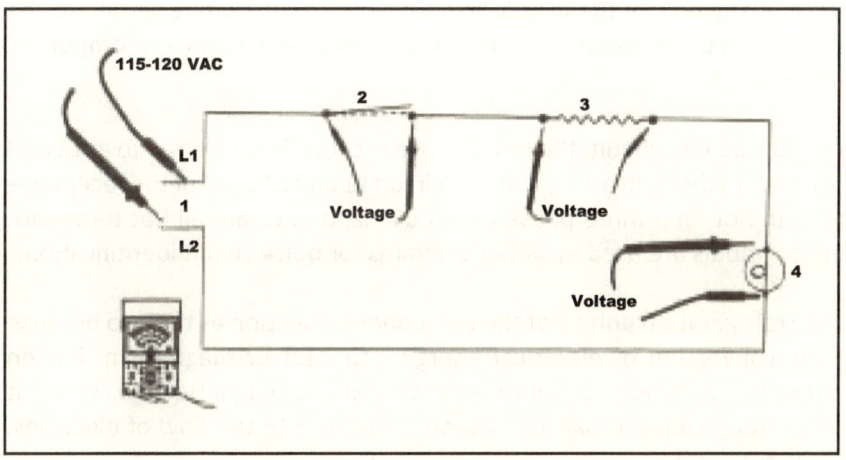

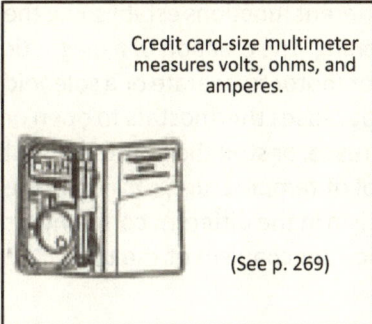

Credit card-size multimeter measures volts, ohms, and amperes.

(See p. 269)

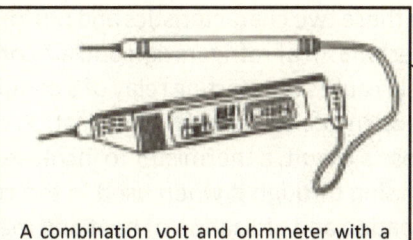

A combination volt and ohmmeter with a digital readout is recommended for ease of use in lieu of the needle-and-scale type illustrated above.

*Courtesy of Beckman Industrial Corporation*

## UNDERSTANDING ELECTRICAL CIRCUITS

Instead of drawing pictures to illustrate the various parts of an electrical circuit, symbols are used in schematic wiring diagrams. Schematics are simpler and easier to read than the pictorial wiring diagrams. They are used on large and complex circuits.

Since these symbols are internationally recognized, they make it easy for any service technician to read the schematic wiring diagrams. Sometimes, some variations of a certain symbol for the same component can be noticed. These are very limited and should not create any problem.

Bear in mind that in order to energize an electrical circuit and make it become operational, current flowing into the circuit from one side of the receptacle must be able to return to the other side of the receptacle. Should any interruption such as a broken wire, loose connection, or a defective switch in any part of the circuit prevent the current from returning to the other side of the receptacle, the circuit becomes inactive and no component in it will work.

In a 110/130 VAC circuit, the hot side of a receptacle is referred to as L1 and the neutral side L2. In a 220/250 VAC circuit L1 and L2 sides of the receptacle are both hot. In a three-phase circuit L1, L2, and L3 are all hot terminals. These symbols are used in wiring diagrams for quick circuit identification.

In any refrigeration unit all of the components function as they do because of the conversion of electrical energy into heat or magnetism. (When current flows through a coil of wire it creates a magnetic field. When it flows through a path that offers some resistance to the flow of electrons, heat is created.)

When a unit is activated, every component in the electrical circuit is affected by these two characteristics and perform different functions establishing the mechanism of refrigeration and air-conditioning. The creation of a magnetic field causes the starting relay of a compressor motor to operate or a solenoid to open or close a valve. The creation of heat causes thermostats to open or close a circuit, a thermister to increase, decrease, or stop the flow of current passing through it when used in the control of temperature, etc. Thus, the creation and utilization of heat and magnetism in the different components in the electrical circuit establish the electrical mechanism of the unit.

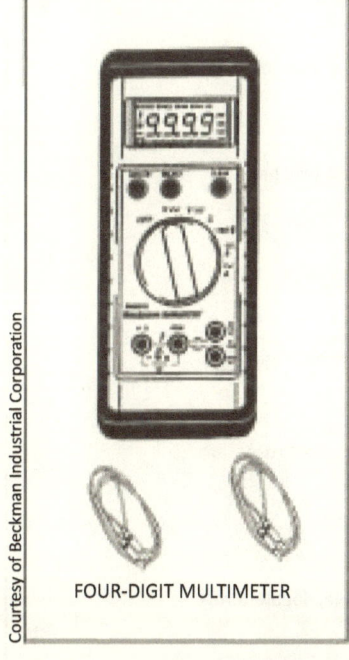

FOUR-DIGIT MULTIMETER

Courtesy of Beckman Industrial Corporation

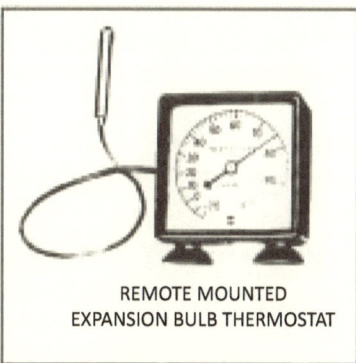

REMOTE MOUNTED
EXPANSION BULB THERMOSTAT

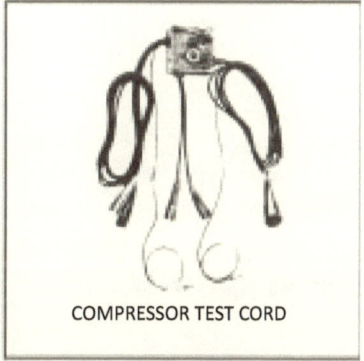

COMPRESSOR TEST CORD

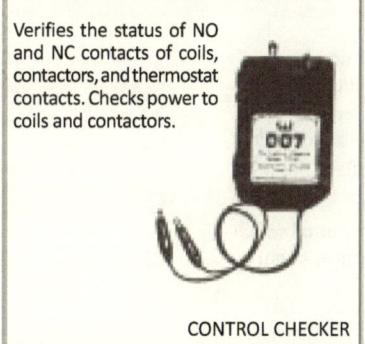

Verifies the status of NO and NC contacts of coils, contactors, and thermostat contacts. Checks power to coils and contactors.

CONTROL CHECKER

Courtesy of Wagner Products Corporation
Miami, Florida

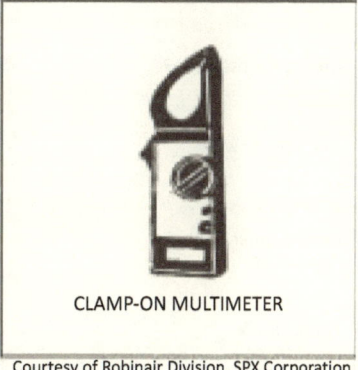

CLAMP-ON MULTIMETER

Courtesy of Robinair Division, SPX Corporation

## SCHEMATIC WIRING DIAGRAM SYMBOLS

|  | |  | |
|---|---|---|---|
| Iron core transformer | | Timer Switch | |
| Transformer | | Main switch | |
| Circuit breaker / Overload protector | | Adjustable thermostat | |
| Three-phase motor | | Connected switches | |
| Fusible link | | Switch | |
| Overload protector | | Thermostat (Closes on heat rise) | |
| Overload protector | | Thermostat (Opens on heat rise) | |
| Variable resistor | | Thermostat | |
| Relay coil | | Single-pole, double-throw switch | |
| Normally open relay | | Resistor | |
| Normally closed relay | | Thermostat (Whirlpool) | |
| Line voltage wires | | Push-button switch (Normally closed) | |
| Defrost timer symbol used by GE | | Push-button switch (Normally open) | |
| Connector | | Two-circuit, push-button switch | |
| Heating element | | Multiposition switch | |
| Overload protector (GE) | | | |

|  | |  | |
|---|---|---|---|
| Transformer | | Split-phase motor | |
| Solenoid coil | | Three-speed motor | |
| Centrifugal switch | | Two-speed motor | |
| Starting relay | | Single-speed motor | |
| Coil | | Neon light | |
| Power supply grounded | | Fluorescent light | |
| Wires connected | | Incandescent light | |
| Light bulb | | Chassis ground | |
| Wires not connected (Crossover) | | Earth ground | |
| Terminal | | Motor | |
| Crossover | | Low-pressure control | |
| Capacitor | | High-pressure control | |
| Thermister | | Three-speed motor | |
| Buzzer | | Two-speed motor | |
| Bell | | | |
| Power supply grounded | | | |

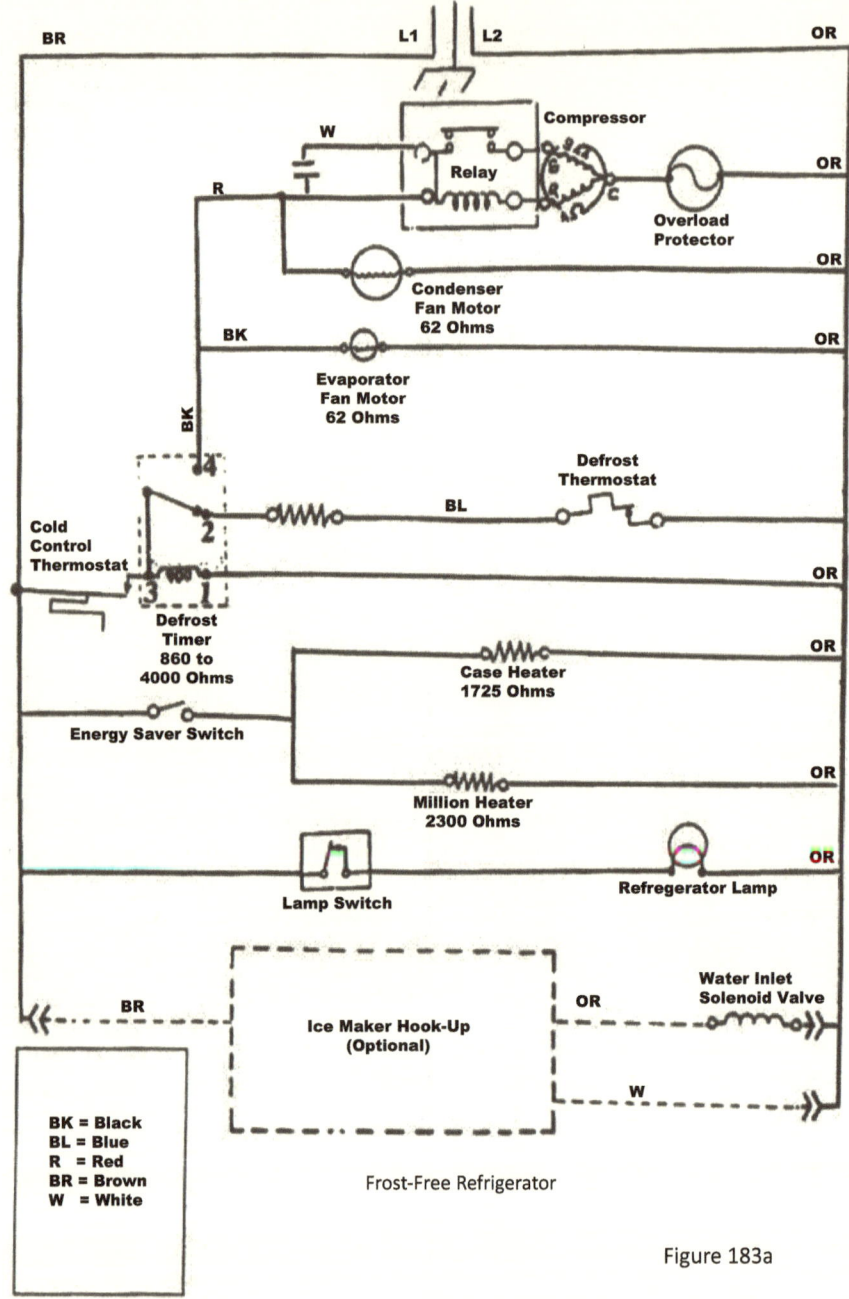

Figure 183a

(See pages 18, 19).

Figures 183a, b, and c show a typical schematic of a frost-free refrigerator. Most refrigerators of this type have a similar illustration.

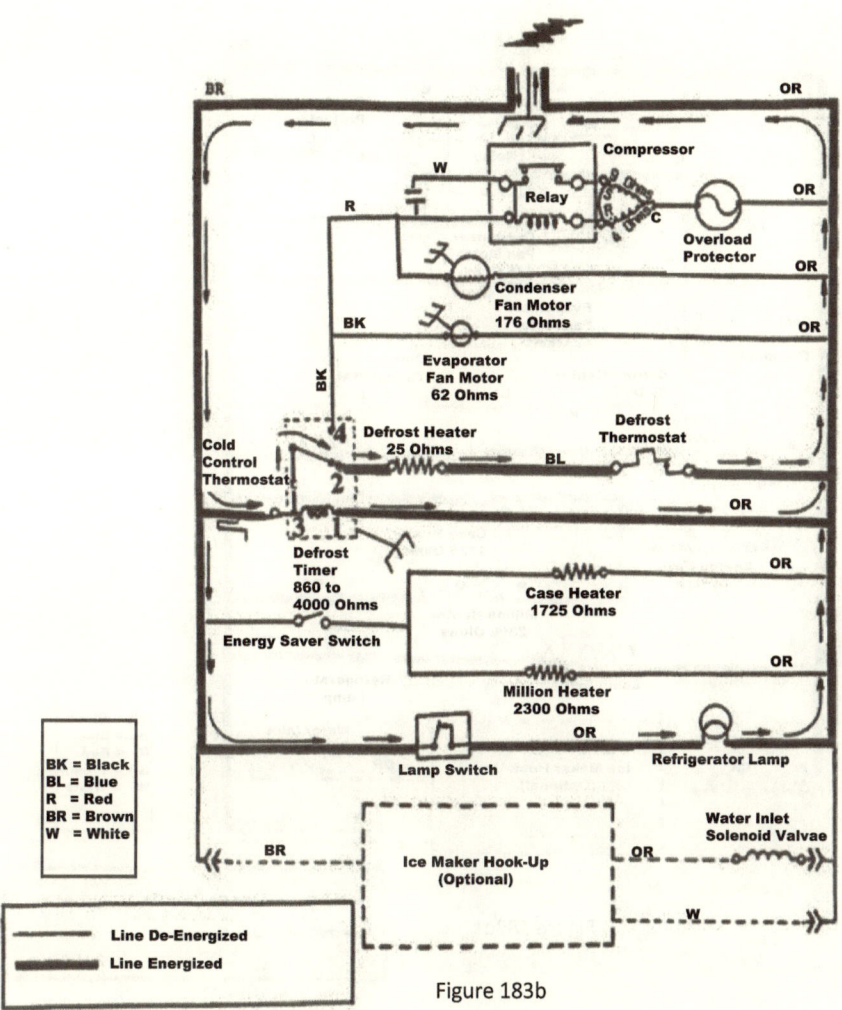

Figure 183b

## A SCHEMATIC WIRING DIAGRAM OF A
## FROST-FREE REFRIGERATOR IN THE DEFROST CYCLE

The timer shifts to number 2 position and takes the unit into the defrost cycle. (This cycle will last no longer than twenty-one minutes.) Pay close attention to the direction of the flow of current. It starts from L1 (the hot terminal in the wall receptacle), flows through the timer terminal number 2, the defrost heater, the defrost thermostat, and then closes its circuit by flowing to L2 (the neutral terminal of the wall receptacle). Power energizes the timer motor through its terminals 1 and 3 as long as the cold control (thermostat) is not satisfied. The refrigerator lamp is always energized as long as the lamp switch is closed (when door is open).

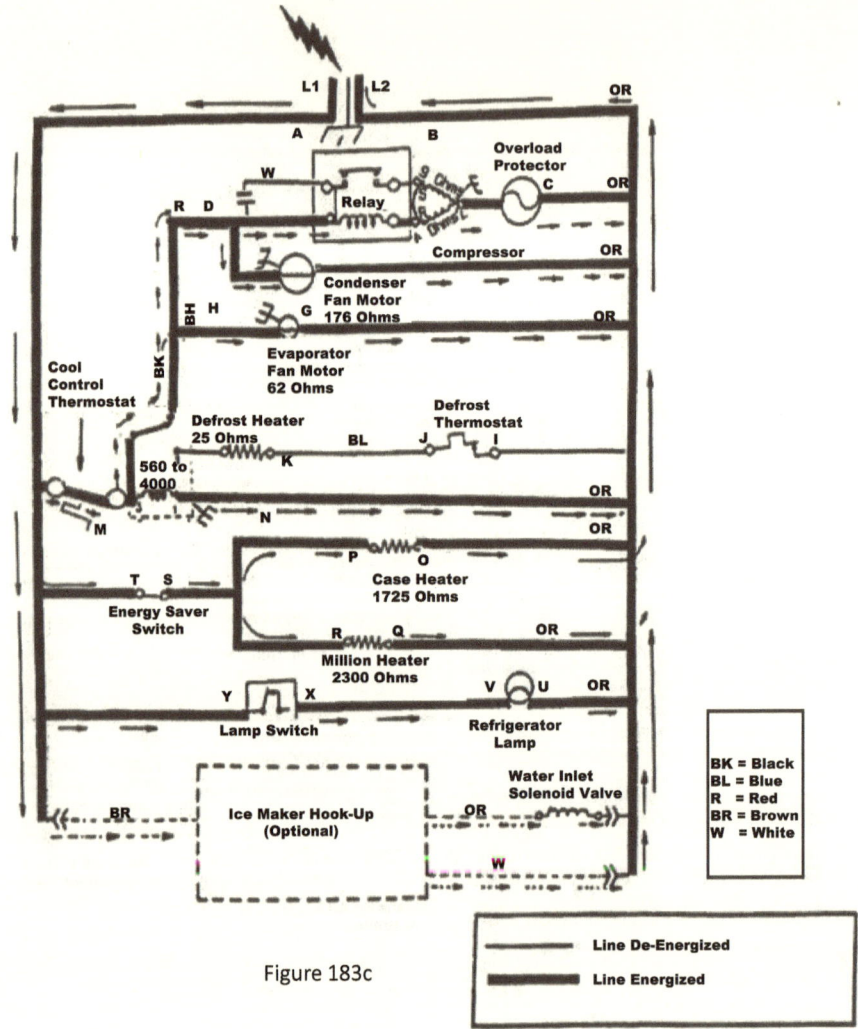

Figure 183c

## THE SAME SCHEMATIC WIRING DIAGRAM SHOWING
## THE FROST-FREE REFRIGERATOR IN THE COOLING CYCLE

The timer moves to position number 4 and takes the unit into the cooling cycle (and the unit stays in this cycle for six, eight, or twelve hours, depending upon the kind of timer used). Pay close attention to the direction of the flow of current. It starts from L1 (the hot terminal in the wall receptacle), flows into the timer from timer terminal 4, then to the compressor, evaporator, and condenser fan motors, and then finds its way back to L2 (the neutral terminal of the wall receptacle) to complete its circuit.

Refer to page 351 for questions and
answers pertaining to this diagram.

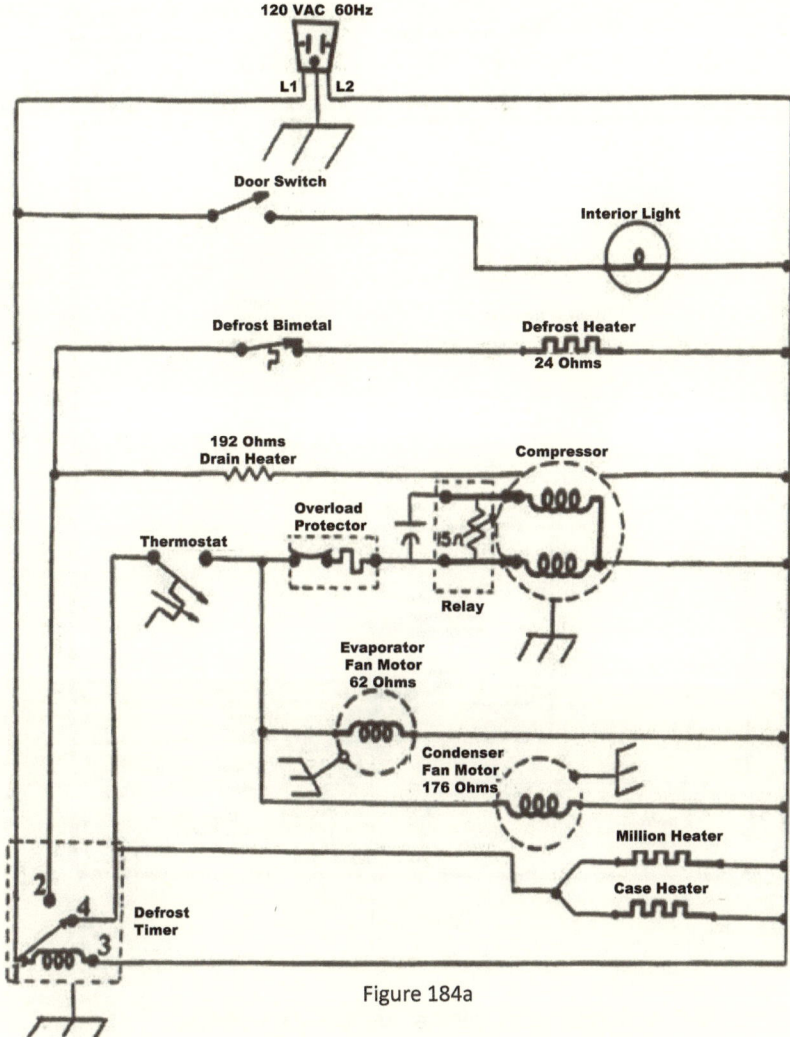

Figure 184a

**A SCHEMATIC WIRING DIAGRAM OF A TYPICAL FROST-FREE (AUTOMATIC DEFROST) REFRIGERATOR**

As the timer motor turns, it moves a cam within itself. The cam changes the direction of current flow through its contacts. This contact position change takes the unit into the defrost or cooling cycle. Cooling cycles last six, eight, or twelve hours, and defrosting periods do not last longer than twenty-one minutes, depending upon the timer used.

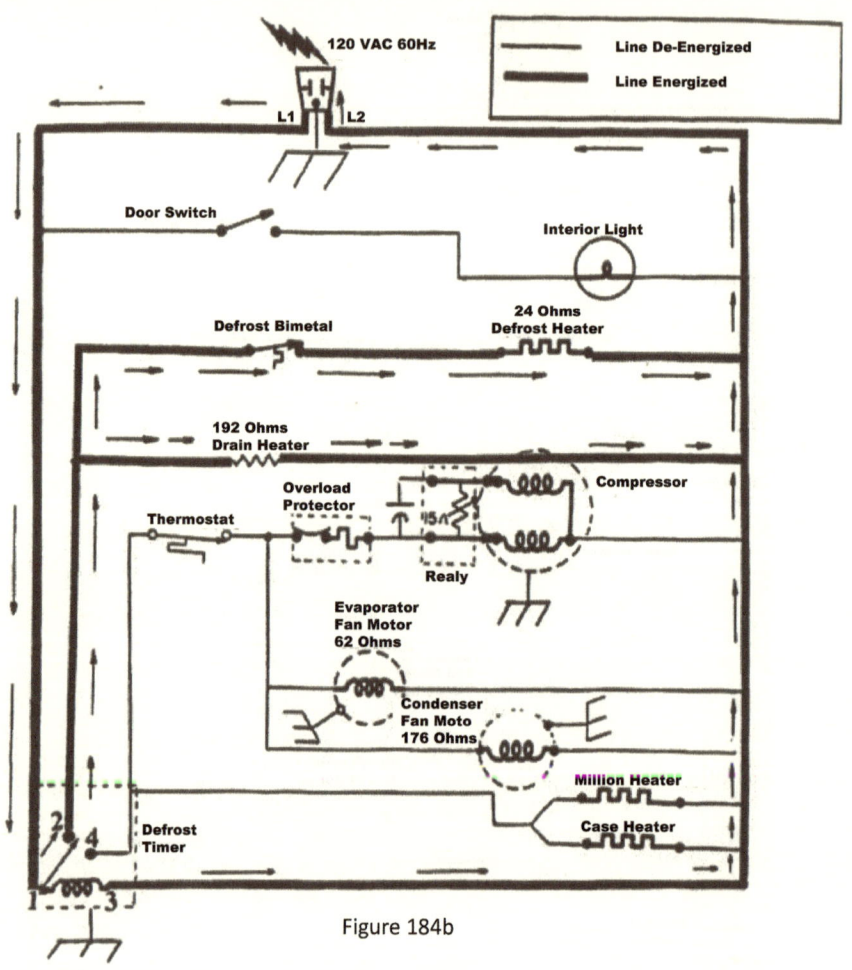

Figure 184b

**A SCHEMATIC WIRING DIAGRAM OF A TYPICAL
FROST-FREE (AUTOMATIC DEFROST) REFRIGERATOR
IN THE DEFROST CYCLE**

The timer shifts to number 2 position and takes the unit into the defrost cycle. (The duration of this cycle is no longer than twenty-one minutes.) Pay close attention to the direction of the flow of electricity. It starts from L1 (the hot terminal in the wall receptacle) and flows into the timer through timer terminal 1. It flows to a defrost heater through a defrost bimetal and also to a drain heater, then back to L2 (the neutral terminal in the wall receptacle) to complete its circuit.

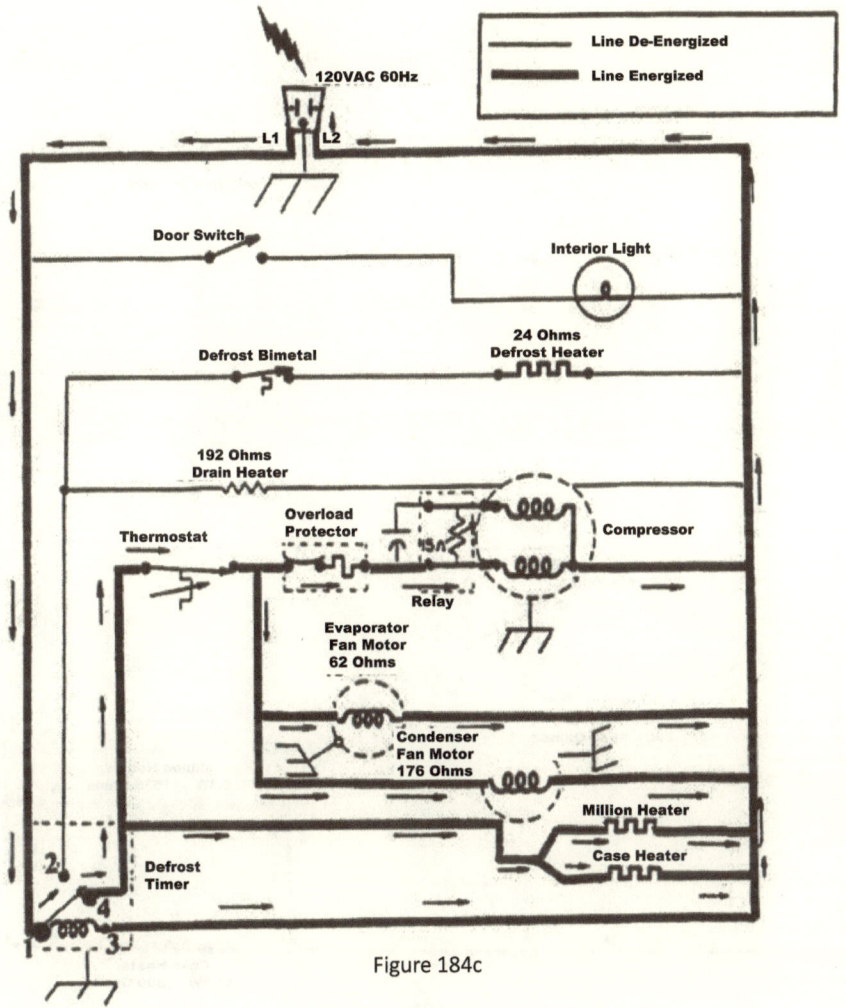

Figure 184c

## AUTOMATIC DEFROST REFRIGERATOR IN THE COOLING CYCLE

The timer cam shifts to position number 4 and takes the unit into the cooling cycle (and stays in this cycle for six, eight, or twelve hours depending upon the kind of timer used). Pay close attention to the direction of the flow of current. It starts from L1 (the hot terminal in the wall receptacle), flows into the timer, and through timer terminal 4, to the compressor, evaporator, and condenser fan motors, through the thermostat and then back to L2 (the neutral terminal of the wall receptacle) to complete its circuit.

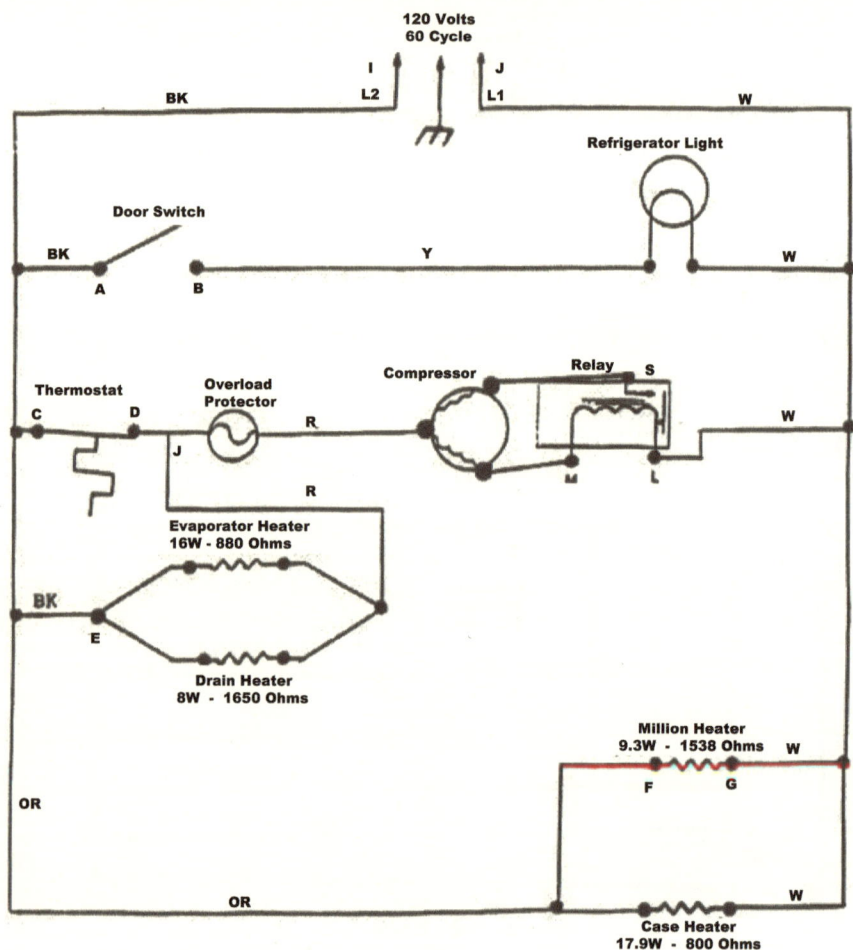

Figure 185a

## A TYPICAL SCHEMATIC WIRING DIAGRAM THAT COMES WITH A CYCLE-DEFROST-TYPE REFRIGERATOR

Color coding is usually indicated by BK—black, W—white, OR—orange, Y—yellow, and R—red. In a more sophisticated unit, an *energy-saver switch* deactivates the mullion and case heaters by cutting off power to them. In this unit, they are always on as long as the power is on. Pay close attention to the direction and path of current energizing or de-energizing the different components that take the unit into the cooling or defrost cycle. (See pp. 315 and 314.)

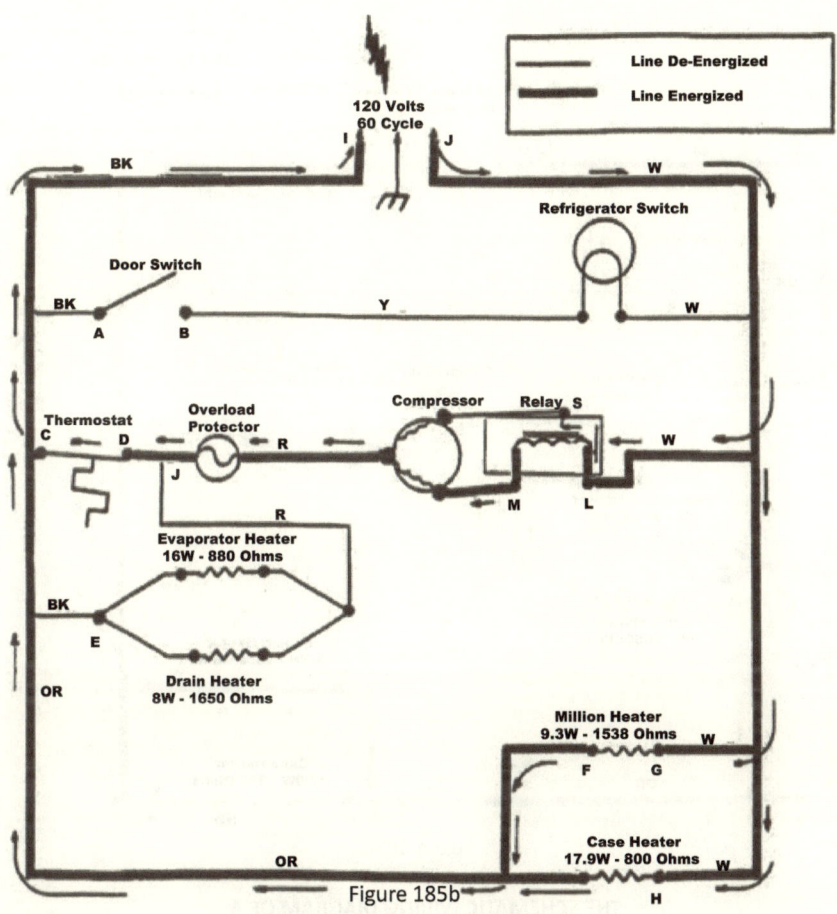

Figure 185b

**SCHEMATIC WIRING DIAGRAM OF A CYCLE-DEFROST REFRIGERATOR IN THE COOLING CYCLE**

As long as the thermostat is not satisfied (closed), the unit stays in the cooling cycle. Note the direction of the current flow. It starts from L1 (the hot terminal in the wall receptacle), flows through the starting relay, the compressor, and the overload protector at point J. Since the resistance of the evaporator and the drain heater is high and because current always follows the path of least resistance, the current flows through the thermostat path (with practically no resistance) and not through the heaters. It completes its path by flowing back to L2, the neutral terminal in the wall receptacle.

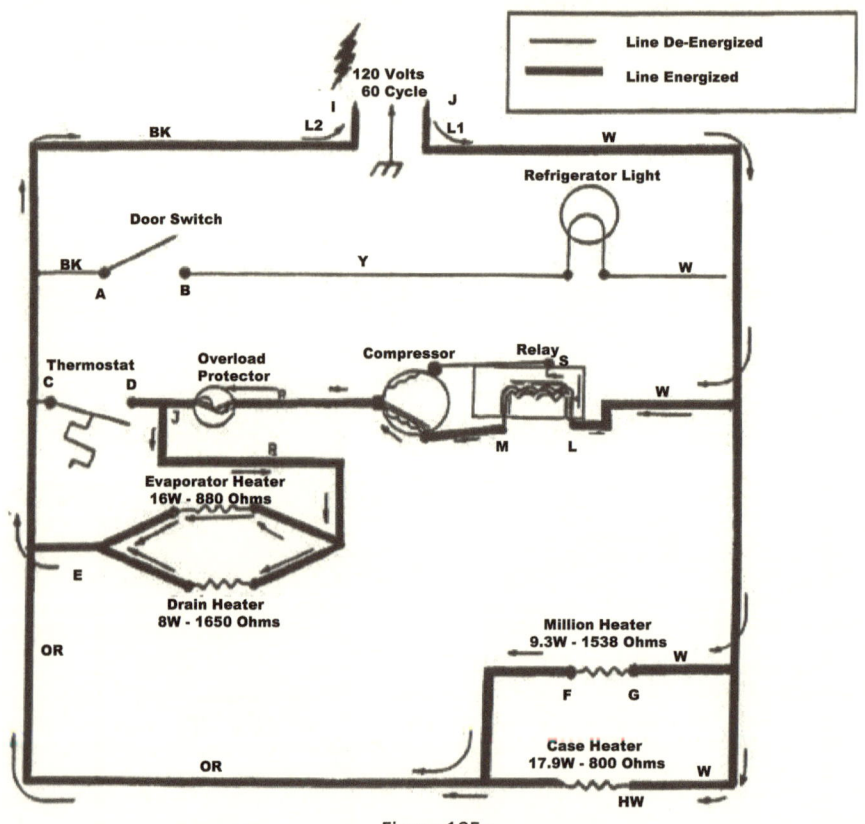

Figure 185c

**THE SCHEMATIC WIRING DIAGRAM OF A
CYCLE-DEFROST REFRIGERATOR IN THE
DEFROSTING CYCLE**

When the thermostat becomes satisfied (opens) the defrost cycle will start. Note the direction of the flow. Power starts from L1 (the hot terminal in the wall receptacle), flows through the starting relay, compressor overload protector, evaporator and drain heater, and back to L2 (the neutral terminal in the wall receptacle) to complete its circuit. Since the resistances of the evaporator and drain heaters are substantially higher than the resistance of the compressor motor in the circuit, the heaters will consume all the power and the compressor will not start (although current flows through it). (See p. 292.)

## A TYPICAL TWO-PIPE, HOT GAS DEFROST SYSTEM

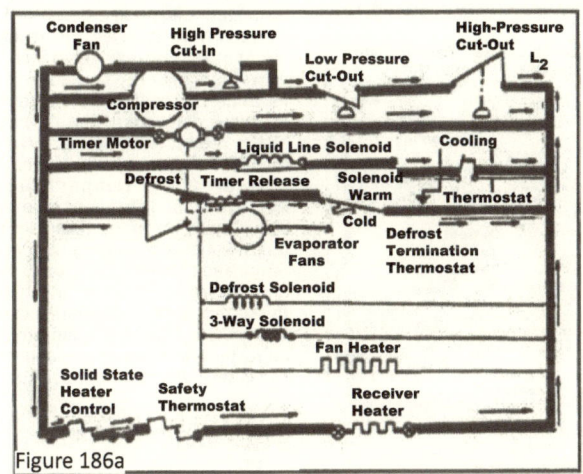

The liquid-line solenoid valve becomes energized and Freon flows through an expansion valve to the evaporator. It is then drawn back into the compressor flowing through a three-way valve and a hold-back valve.

**Refrigeration**
Normally Closed

**Defrost**
Normally Open

Figure 186a

### Unit in Cooling Cycle

Figure 186a displays a momentary position of the termination thermostat on warm position in the beginning of each refrigeration cycle. Timer-release solenoid is energized, and timer defrost contact is released. Due to low evaporator temperature, termination thermostat is moved to cold position and current flows through the evaporator fan and the refrigeration cycle continues.

Figure 186b shows the direction of refrigerant flow in the refrigeration cycle. Compressor forces out hot vapor to the condenser. It is then forced to flow through a receiver.

— Line De-Energized
■ Line Energized

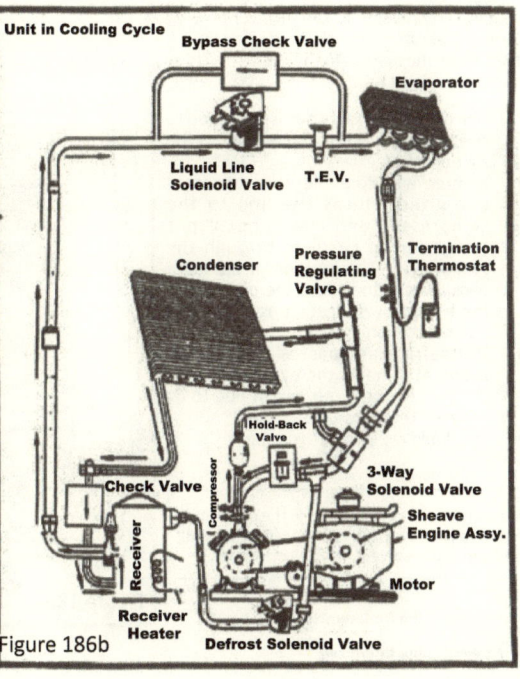

Figure 186b

A typical two-pipe hot gas defrost system

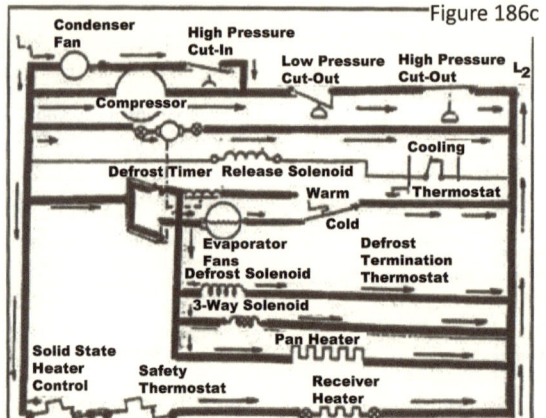

Figure 186c

the TEV through a check valve and through the liquid line flows to the receiver. There is an electric heating element in the receiver which is activated in this cycle to vaporize the returning refrigerant in the receiver for more defrosting. The purpose of the pressure-regulating valve is to sustain favorable hot refrigerant pressures and temperatures. Defrosting of the evaporator in this system normally takes six to ten minutes. The sensing bulb of a termination thermostat is mounted on the evaporator outlet. It returns the system to the refrigerating cycle when the ice on the evaporator is defrosted and it senses an adequate rise in the evaporator temperature.

### Unit In Defrost Cycle

In figure 186c and d, hot gas is circulated from the compressor to the receiver flowing to the evaporator through the suction line. A timer takes the system to the defrost cycle at a predetermined time.

1. Defrosting solenoid valve is energized to open the line from the receiver to the suction line.
2. A three-way solenoid valve becomes energized, closes the line to the compressor, and allows hot vapor refrigerant to circulate through the suction line. (The function of the hold-back valve is to reduce the pressure of the hot vapor refrigerant as it is drawn into the compressor).
3. Hot refrigerant passes to the cold evaporator, heats the evaporator, and changes to liquid refrigerant due to its lowered temperature.
4. The liquid refrigerant bypass

The purpose of a check valve in the condenser outlet is to prevent refrigerant from backing up from the receiver into the condenser.

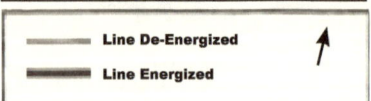

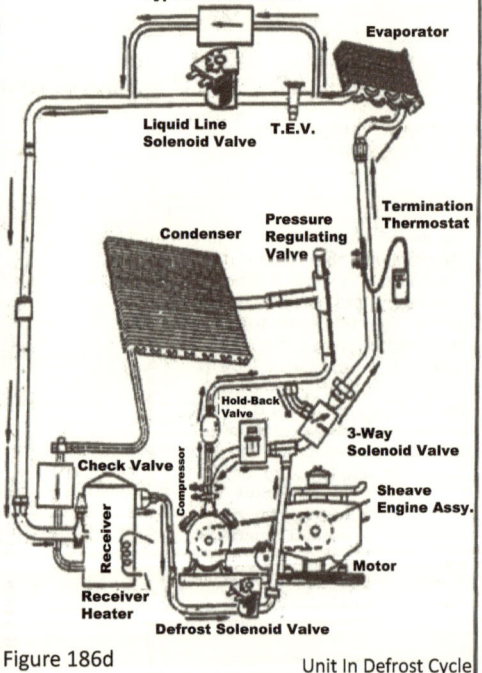

Figure 186d

Unit In Defrost Cycle

# AIR CONDITIONING AND REFRIGERATION REPAIR MADE EASY

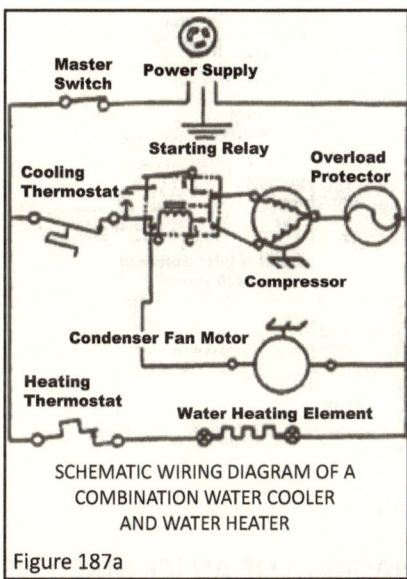

SCHEMATIC WIRING DIAGRAM OF A COMBINATION WATER COOLER AND WATER HEATER

Figure 187a

## Schematic Wiring Diagram of a Combination Water Cooler and Water Heater

The cooling and heating operations of the water dispenser are independent of each other.

When the cooling thermostat is satisfied, the unit will no longer cool.

When the heating thermostat is satisfied, the unit will stop heating the water until the water temperature drops to a predetermined point.

See pages 203 through 206.

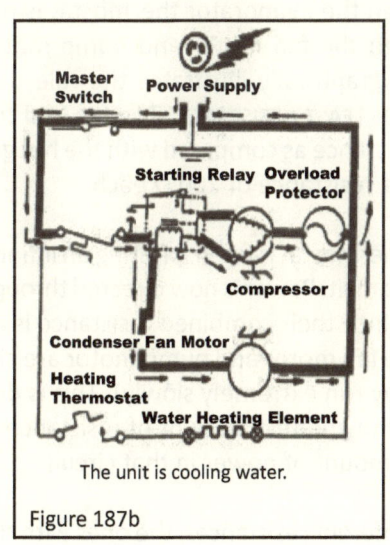

The unit is cooling water.

Figure 187b

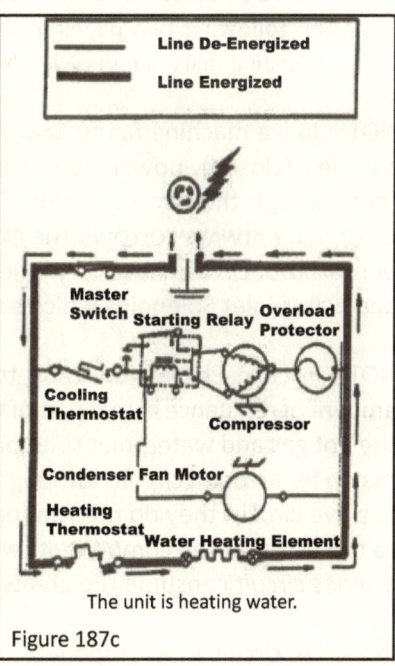

The unit is heating water.

Figure 187c

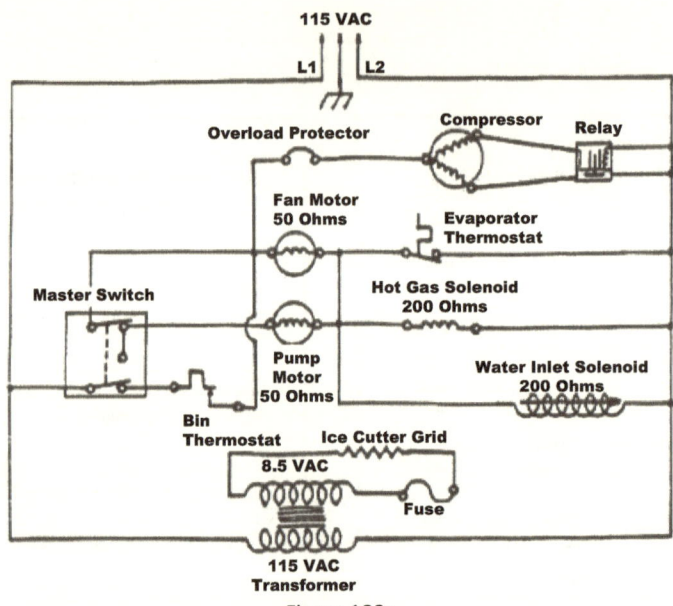

Figure 188a

## A TYPICAL SCHEMATIC WIRING DIAGRAM OF AN ICE MACHINE

The following two pages illustrate this circuit in the cooling and defrosting (harvesting) cycles. Note the path of current in each cycle.

NOTE: In ice machine figure 188c, when the evaporator thermostat is not satisfied (closed), power flows through the fan motor and pump motor (not through the two solenoids). This graphically illustrates the rule that ELECTRICITY ALWAYS FOLLOWS THE PATH OF LEAST RESISTANCE. The fan and the pump motor circuits have very little resistance as compared with the hot gas and water-inlet Solenoids which have a resistance of 200 Ω each.

NOTE: In ice machine figure 188b, the thermostat is open, offering an infinite amount of resistance in that part of the circuit. Power is now directed through the hot gas and water-inlet solenoids since their combined resistance is no match for an open circuit. Although the fan motor and pump motor are still in a live circuit, they do not run (or they run extremely slowly). This is due to the fact that the components with the greatest amount of resistance in a *series circuit* consume the greatest amount of power in that circuit.

In a series circuit with components of different resistances, the ones with the highest amount of resistance consume most of the power in being activated and do not leave enough voltage for the other components with less resistance to become activated even though they are still in the circuit. (See pp. 186 through 202.)

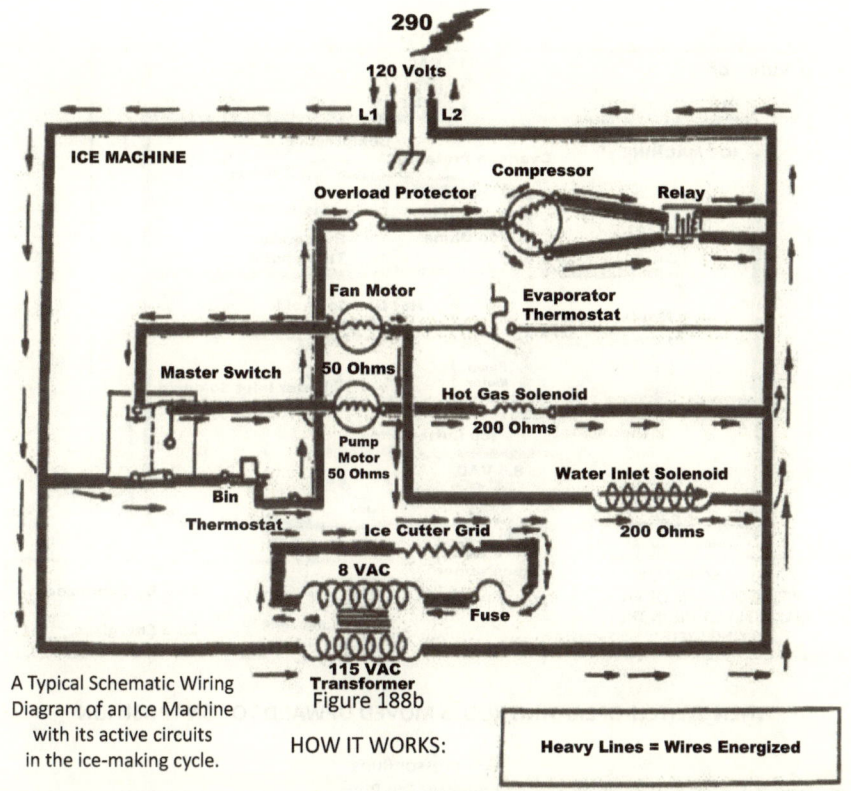

A Typical Schematic Wiring Diagram of an Ice Machine with its active circuits in the ice-making cycle.

Figure 188b

HOW IT WORKS:

Heavy Lines = Wires Energized

**WHEN SWITCH OPERATING ROD IS MOVED UPWARD TO "ON" POSITION**

Compressor Runs
Condenser Fan Runs
Water Pump Runs and Circulates Water
Cutter Grid is Warm

Since evaporator thermostat is not satisfied (ice has not reached desired thickness), it keeps the circuit closed. Since current always follows the path of least resistance, it will not flow through the hot gas or water inlet solenoid as each have 200 Ω of resistance. So it goes through the thermostat with virtually no resistance. (See p. 295, 296)

**WHEN ICE SLAB REACHES PRESET THICKNESS, THE THERMOSTAT OPENS, AND THE HARVEST CYCLE BEGINS.**

The transformer is always energized as it is connected directly to L1 and L2. It converts 115 VAC to 8.5 VAC to keep the grid warm as long as the unit is plugged in.

**UNIT RESTARTS FREEZING CYCLE WHEN SLAB IS RELEASED FROM EVAPORATOR, AND CUTTING PROCESS BEGINS.**

Cutter Grid Remains On

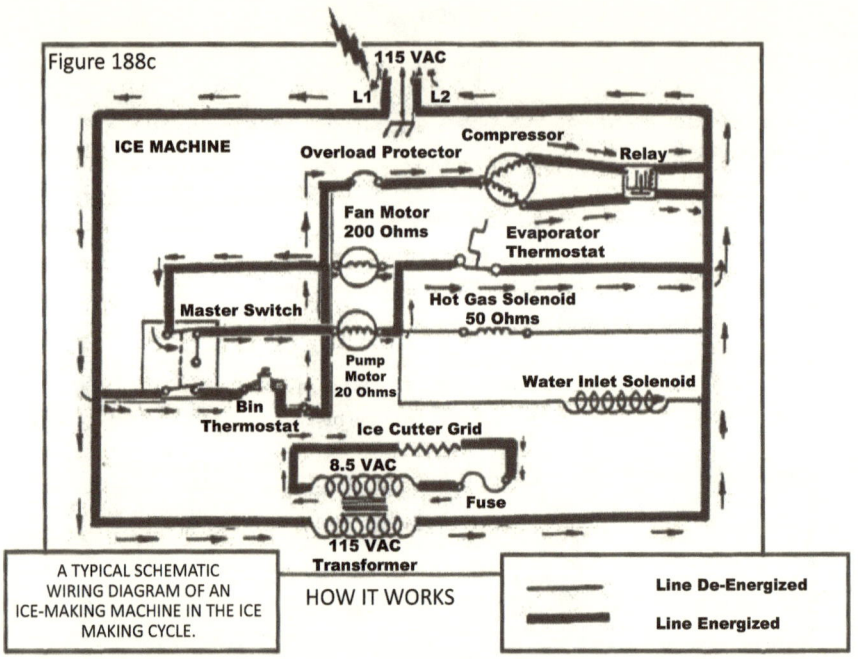

**WHEN SWITCH-OPERATING ROD IS MOVED UPWARD TO "ON" POSITION**

Compressor Runs
Condenser Fan Runs
Water Pump Runs and Circulates Water
Cutter Grid is Warm

**WHEN ICE SLAB REACHES PRESET THICKNESS, HARVEST CYCLE BEGINS AND THE FOLLOWING HAPPENS:**

Compressor Keeps Running
Evaporator Thermostat Is Satisfied
Condenser Fan Stops or Slows
Water Pump Stops
Hot Gas Solenoid Opens
Cutter Grid is Warm
Harvest Cycle Lasts One to Two Minutes

When the evaporator thermostat opens, current flow has no easier path than through the hot gas and water-inlet valves (see page 292). Since the heaviest loads in the circuit consume most or all of the available voltage, the power remaining in the line will not be enough to activate the condenser fan motor and the water pump motor as each have only a 50 Ω value.

UNIT RESTARTS FREEZING CYCLE WHEN SLAB IS RELEASED FROM EVAPORATOR AND CUTTING PROCESS BEGINS. WHEN STORAGE BIN IS FULL, BIN THERMOSTAT OPENS.

Cutter grid remains on.

# Air conditioning and Refrigeration Repair Made Easy

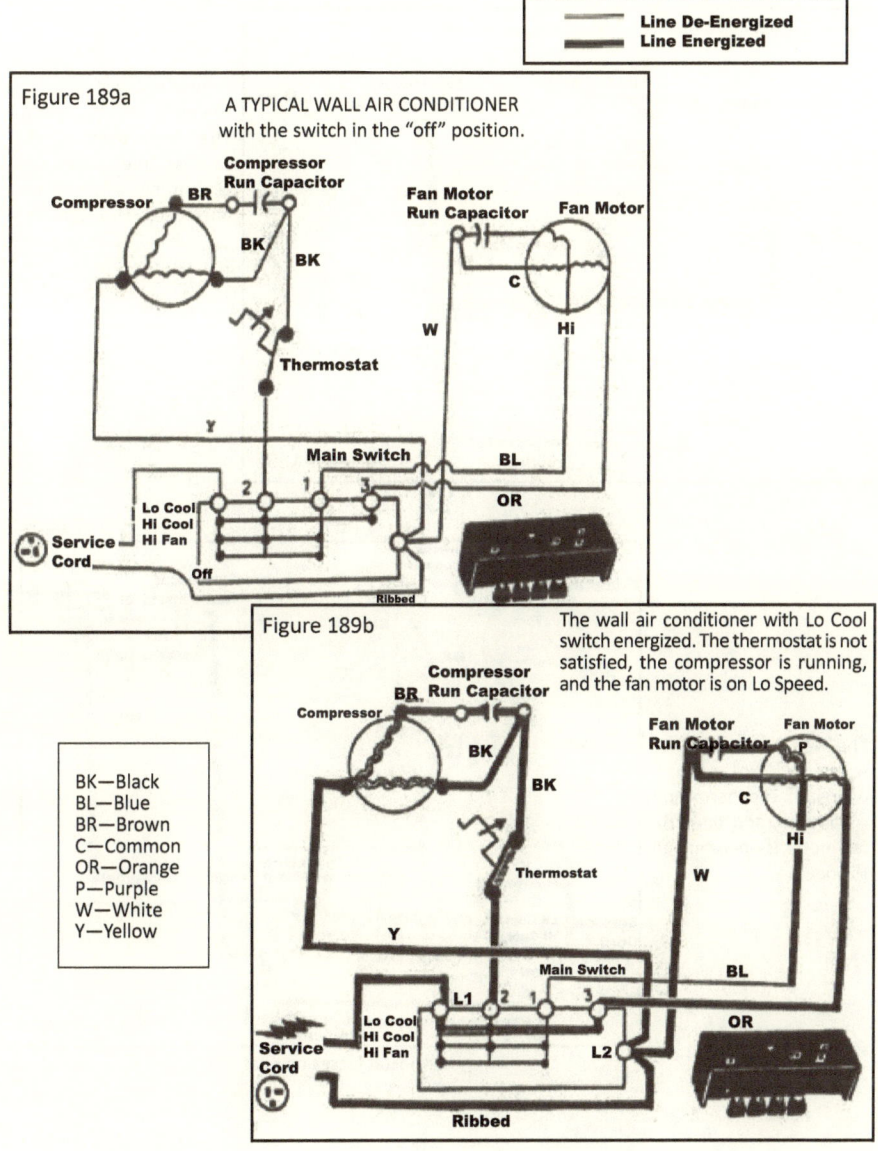

Figure 189a — A TYPICAL WALL AIR CONDITIONER with the switch in the "off" position.

Figure 189b — The wall air conditioner with Lo Cool switch energized. The thermostat is not satisfied, the compressor is running, and the fan motor is on Lo Speed.

BK—Black
BL—Blue
BR—Brown
C—Common
OR—Orange
P—Purple
W—White
Y—Yellow

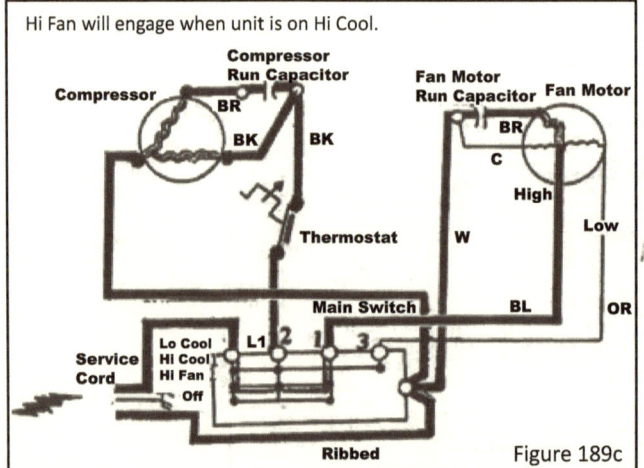

The wall air conditioner with Hi Cool switch energized. The thermostat is not satisfied, the compressor is running, and the fan operates on Hi Speed.

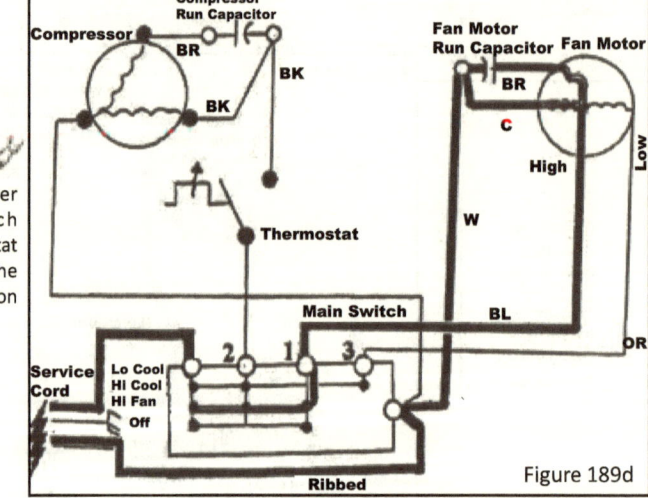

The wall air conditioner with Hi Cool switch energized. The thermostat is satisfied, and only the fan motor is operating on Hi Speed.

Thermostat closes on heat rise.

# Air Conditioning and Refrigeration Repair Made Easy

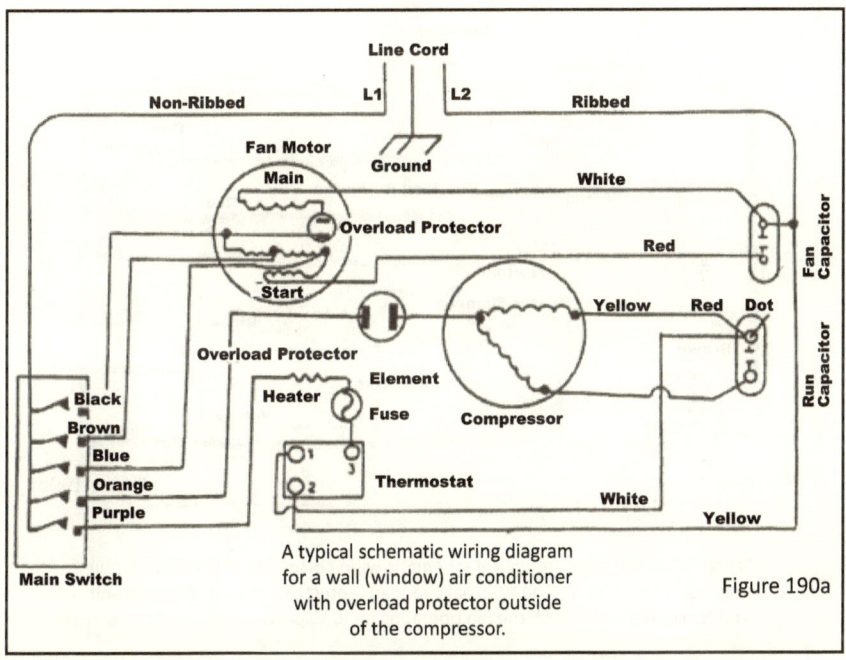

A typical schematic wiring diagram for a wall (window) air conditioner with overload protector outside of the compressor.

Figure 190a

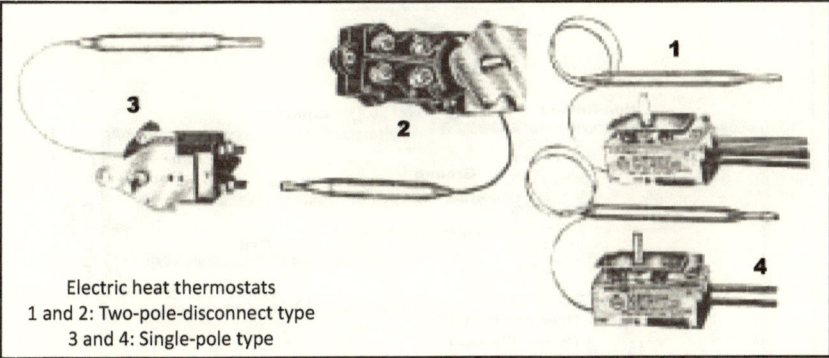

Electric heat thermostats
1 and 2: Two-pole-disconnect type
3 and 4: Single-pole type

*Courtesy of White-Rodgers, Division of Emerson Electric Company*

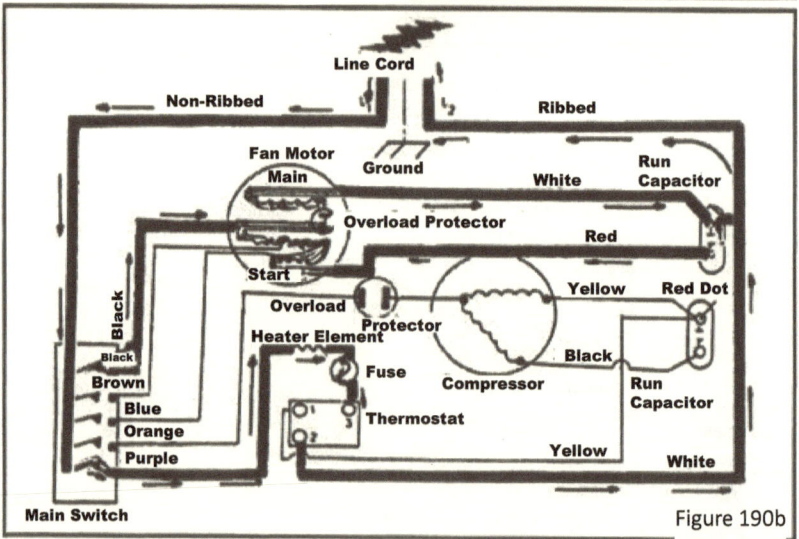

Figure 190b

Figures 190b and 190c are typical schematic wiring diagrams of a wall (window) air-conditioning unit with an external overload protector. In figure 190b, the unit is in the cooling mode with the fan operating on low speed.

In figure 190c, the unit is in the heating mode with its fan operating on high speed. This unit uses an electric element for heating. These units cost less to buy but are more expensive to operate than heat pumps.

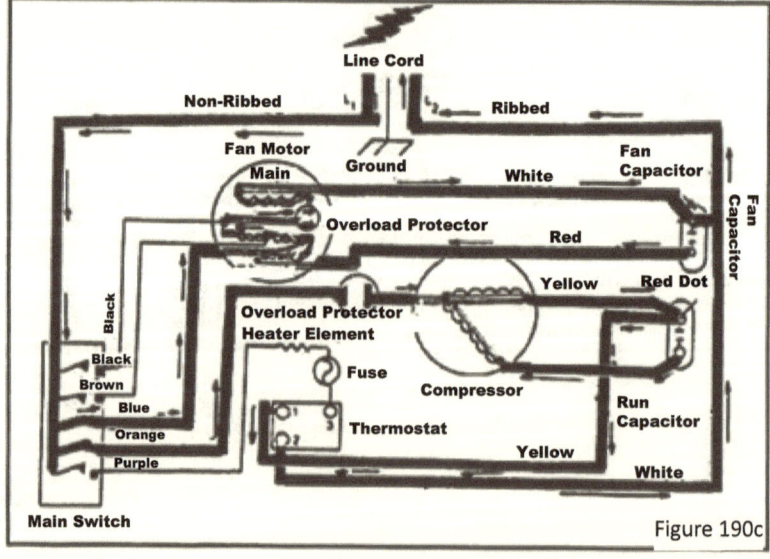

Figure 190c

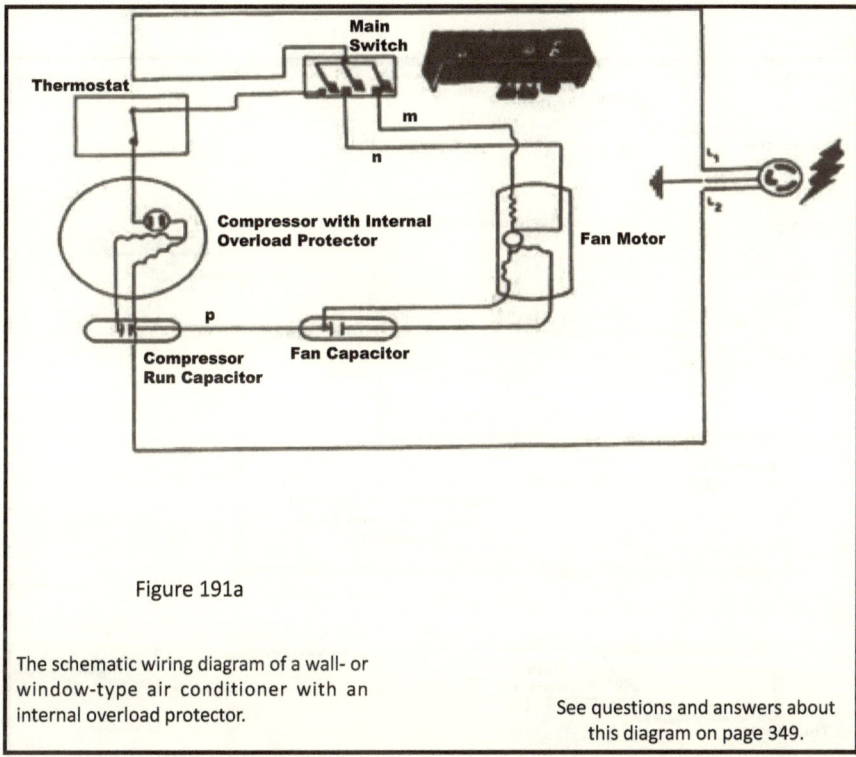

Figure 191a

The schematic wiring diagram of a wall- or window-type air conditioner with an internal overload protector.

See questions and answers about this diagram on page 349.

Before you begin servicing an electrical circuit make certain the electrical supply at the outlet is correct. Check the voltage specifications of the unit to be serviced. Check the voltage again with the unit in operation (see fig. 34). There should be about 5-volt decrease with the unit running. If the voltage drops 10 volts or more, there could be (a) bad wiring to the wall outlet, (b) an overload such as an overcharged or restricted system, or (c) a bad motor winding. If the compressor fails to start with sufficient line voltage reaching the motor, check the overload protector and the starting relay (if applicable). If they check OK, disconnect the wiring from the compressor motor and check it with a compressor test cord (see fig. 35). See figure 36 for testing a capacitor-start-capacitor-run motor or a capacitor-start-induction-run motor. If the compressor fails to operate when tested, it must be replaced.

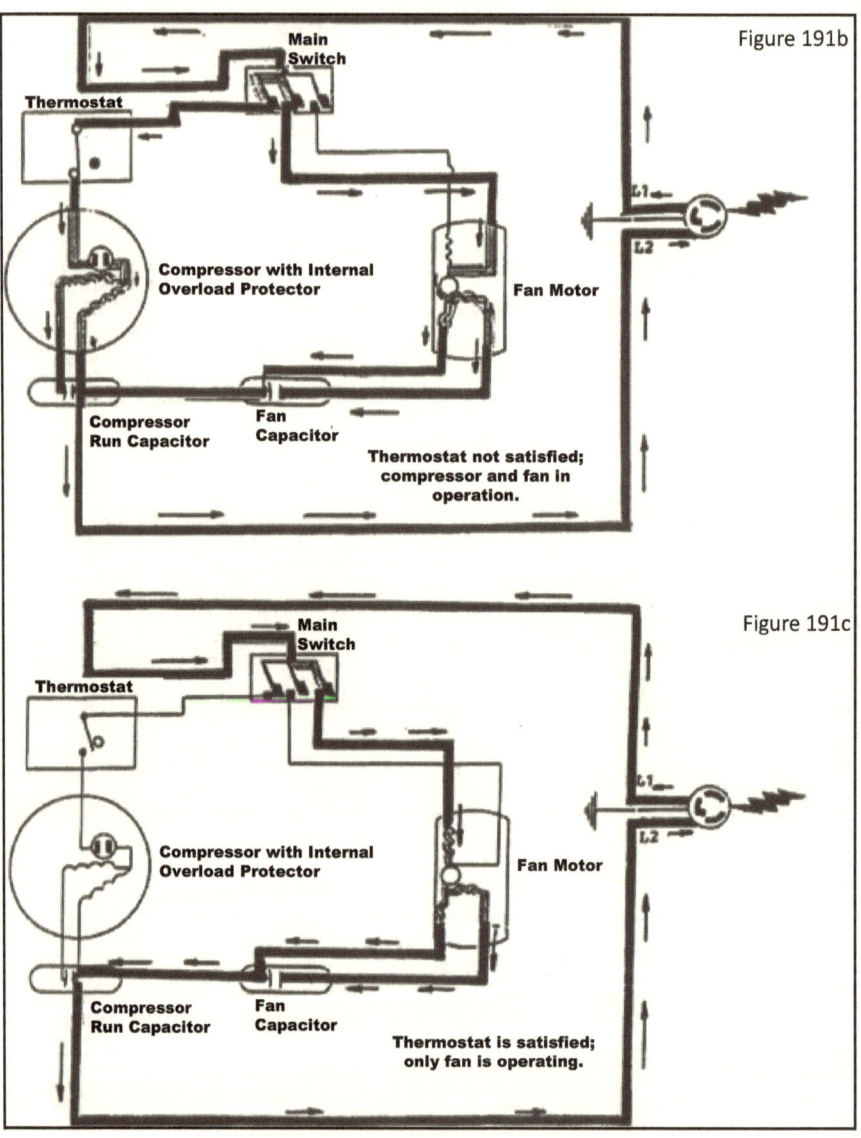

Figure 191b — Thermostat not satisfied; compressor and fan in operation.

Figure 191c — Thermostat is satisfied; only fan is operating.

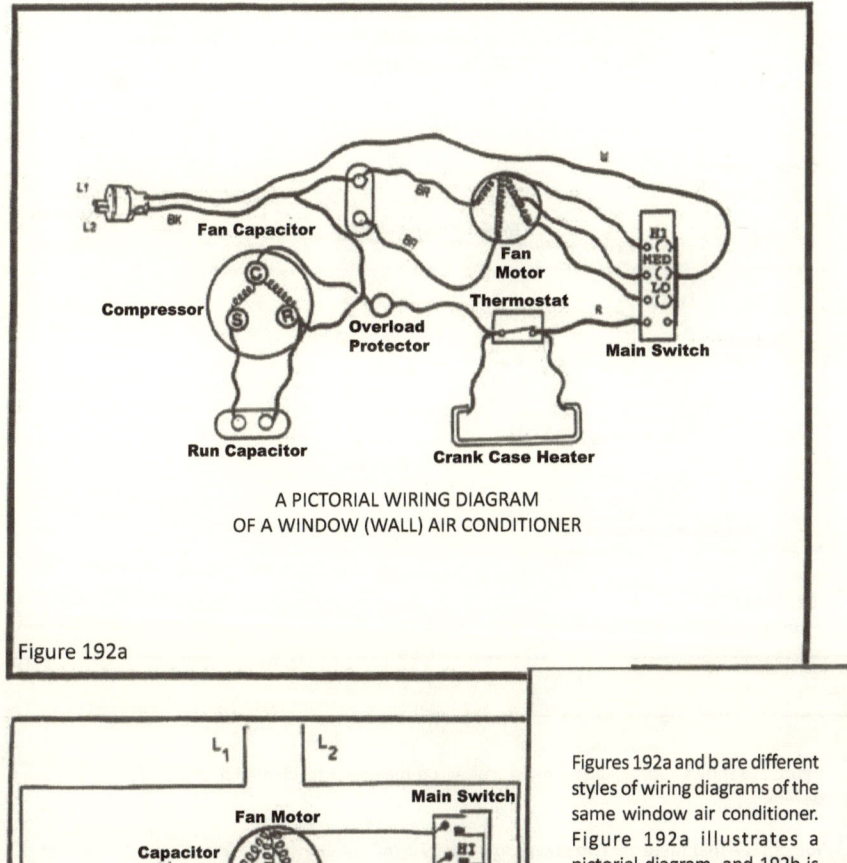

A PICTORIAL WIRING DIAGRAM
OF A WINDOW (WALL) AIR CONDITIONER

Figure 192a

Figure 192b

Figures 192a and b are different styles of wiring diagrams of the same window air conditioner. Figure 192a illustrates a pictorial diagram, and 192b is a schematic wiring diagram. It is much easier to read a schematic diagram than a pictorial one.

A TYPICAL SCHEMATIC WIRING DIAGRAM OF A WINDOW (WALL) MOUNT AIR CONDITIONER

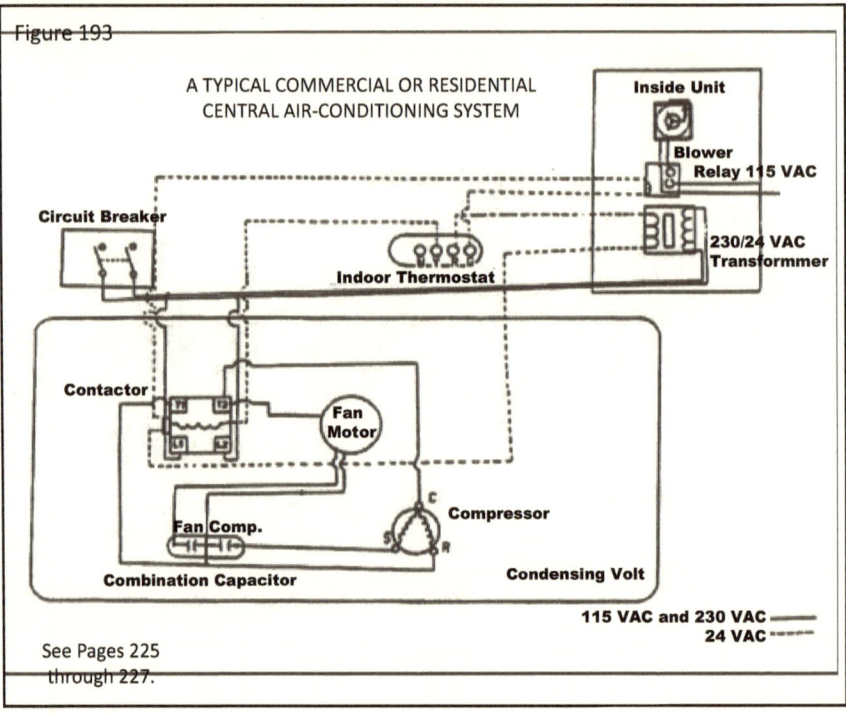

Figure 193

Figure 193 shows a wiring diagram of a typical commercial or residential central air-conditioning system.

You can see that there are two power sources: 220 VAC coming directly from the circuit breaker box next to the condensing unit outside the structure, and 115 VAC which would come from a wall outlet to power the fan relay inside the plenum chamber.

The low voltage created by the step-down transformer goes through the indoor wall thermostat to energize the contactor inside the condensing unit (see p. 244). When the contactor coil is energized, it creates a magnetic field and overcomes the spring pressure which draws down the armature causing the points to connect $T_1$ and $T_2$ with $L_1$ and $L_2$. Since the 230-line voltage is also connected to $L_1$ and $L_2$, the 220 VAC circuit to energize the compressor and the condenser fan motor(s) is created. The circuit will remain in this state until the wall thermostat is satisfied and disconnects the low voltage from the contactor. At this time, the coil loses its magnetic field and spring pressure takes over to return the armature to its upper (disconnect) position. Compare this schematic with figure 163 on page 254 to get a better understanding.

Turning on air-conditioning units requires closing or opening many different electrical circuits to operate the unit.

To make this possible, a control relay which operates on low voltage (or sometimes on line voltage, depending on the unit) becomes energized causing its contacts to snap open or close, thus energizing or de-energizing various circuits. (See fig. 151.)

The main circuits run through the relay contacts. When the relay coil is energized, it creates a magnetic field moving an armature which in turn shifts the contact points. As these points come into contact or separate from one another, by the movement of the armature, individual circuits are interrupted or connected.

In other words, a relay that can operate on low voltage power controls other circuits that are connected to higher voltage lines.

In the schematic wiring diagrams, each contact of a relay is shown on the line voltage circuits which is intersected by that particular contact.

The relay contacts that are normally open, close when being energized, and those that are normally closed, open when energized.

Refer to figures 194a through 194c. Note the representation of relay contacts on the schematic wiring diagram.

Courtesy of OMRON Electronics Inc.

1. *Three-phase-motor protector.* It protects the compressor by shutting off power to the unit in case of (a) energy surge, (b) low voltage, (c) phase loss, (d) phase imbalance, (e) phase reversal, or (f) short-cycling. It indicates the specific problem(s) and in which order they occurred. Six indicator lights display line status and faults in memory.
2. *Five-seconds-to-eight-minutes adjustable delay timer.* After power interruption, time period starts when power is restored and thermostat closes. The two terminals are connected in series with the load.
3. This timer will prevent short-cycling by delaying the start-up of a compressor for five minutes after a power interruption. Its terminals are connected in series with the load.
4 and 5. Different styles of bayonet-type relays for solid-state circuit boards. They are widely used in ice machines.

Courtesy of Wagner Products Corp.
Miami, Florida

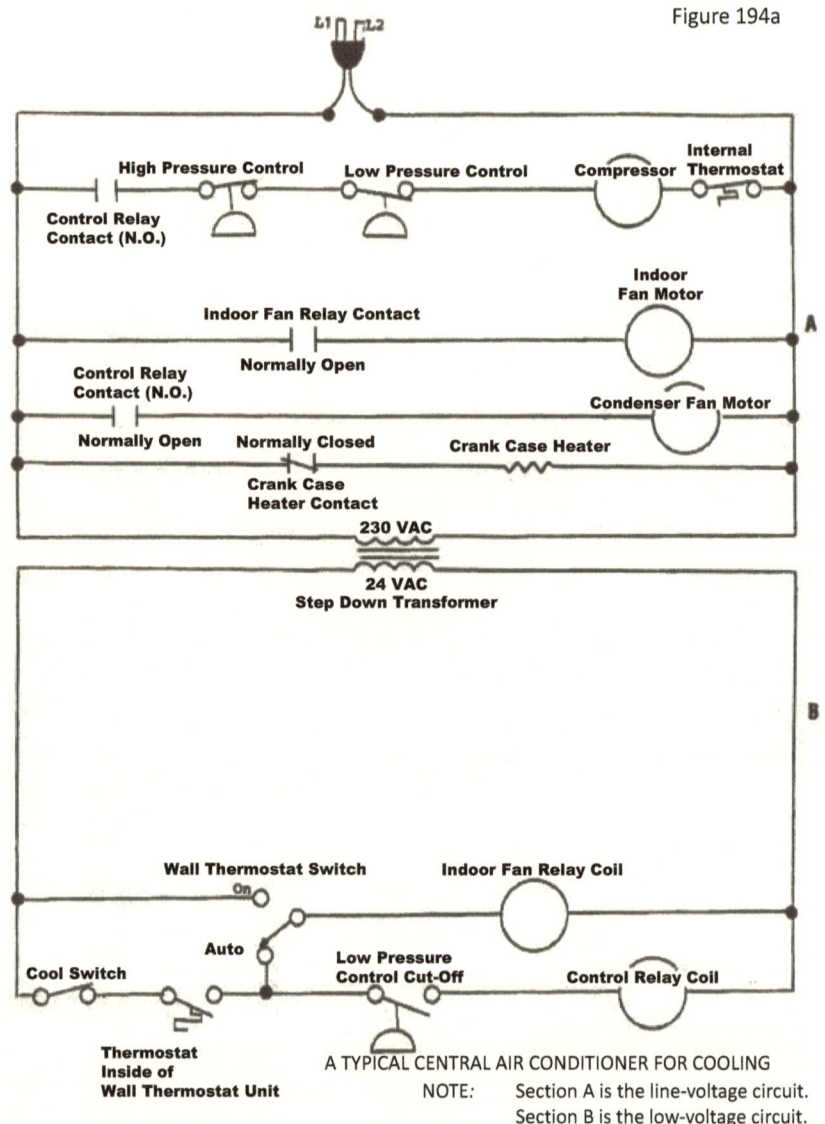

Figure 194a

A TYPICAL CENTRAL AIR CONDITIONER FOR COOLING

NOTE: Section A is the line-voltage circuit.
Section B is the low-voltage circuit.

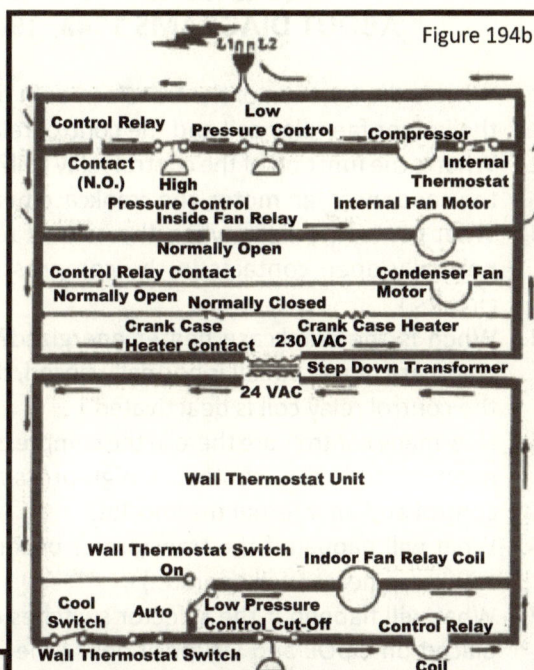

**Figure 194b**

A central air conditioner with the thermostat on "Auto" position. A 230 VAC is decreased to 24 VAC to activate the *indoor fan relay coil*, which in turn activates the *compressor*, the *internal fan motor* and the *condenser fan motor*. In order for this to happen, the *cool switch* and *thermostat* must be closed.

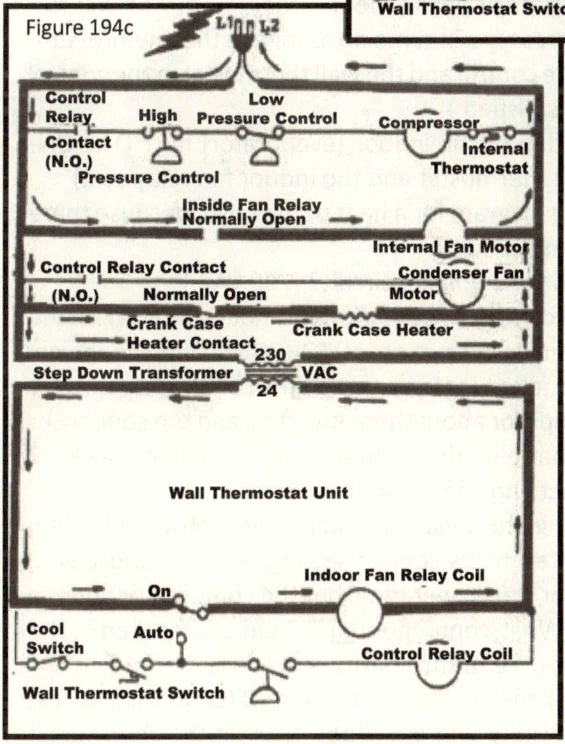

**Figure 194c**

**HEAVY LINES-WIRES ENCIRCLED**

**Figure 194c**

Note the different paths in which the current travels in each of the diagrams.

When the thermostat switch is ON, the indoor fan relay coil is energized and activates the internal fan motor.

NOTE: The normally open (NO) contacts close, and the normally closed (NC) contacts open in the 220 VAC circuits when the control coil(s) is (are) energized in the 24 VAC circuits.

# QUESTIONS AND ANSWERS
## ABOUT DIAGRAMS 194a, 194b, and 194c

1. What is the function of the transformer in these diagrams? (To activate the indoor fan relay coil and the control relay coil.)
2. What is the function of the control relay coil? (To control the operation of the condenser fan motor, the crankcase heater and the compressor.)
3. What contact(s) close when the control relay coil is energized? (The normally open contacts in the compressor and the condenser fan circuits.)
4. When is the crankcase heater energized? (Since the contact in the crankcase heater circuit is normally closed, the heater is energized when the control relay coil is deactivated.)
5. How many controls are there in the compressor circuit? (Four: a normally open control relay contact, a high-pressure control, a low-pressure control and an internal thermostat.)
6. What will happen if the transformer becomes inoperative? (Only the crankcase heater will operate.)
7. What will happen if the selector switches on the wall thermostat are placed on COOL and ON positions? (The internal fan motor will run constantly and the compressor will come on as long as the low-pressure control, the high-pressure control and the wall thermostat in the control relay coil circuit are not satisfied.)
8. What causes the operation of the indoor (evaporator) fan? (The ON-AUTO switch on the wall thermostat and the indoor fan relay coil.)
9. Is this a schematic wiring diagram for a heat pump? (No, because there is no reversing valve in the circuit.)
10. What is the voltage in the line voltage circuit? (230 VAC.)
11. On a hot summer day, a central air conditioner is not cooling. With the controls on the wall thermostat set on AUTO, COOL, and the lowest temperature setting, the following is observed: the compressor runs for about three minutes, stops for about three minutes, and the condenser fan is not operating. What electrical components should be checked? (The control relay contact and condenser fan motor.)
12. In answering a "no cooling" complaint, the following is observed: when the thermostat is adjusted to its coolest setting, + AUTO and COOL, only the Indoor fan motor starts operating, but the compressor and the condenser fan do not. What component(s) should be checked? (The low-pressure control cut-off and the control relay coil.)
13. After adjusting the wall thermostat to COOL, AUTO and its lowest temperature setting, nothing worked. Upon turning the thermostat

control from AUTO to ON, the indoor fan starts running, but nothing else works. What conclusion can be drawn and what component(s) may be defective? (It may be concluded that the wall thermostat is defective; check the COOL switch and the mercury switch and refer to page 236, "Testing a Wall Thermostat." Replace the wall thermostat if necessary.)
14. In answering a "no cooling" complaint for a central air conditioner and after setting the wall thermostat controls to the coolest position, nothing works, the fan and compressor remain inoperative. The position of the switch is changed from AUTO to ON while the switch remains on COOL. Nothing happens, the fans and compressor fail to start. What components should be checked? (Check the transformer primary lines for 230 VAC. Check the transformer secondary lines for 24 VAC, and check for a tripped circuit breaker or blown fuse.)
15. In answering a "no cooling" complaint on a central air conditioner; after the Wall Thermostat is adjusted to its coolest setting, the levers on AUTO and COOL, the unit starts operating but the compressor short-cycles (operates for four or five seconds then stops, then restarts after two or three minutes for only four or five seconds). What electrical components should be checked? (Check the compressor for a grounded or disconnected run winding or for low voltage at the power source.)

## QUESTIONS AND ANSWERS ABOUT THE SCHEMATIC WIRING DIAGRAM FIGURE 194a

1. How many controls are there in the compressor circuit? (Five: control relay contact, high-pressure control, low-pressure control, internal thermostat and the compressor.)
2. If the control relay becomes inoperative, how many components will not function? (Two: the compressor and the condenser fan motor)
3. What controls the compressor crankcase heater? (The control relay coil.)
4. What controls the indoor fan motor? (Indoor fan relay coil)
5. If the low-pressure control opens, what components will become inoperative? (Compressor and condenser fan motor)
6. What happens if the transformer becomes inoperative? (Nothing will work except the crankcase heater.)
7. What happens if the wall thermostat switch is turned to ON position? (The indoor fan motor will run all the time.)

# Air Conditioning and Refrigeration Repair Made Easy

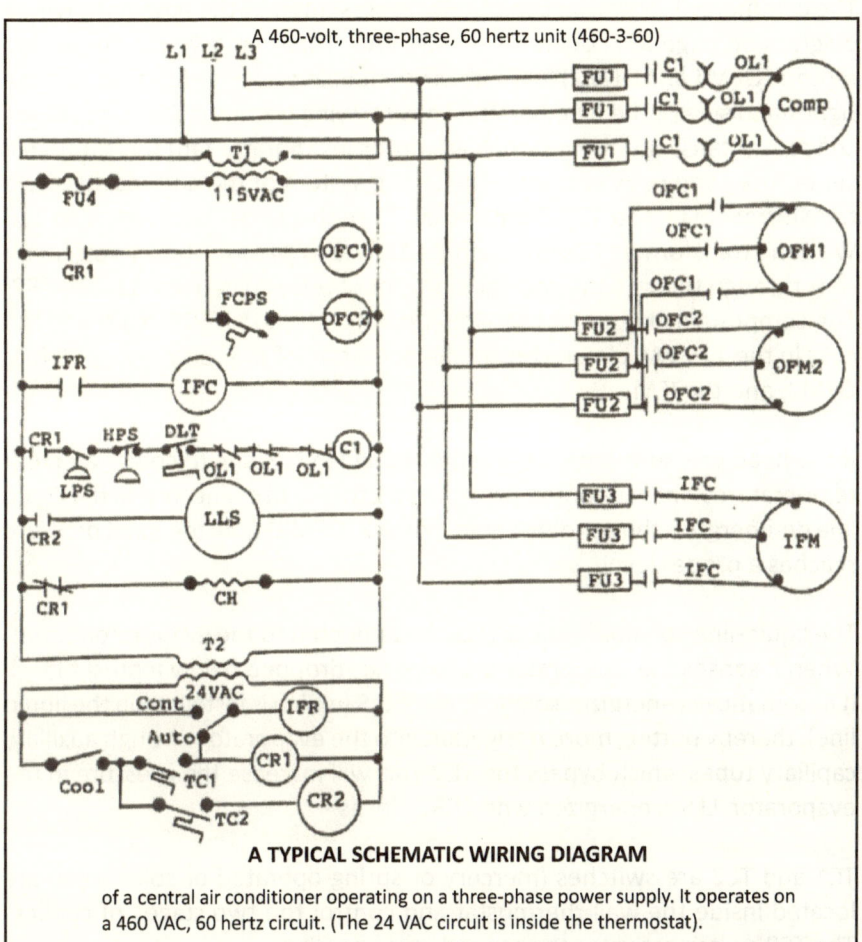

**A TYPICAL SCHEMATIC WIRING DIAGRAM**

of a central air conditioner operating on a three-phase power supply. It operates on a 460 VAC, 60 hertz circuit. (The 24 VAC circuit is inside the thermostat.)

LEGEND:

| | | |
|---|---|---|
| CR | — | Control Relay |
| IFR | — | Internal Fan Relay |
| TC1 | — | Thermostat Contact (Cooling Stage 1) |
| CONT | — | Internal Fan Operating Continuously |
| TC2 | — | Thermostat Contact (Cooling Stage 2) |
| IFC | — | Internal Fan Coil |
| CH | — | Crankcase Heater |
| OFC1 | — | Outside Condenser Fan 1 |
| IFM | — | Internal Fan Motor |
| OFC2 | — | Outside Condenser Fan 2 |
| FCPS | — | Fan Control Pressure Switch |
| LLS | — | Liquid Line Solenoid |
| HPS | — | High Pressure Switch (Control) |
| DLT | — | Time-Delayed Thermostat |
| LPS | — | Low-Pressure Switch (Control) |
| FU1 through FU4 | — | Fuse 1 to fuse 4 |

## EXPLANATION OF THE DIAGRAM ON PAGE 337

The commercial central air conditioner represented by the schematic wiring diagram on page 337, operates on 460 VAC. There are two condenser fan motors (OFM1 and OFM2) for more efficient heat exchange (cooling). On the right-hand side of the diagram, the lines carrying power to the compressor, condenser fans, and the internal fan motor (evaporator fan) run through a set of fuses and relay contacts. The wiring system of the unit includes two transformers (T1 and T2). Transformer T1 reduces the line voltage to 115 VAC, and transformer T2 reduces that 115 VAC to 24 VAC. The 24-volt circuit runs through the wall thermostat and a set of relay coils, IFR, CR1, and CR2. The components in the 24 VAC circuit control the 115 VAC circuit; and the coils in the 115 VAC circuit control the operation of the compressor, OFM1, OFM2, and the IFM.

If the head pressure drops to an unacceptable level due to cooler ambient temperatures, the fan control pressure switch (FCPS) connected to the liquid line de-energizes the condenser fan motor 2 (OFM2) until the head pressure reaches a proper level.

The liquid-line solenoid control (LLS) is connected to the evaporator outlet. When it senses the evaporator pressure has dropped below a preset level, it automatically energizes solenoid valve LLS (which is installed on the liquid line), thereby putting more refrigerant into the evaporator through auxiliary capillary tubes which bypass the TEV This will increase the pressure in the evaporator. LLS is energized when CR2 closes.

TC1 and TC2 are switches (mercury or spring operated or solid-state) are located inside the wall thermostat and control the two stages of cooling. The TC2 switch activates the second-stage cooling.

When C1 coil in the 115-volt circuit is activated, it closes the C1 contacts energizing the compressor.

When OFC1 coil is activated, it closes the OFC1 contacts to energize the outside number 1 fan motor (OFM1).

When OFC2 coil is activated, it closes the contacts to condenser fan motor number 2 (OFM2).

When IFC coil is energized, it closes the IFC contacts to energize the indoor fan motor.

## QUESTIONS AND ANSWERS
## ABOUT THE SCHEMATIC WIRING DIAGRAM ON PAGE 337

1. How many transformers are there in the entire circuit? (Two: one transformer converts line voltage to 115 VAC; the second one converts 115 VAC to 24 VAC.)
2. What happens if the high-pressure control is activated? (The compressor becomes energized.)
3. What happens if fuse 4 burns out? (Nothing in the entire system will become energized.)
4. What happens if transformer T2 burns out (or becomes deactivated)? (The internal fan relay and control relays 1 and 2 will become deactivated; consequently, the internal fan motor, compressor, and the condenser fan motors will become de-energized.)
5. What happens if one of the number 1 fuses burns out? (The compressor will become inoperative.)
6. What happens if the relay coil, CR1, burns out? (The compressor and outside fan motors 1 and 2 will become inoperative. The crankcase heater will remain warm all the time.)
7. On a service call for not enough cooling, the thermostat is adjusted to its lowest setting and the following is observed:

    a. The compressor operates for about three minutes and shuts off for about four minutes.
    b. The inside fan motor is operating.
    c. The outside fan motors 1 and 2 are not operating.

    Name a specific component that could cause this problem. (The CR1 contacts responsible for energizing OFC1 and OFC2 have burnt out.)

8. On a service call for insufficient cooling, the wall thermostat is set on COOL and AUTO, and it is on its lowest setting. During the operation of the unit, it is observed that

    a. the IFM is operating,
    b. OFM2 is not operating,
    c. OFM1 is operating.

    What three components could cause OFM2 to become nonoperational? (Either a defective FCPS switch or OFM2 has shorted out or one of the FU2 fuses has blown.)

9. On a hot summer day, a service call for insufficient cooling has produced the following with the thermostat set on COOL, AUTO, and its lowest temperature:

   a. The compressor is operating.
   b. The IFM is operating.
   c. OFM1 and OFM2 are both operating and some cooled air is delivered.

   List the electrical components that are likely to cause insufficient cooling. (Explanation: LLS, the liquid-line solenoid, is responsible for opening the liquid line to allow more refrigerant to circulate in the system as the ambient temperature increases in the conditioned area. This is activated by the TC2 switch in the wall thermostat). Answer: TC2, CR2, or LLS.

10. On a service call for no cooling, the following is observed:

    a. With the wall thermostat set on CONT or AUTO position, and on its lowest temperature, no motor will operate.
    b. The crankcase heater is warm.

    What could cause this problem? (The T2 Transformer creating the 24 VAC circuit must have burned out. Explanation: when coil CR1 is not activated, contact CR1 for the crankcase heater is closed.)

11. What reading should be expected at T1, secondary terminals? (115 VAC)
12. What reading should be expected at T2, primary terminals? (115 VAC)
13. How many cooling stages are there in this unit? (Two)
14. How many heating stages are there in this unit? (None)
15. If everything works except the number one outside fan motor, what electrical components are to be checked? (OFC1 and OFM1)
16. What electrical components were likely to be responsible if the internal fan motor ran, but the compressor, the number 1 and number 2 outside fan motors did not operate with the thermostat set on AUTO, COOL, and lowest temperature? (Check CR1)
17. If all of the fuses are good but none of the components are energized, what must be checked first? (Circuit breaker, T1, and T2 transformers)
18. What kind of motors do the compressor, OFM1, OFM2, and IFM have? (Three-phase motors because three hot wires are connected to each one of them.)

# Air Conditioning and Refrigeration Repair Made Easy

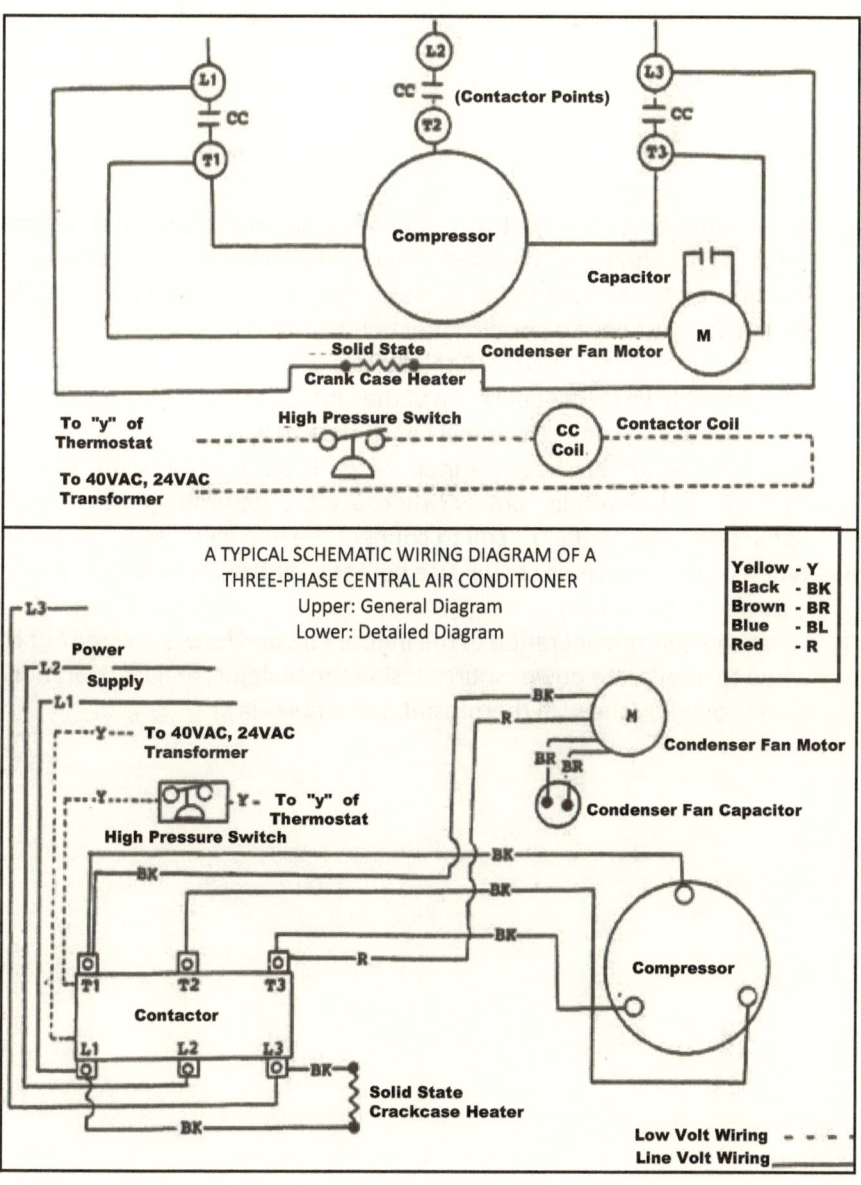

## QUESTIONS AND ANSWERS CONCERNING THE SCHEMATIC WIRING DIAGRAM ON PAGE 341

1. What color wires activate the contactor coil? (Yellow)
2. Which contactor terminals are connected to the fan motor? (T1 and T3)
3. Which contactor terminals are connected to the compressor? (T1, T2, and T3)
4. Which contacts in the contactor are connected to the power supply lines? (L1, L2 and L3)
5. If the compressor runs but the condenser fan motor does not, which components should be checked as most likely to cause this problem? (Fan motor, fan capacitor, fan wires connected to contactor terminals T1 and T3 must have a good connection.)
6. If a fuse blows in the power supply line to contactor terminal L2, what effect will it have on the operation of the unit? (The condenser fan motor will run, but the compressor will fail to start.)
7. Why will the system become inoperative if the high-pressure switch cuts off the low-voltage power? (Because the low-voltage power will not activate the contactor coil to connect the line voltage.)
8. When is the compressor crankcase heater energized? (As long as there is power to L1 and L3)
9. What controls the operation of the indoor fan, and how is it wired? (It is wired to a separate power source inside the building, and its operation is controlled by the wall thermostat and a fan relay.)

The following is a typical example of troubleshooting by the voltmeter method. It illustrates the expected readings as the probes touch different parts of the circuit. Any interruption or short can be detected when compared with the expected readings.

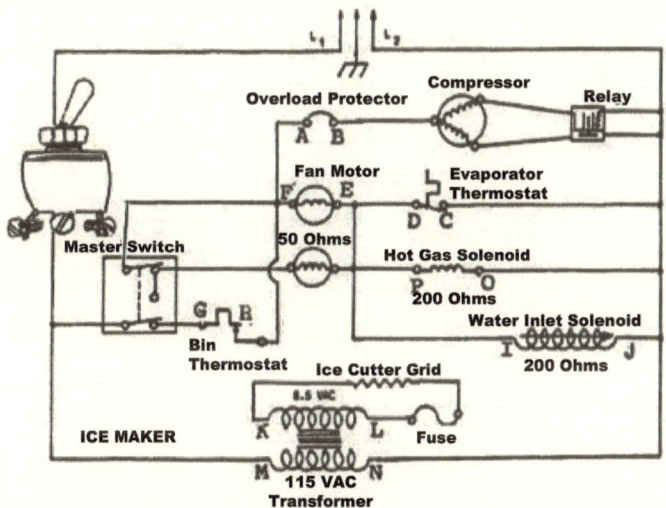

FOR THESE READINGS, ASSUME THAT THE MASTER SWITCH, BIN, AND EVAPORATOR THERMOSTATS ARE CLOSED UNLESS OTHERWISE INDICATED.

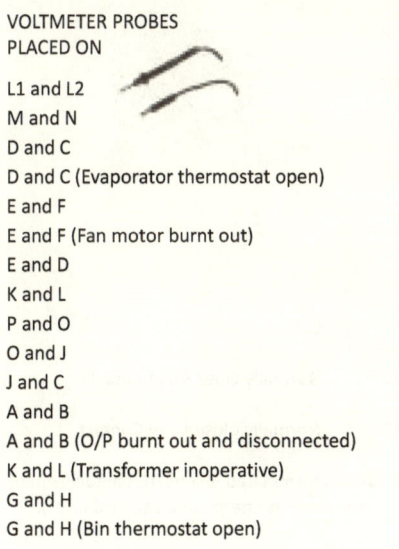
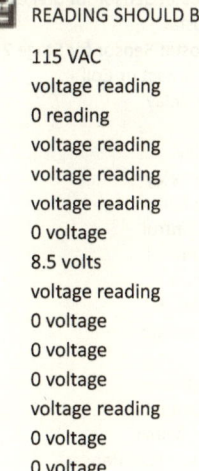

| VOLTMETER PROBES PLACED ON | VOLTMETER READING READING SHOULD BE |
|---|---|
| L1 and L2 | 115 VAC |
| M and N | voltage reading |
| D and C | 0 reading |
| D and C (Evaporator thermostat open) | voltage reading |
| E and F | voltage reading |
| E and F (Fan motor burnt out) | voltage reading |
| E and D | 0 voltage |
| K and L | 8.5 volts |
| P and O | voltage reading |
| O and J | 0 voltage |
| J and C | 0 voltage |
| A and B | 0 voltage |
| A and B (O/P burnt out and disconnected) | voltage reading |
| K and L (Transformer inoperative) | 0 voltage |
| G and H | 0 voltage |
| G and H (Bin thermostat open) | voltage reading |

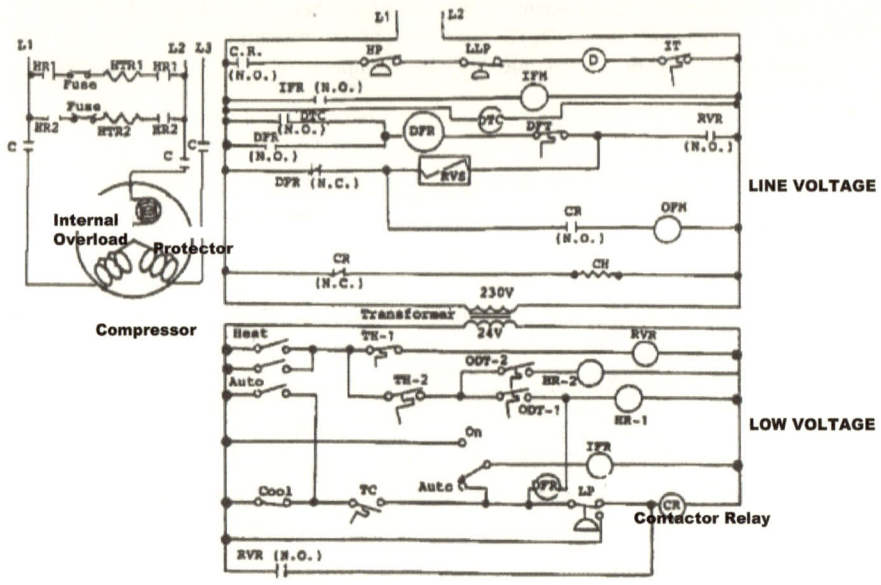

LEGEND:

CR— Relay Contactor
CH— Crankcase Heater
TH-1 Thermostat (for heater)
TH-2 Thermostat (second stage heating in wall thermostat)
DTC— Defrost timer coil
ODT-1 Outside Thermostat Sensor for Stage 1 Heating
DFT— Defrost Thermostat
ODT-2 Outside Thermostat Sensor for Stage 2 Heating
DFR— Defrost Relay, Contact or Coil
RVR— Reversing Valve Relay
RC— Relay Coil
IFR— Internal Fan Relay
HR-1 Stage 1 Heating Relay
HR-2 Stage 2 Heating Relay
HP— High-Pressure Control
LLP— Low-Pressure Control
IFM— Internal Fan Motor
IT— Internal Thermostat
DTC— Defrost Timer Contact
NO— Normally Open
NC— Normally Closed
DFR— Defrost Relay Contact
RVS— Reversing Valve Switch
OFM— Outside Fan Motor (Condenser)

○ Coil

NO ‖ Normally Open Coil Contact

NC ⋈ Normally Closed Coil Contact

(Defrost Cycle is terminated when DTC contacts open. The defrost timer is energized for ten minutes).

A TYPICAL SCHEMATIC WIRING DIAGRAM
OF AN AIR-CONDITIONING UNIT USING
A HEAT PUMP

## EXPLANATION OF THE DIAGRAM ON PAGE 344

The diagram on page 344 is a typical schematic wiring diagram of a heat pump. It operates on three different circuits. The compressor works on a three-phase, 460 VAC, 60 Hz circuit, and its operation is controlled by a contactor operating on 230 VAC. The 24 VAC circuit is controlled by the wall thermostat.

When there is a need for heating, the thermostat is set on HEAT and AUTO. The RVR (reversing valve relay) is energized, and its normally open contacts close. This will cause the reversing valve solenoid to become energized in the 220 VAC circuit and the control relay to become energized in the 24 VAC circuit. When the reversing valve solenoid is energized, the outside coil will act as an evaporator, and the inside coil will become a condenser to heat the conditioned area.

The two thermostats (TH-1 and TH-2) are within the wall thermostat unit and are designed to perform two different stages of heating. TH-1 is manually controlled by the user. When TH-2 senses a need for additional heating, it will close the circuits to the outside heaters, HR-1 and HR-2, which in turn are controlled by the two thermostats clipped to the outside coil, ODT1 and ODT2, thus commencing the second stage of heating. (HR-1 and HR-2 are installed on the outside coil). ODT1 and ODT2 are both affected by outside temperatures.

Since the outside coil temperature is set about 20°F below the average ambient temperature, this temperature difference is heat which is transferred to the inside coil, now acting as a condenser.

Considering the low temperatures in winter, the outside coil temperature tends to drop very rapidly. The heating of the outside coil will help alleviate this problem and increase the heating efficiency of the unit.

When the wall thermostat is set on COOL, the TC cooling thermostat closes a circuit to energize IFR and CR relays and takes the unit into the cooling cycle. The RVR (reversing valve relay) stays de-energized and the unit cools. (The outside coil now acts as a condenser and the inside coil acts as an evaporator; hence there is no need for heating the outside coil). When the unit is in the heating mode, every ninety minutes the DTC defrost timer takes the unit into the defrost cycle for a maximum time of ten minutes.

## QUESTIONS AND ANSWERS
## ABOUT THE WIRING DIAGRAM ON PAGE 344

1. How is the reversing valve solenoid energized? (When the reversing valve relay closes its normally open contacts.)
2. Assuming all switches on the C (contactor coil) and CR (control relay coil) are operational, the wall thermostat is set on COOL and the TC thermostat is adjusted to its lowest setting; the compressor fails to start. What components will require checking? (CR, C, HP, LLP and the compressor)
3. What switch(es) must be closed to energize the heater relay HR-1? (Heat switch on the wall thermostat, TH-2 and ODT1)
4. What indicates that the transformer is burnt out and no longer functional? (Nothing in the system will work and the crankcase heater will stay on all the time.)
5. When the thermostat TH-1 closes, what determines the number of supplemental heater strips that are energized? (The inside temperature)
6. Assuming all the loads are operating and the wall thermostat switches are correctly set, can the second stage of the supplemental heat (HR-1 or HR-1 and HR-2) be energized without the compressor running? (No. The compressor is energized when RVS is energized.)
7. Assuming the AUTO switch is closed, what other switches on the wall thermostat determine if the heat pump operates in the heating or cooling mode? (The HEAT or COOL switch and the thermostat setting.).
8. How many line voltage load circuits are shown in the diagram? (Eight: CH, OFM, RVS, IFM, and C circuits, compressor, and two outside heater circuits.)
9. Thermostat TH-1 and the HEAT switch are closed and the fan switch is set on AUTO; name the switches that must be closed to energize the indoor fan motor, in that case what other loads will be energized in the circuit(s)? (LP.) RVR relay and its contacts within the relay. Other components energized with the thermostat at this setting are CR, RVS, OFM, C and the compressor.)
10. Thermostat TC and the COOL switch are closed and the fan switch is set on AUTO position. Name all the other switches that must be closed to energize the (CR) control relay. (LP)
11. Thermostat TC and the COOL switch are closed and the fan switch is set on AUTO position; name the electrical components that are energized assuming that all pressure controls and the IT thermostat are closed. (CR, IFR, OFM, IFM, C and the compressor.)
12. Manual thermostat TH-1, the HEAT and AUTO switches are closed and assuming all electrical components are functional; explain the step-by-

step electrical operation in the circuit. (24 VAC power flows in the circuit through the HEAT switch, TH-1, RVR, then back to the other side of the line. With the RVR contact (at the bottom of the diagram) closed, power flows through the circuit to the other side of the line energizing the CR (coil relay). A 24 VAC also flows through the IFR (internal fan relay) and to the other side of the line. With the CR, IFR, and the RVR (reversing valve relay) energized, the contacts within the relays will become activated and the 230 VAC circuitry will close the RVS, OFM, IFM, and C causing the reversing valve, outside fan motor, inside fan motor, and the compressor to become energized and commence operating. The CH will become de-energized).

13. Can the heater relay, HR-1 (second-stage supplemental heat) become energized without the compressor operating? (No)
14. What purpose does a defrost thermostat (DFT) serve? (By sensing the temperature on the outside coil, it initiates or terminates the defrosting.)
15. How and when is the defrost cycle started? (It is automatically started every ninety minutes by the defrost timer.)
16. How does the compressor start? (When the inside thermostat (IT) (or the TH-1) is closed, the heat or cool switch is closed and when the contactor relay (CR) is energized.)
17. What are the two ways the unit is taken out of the defrost cycle? (The defrost timer contacts [DTC] open after a maximum of ten minutes every ninety minutes, or the defrost thermostat [DFT] opens.)
18. What contact(s) operate in the reversing valve solenoid (RVS) circuits, and what is/are the normal position(s) of the switch(es)? (DFR contact normally closed; RVR contact normally open.)
19. If the control relay (CR) in the 24 VAC circuit is energized and the compressor does not start, name all the switches in the 220 VAC circuit responsible for this failure that should be checked. (CR, HP, LLP, IT, and C.)
20. Name the switches in the defrost relay coil (DFR) circuit when the heat pump is in the cooling cycle. (Cool switch, TC thermostat, AUTO (or ON) switch, and LP)
21. How does the first and second stage heating start? (The first stage heating starts when the TH-1 is manually closed on the wall thermostat. The second stage begins when TH-2 on the wall thermostat senses the need for more heating and if ODT1 and/or ODT2 thermostat closes [which is/are, clipped to the outside coil], the second stage heating will begin.)
22. How many switches are there in the contactor relay? (Three: two normally open and one normally closed)
23. Which switches on the wall thermostat are manually adjustable? (HEAT, AUTO, COOL, ON, OFF, TH-1, and TC.)

24. The wall thermostat is set on the highest temperature, the compressor is running, but little heating is produced. Which electrical components are most likely responsible? (ODT1, ODT2, HR-1, HR-2, reversing valve, and the compressor.)
25. Assume the compressor is OK and there is sufficient refrigerant in the sealed system. When the thermostat is set on COOL, the compressor starts running, but there is no cooling. When set on HEAT, the compressor starts operating and heating is produced. What electrical components are most likely responsible for this and must be checked? (The reversing valve solenoid or the reversing valve relay.)

## QUESTIONS AND ANSWERS
## ABOUT THE DIAGRAM ON PAGE 327

1. What type of fan is operating in this unit? (PSC type. It is not a shaded pole type because it uses a capacitor.)
2. How many speeds does this fan have? (Two)
3. Does the compressor use an internal or an external overload protector? (Internal)
4. How is the start winding removed from the compressor circuit? (There is no starting relay in this type of compressor.)
5. How can the fan motor be directly tested? (With an ohmmeter.) (See pages 82 through 85.)
6. Does the fan motor cycle off with the compressor? (No)
7. If the fan motor ran on low speed but not on high speed, would it be defective? (No, the switch would be defective.)
8. If the unit normally draws 9 A while running, how many amperes will it draw in starting? (At least three times the running amperes.)
9. What type of fuse should be used for this unit? (A 20 A time-delay type to tolerate the initial surge that start-up draws.)
10. If power reaches the unit but the compressor fails to start, what electrical components are to be checked? (The capacitor, thermostat, switch and compressor.)
11. How is the compressor checked? (By direct testing. Connect the L1 and L2 of the 220 VAC to the compressor common terminal and the capacitor terminal for the run winding, bypassing the switch. Make sure the capacitor is not defective. Observe all safety precautions.)
12. How is the fan motor tested? (Bypass the selector switch and the compressor. (a) Disconnect the unit from the power supply. (b) Disconnect m, n and p wires from the switch and the compressor run capacitor. (c) With a jumper cable, connect L1 and L2 directly to wires n and p with the m wire well insulated. When power is reconnected, fan should start. (d.) Disconnect L1 and L2 from power. (e) Connect L1 and L2 to wires m and p with n well insulated. (f) Connect power to L1 and L2. The fan motor should start with either of these methods, if not, it is defective. Make sure the fan capacitor is not defective.)

**WARNING! THE ABOVE TWO PROCEDURES HAVE A POTENTIAL ELECTRICAL SHOCK HAZARD. BE SURE TO USE COMMON SENSE AND OBSERVE ALL SAFETY PRECAUTIONS.**

350

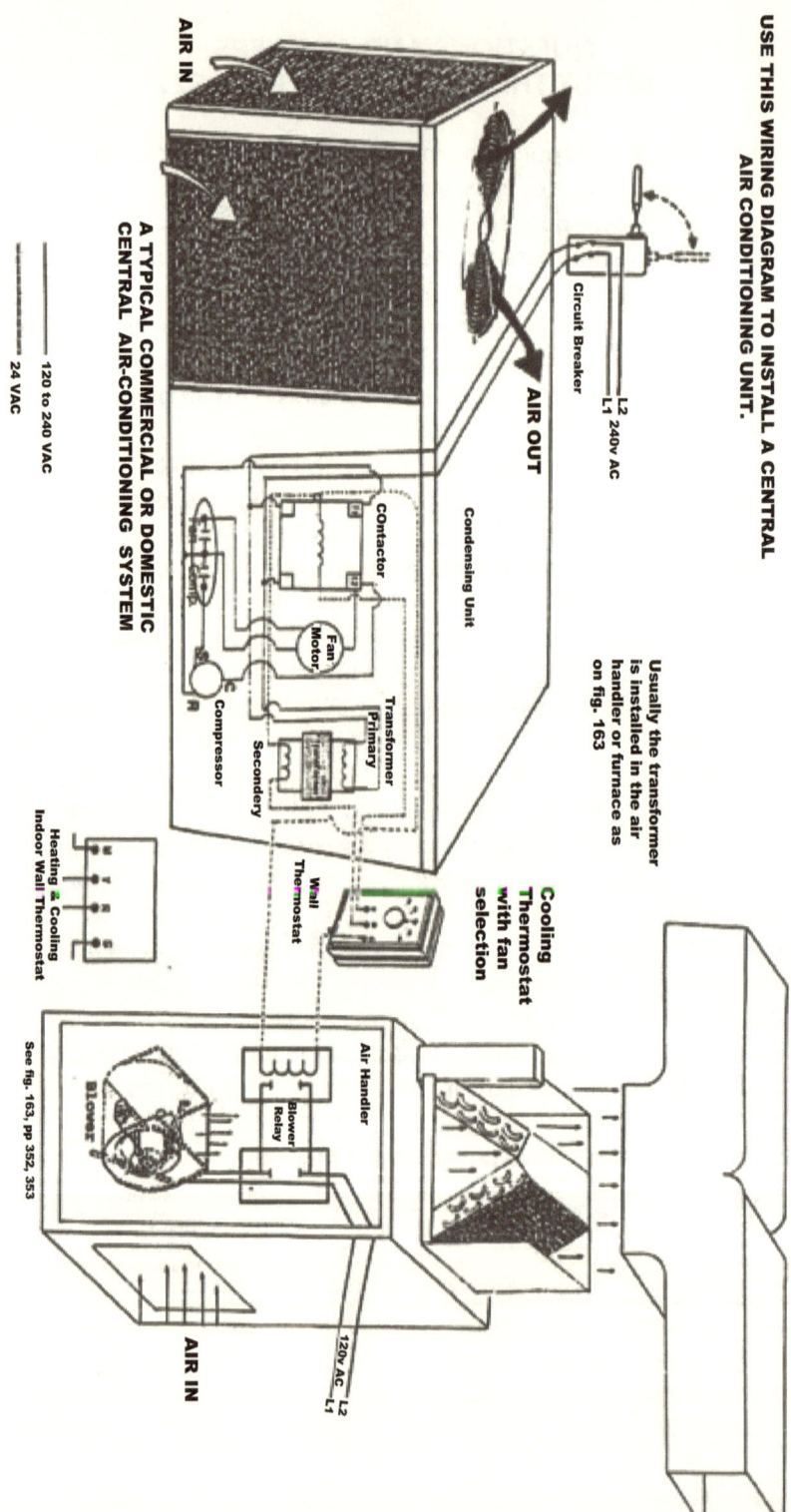

## QUESTIONS AND ANSWERS
## ABOUT THE DIAGRAM ON PAGE 310

1. Is this a frost-free refrigerator? (Yes, because a defrost timer is being used in the wiring.)

2. On what scale should the ohmmeter be set in order to get the correct ohm reading (or continuity) in the following:

    a. The compressor. (The lowest scale.)
    b. The evaporator or condenser fan motor. (The lowest scale.)
    c. The capacitor. (On the highest scale.)
    d. The defrost heater. (On the lowest scale.)

3. What voltage reading should be expected between points A-M? (0 VAC)
4. Is the compressor grounded to the chassis? (Yes)
5. What voltage reading should be obtained between points N-O? (0 volts)
6. Is the unit in a cooling or defrosting cycle? (Cooling)
7. What three components in the compressor circuit control the operation of the compressor? (The defrost timer, thermostat, and overload protector.)
8. In the defrost heater circuit, what component controls the operation of the heater? (The defrost thermostat.)
9. What controls the operation of the light bulb? (Y-X, light switch)
10. What components in the entire diagram are grounded? (four: the compressor, condenser fan motor, evaporator fan motor, and the defrost timer)
11. Supposing the timer motor has a resistance of 4000 Ω. By placing a flux-type ammeter between points M-N, what ampere reading should be expected to determine that the timer motor is *not* defective? (I = E/R = 120/4000 = about 0.3 A)
12. How many components are controlled by the cold control (thermostat)? (Four: the compressor, evaporator fan motor, condenser fan motor and the defrost heater)
13. Since the resistance of the defrost heater is 25 Ω, what amperage should be expected between the timer terminal 2 and point L? (I = E/R = 120/25 = about 4.8 A)

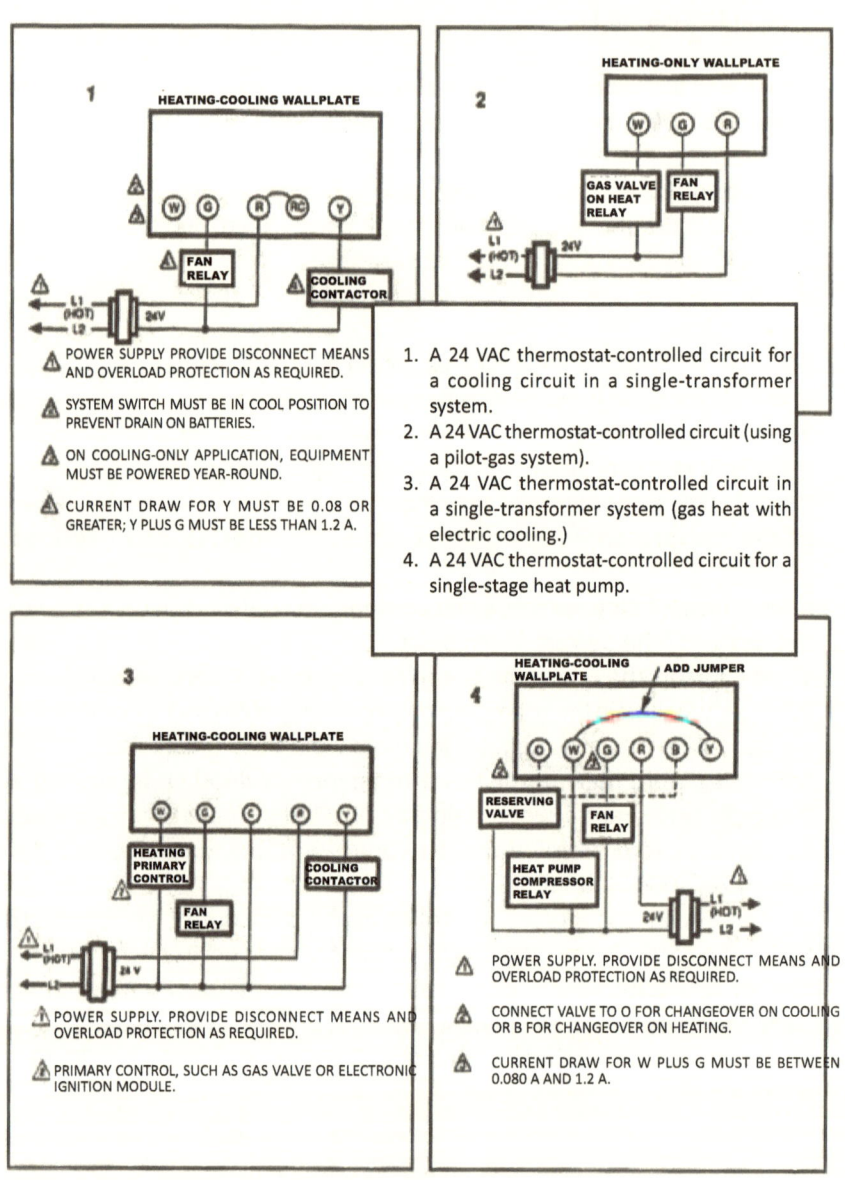

# Air Conditioning and Refrigeration Repair Made Easy

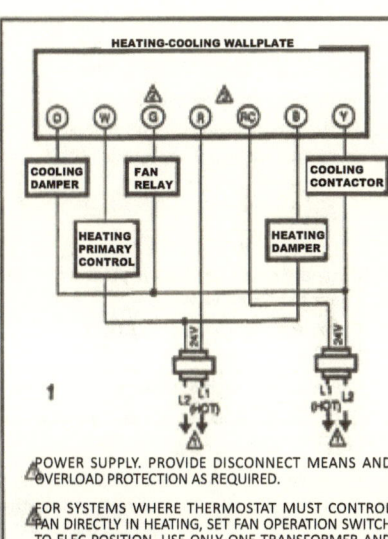

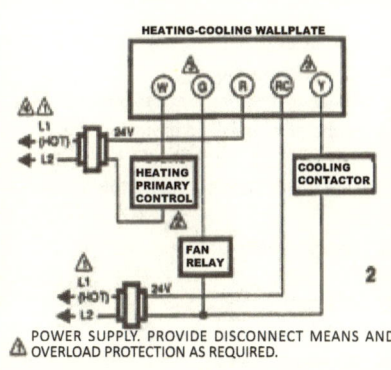

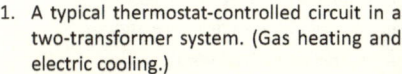

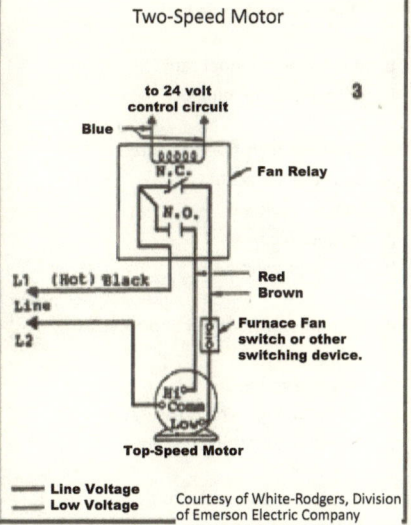

1. A typical thermostat-controlled circuit in a two-transformer system. (Gas heating and electric cooling.)

2. A typical thermostat-controlled heating/cooling circuit in a two-transformer system (one for heating, one for cooling).

3. A typical relay connection (operating a two-speed blower fan in the furnace).

# COMPRESSOR TROUBLESHOOTING CHART
## (Commercial Units)

**Problem and Possible Cause** — **Remedy**

1. Unit Won't Start—No Hum

   a. No power.
   b. Open circuit in motor.
   c. Timer or pressure contacts open.
   d. Inoperative overload protector.

   a. Check fuses, circuit breakers, lines.
   b. Replace motor or hermetic unit.
   c. Check timer controls and pressures.
   d. Check and replace if necessary.

2. Unit Won't Start But Hums Intermittently

   a. Excessive high-side pressure.
   b. Inoperative compressor.
   c. Bad or weak start capacitor.
   d. Wired incorrectly.
   e. Open stator wiring (blows fuse).
   f. Low line voltage.
   g. Start winding open or shorted.
   h. Relay contacts not closing.

   a. Eliminate cause or wait for pressure to drop. Check for closed valves.
   b. Check and replace if necessary.
   c. Replace capacitor.
   d. Check wiring diagram, rewire.
   e. Check leads. If OK, replace compressor.
   f. Check for voltage drop.
   g. Check leads. If OK, replace compressor.
   h. Operate manually. Replace relay if defective.

3. Compressor Starts But Start Winding Does Not Disengage

   a. Start or run winding bad.
   b. Run capacitor inoperative.
   c. Compressor binding.
   d. Excessive high-side pressure.
   e. Bad start capacitor.
   f. Low line voltage.
   g. Inoperative relay.
   h. Improper wiring.

   a. Check resistance. Replace compressor if start or run winding faulty.
   b. Check and replace if necessary.
   c. Check for worn bearings, low oil level. Replace faulty compressor.
   d. Check and correct as necessary.
   e. Check and replace if necessary.
   f. Check for voltage drop.
   g. Check and replace as necessary.
   h. Check against wiring diagram.

4. Compressor Runs But Short-Cycles

   a. Low line voltage.
   b. Excessive high-side pressure.
   c. Excessive low-side pressure.
   d. Weak overload protector.
   e. Defective run capacitor.
   f. Defective discharge valve.
   g. Compressor binding.
   h. Insufficient motor cooling.

   a. Check for voltage drop.
   b. Check for restrictions or overcharge.
   c. Check for open TEV, insufficient compressor, system overcharge.
   d. Check and replace as necessary.
   e. Check capacitance, replace if defective.
   f. Check and repair or replace.
   g. Repair or replace compressor.
   h. Check and correct cooling system.

| Problem and Possible Cause | Remedy |
|---|---|
| i. Windings shorted or grounded. | i. Check windings with ohmmeter, replace motor if defective. |
| j. Overload protector wired wrong. | j. Make sure components are not wired to the wrong side of the protector. |
| **5. Run Capacitors Burn Out** | |
| a. Excessive voltage. | a. Check line voltage, should not be over 10% above motor rating. |
| b. Water shorts capacitor terminals. | b. Protect capacitors from moisture. |
| c. Low voltage rating on capacitor. | c. Install capacitors with correct voltage rating. |
| **6. Relays Burn Out** | |
| a. Line voltage too low. | a. Correct line voltage to not less than 10% below motor rating. |
| b. Line voltage too high. | b. Correct voltage to no more than 10% above motor rating. |
| c. Relay vibrates. | c. Check relay operation, mount firmly. |
| d. Improper relay. | d. Install relay recommended for unit. |
| e. Improper run capacitor. | e. Install capacitor with correct mfd rating. |
| f. Unit short-cycles. | f. Correct cause of short-cycling. |
| **7. Start Capacitor Burns Out** | |
| a. Improper capacitor. | a. Install capacitor with recommended voltage and mfd rating. |
| b. Water shorts out terminals. | b. Protect capacitor terminals from moisture, or relocate capacitor. |
| c. Voltage rating too low. | c. Install capacitor with recommended voltage and mfd rating. |
| d. Sticking contacts on relay. | d. Clean contacts or replace. |
| e. Operates too long on start winding. | e. Replace faulty relay, reduce start-up load, increase voltage if too low. |
| f. Unit short-cycles. | f. Correct cause of short-cycling. |
| g. Faulty relay. | g. Replace relay. |

# TROUBLESHOOTING REFRIGERANT FLOW CONTROLS

This section covers refrigerant flow controls, consisting of a troubleshooting guide for the fluid flow problems, troubleshooting the solid-state expansion valve (electric valve), and servicing cooling towers.

The author wishes to express his deep appreciation for the most generous contributions of ALCO Controls, Division of Emerson Electric Company, and Eaton Corporation, Appliance and Specialty Controls Division, without which, the preparation of this portion of the book would not have been possible. The material in this section represents the latest available research data and technology for troubleshooting refrigerant-flow problems, including the electric valve and water-cooled condensing equipment.

## I  CAUSES OF HIGH SUPERHEAT

1. Flash gas
2. Liquid-line restriction
3. Improper piping design
4. Inadequate subcooling
5. Low head pressure
6. Capillary tube or TEV distributor restricted
7. Excessive load on the evaporator above design conditions
8. System contamination
9. Results of using an undersized TEV
10. Using an internally equalized TEV
11. Gas-charged TEV
12. Sensing bulb failure or loss of charge
13. Results of using the wrong thermostatic charge
14. Measuring and adjusting operating superheat
15. Oversized evaporator or undersized compressor
16. Superheat adjustment too high

## II  CAUSES OF LOW SUPERHEAT

1. Overcharge of refrigerant and/or oil
2. Compressor oversized
3. Uneven or inadequate evaporator loading
4. Excessive accumulation of oil in evaporator
5. Poor bulb and equalizer location
6. External equalizer line plugged or crimped
7. Cracked diaphragm or bellows (automatic expansion valves)
8. System contamination
9. Evaporator fan blade on backward, motor running backward
10. Oversized condenser
11. Excessive subcooling
12. Poor distribution through evaporator circuits
13. Faulty TEV or wrong charge in sensing bulb
14. Interrupted pump-down
15. Oversized TEV
16. Cold suction line or compressor location encourages liquid migration to the low side during the off cycle
17. TEV seat leaks, or liquid line solenoid seat leaks, or the compressor discharge valve leaks during the off cycle
18. Excessive coil frosting
19. A long, free-draining suction line to the compressor
20. TEV superheat setting too low

## III  CAUSES OF DISCHARGE PRESSURE PROBLEMS

1. High discharge pressure
2. Low discharge pressure
3. Fluctuating discharge pressure

## IV  CAUSES OF SUCTION PRESSURE PROBLEMS

1. High suction pressure—high superheat (evaporator outlet)
2. High suction pressure—low superheat (evaporator outlet)
3. Low suction pressure—high superheat (evaporator outlet)
4. Low suction pressure—low superheat (evaporator outlet)
5. Fluctuating suction pressure

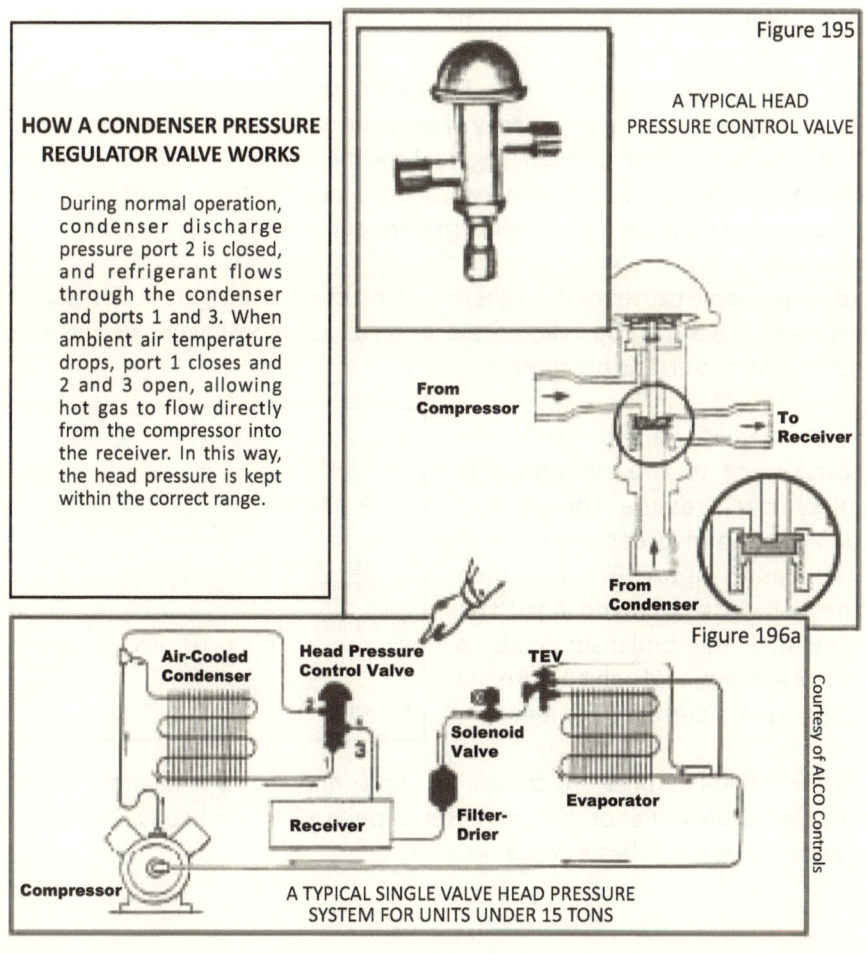

**HOW A CONDENSER PRESSURE REGULATOR VALVE WORKS**

During normal operation, condenser discharge pressure port 2 is closed, and refrigerant flows through the condenser and ports 1 and 3. When ambient air temperature drops, port 1 closes and 2 and 3 open, allowing hot gas to flow directly from the compressor into the receiver. In this way, the head pressure is kept within the correct range.

Figure 195 — A TYPICAL HEAD PRESSURE CONTROL VALVE

Figure 196a — A TYPICAL SINGLE VALVE HEAD PRESSURE SYSTEM FOR UNITS UNDER 15 TONS

Courtesy of ALCO Controls

## V    MISCELLANEOUS PROBLEMS

1. Compressor starts, but motor cycles off on overload protector.
2. Unit starts, but short-cycles.
3. Unit runs continuously—adequately sized, inadequate cooling.
4. Unit will not start.
5. Faulty controls.
6. High amperage draw.
7. Loss of oil, loss of oil pressure, or oil-pressure control trip-out.

## CAUSES OF HIGH SUPERHEAT

### I-1.    Flash Gas and High Superheat

That portion of the liquid refrigerant which evaporates instantly (flashes) and turns into a vapor as it passes through the orifice of a refrigerant control is referred to as "flash gas."

Since refrigerant controls (TEVs, AEVs) are designed to control liquid and not gas, flash gas reduces the efficiency of the control. Flash gas at the inlet of the TEV obstructs the flow of liquid refrigerant and therefore starves the evaporator. If flash gas enters the TEV, the valve will operate inefficiently, causing the system to lose some of its cooling efficiency and the superheat to increase.

Three probable causes of flash gas are (a) insufficient refrigerant charge, (b) excessive pressure drop in liquid line, or (c) a lack of positive head pressure control causes erratic head pressures.

The remedy for a and b is to add refrigerant to the system; this raises head pressure. For c, install positive head pressure control valve as in figure 97d in which the valve stays closed until the pressure in the condenser builds up to a predetermined point before the refrigerant is permitted to leave.

> If there is no sight glass in the liquid line at the TEV inlet,
>
> (a) Flash gas can sometimes be detected by a characteristic steady whistling sound at the TEV.
> b) Flash gas can also be detected by installing a pressure tap at or near the TEV inlet.
>
> Take a pressure and temperature reading at this location. If the temperature measured is above its saturation temperature (taken from the pressure/temperature chart), then flash gas is present.

Or the liquid-line pressure control can be connected to a solenoid valves installed at the outlet of the condenser to open when the head pressure (in the condenser) reaches a desirable level.

## I-2. Liquid Line Restriction and High Superheat

A probable cause is a partially plugged liquid-line filter-drier. If the temperature at its inlet is higher than at its outlet, the drier is restricted. Install a new filter-drier as previously outlined.

An effective way of eliminating moisture from the system is to properly vacuum the system after installing a new filter-drier before recharging. The minimum pump running time increases as the size of the unit being serviced increases. Be sure your lines, gauges and hoses are as dry as possible before beginning. Also, keep your refrigerant oil container sealed from the atmosphere at all time. Refrigerant oil has an attraction for moisture. When it is left open, the oil absorbs moisture rapidly.

In systems using capillary tube(s), a strainer is installed between the condenser and the capillary tube on the liquid line (see fig. 142). The purpose of the strainer is to prevent any foreign matter in the sealed system from entering the capillary tube and restricting the flow of refrigerant. Often, a sleeve-shaped screen is installed in the last part of the liquid line just before the capillary tube connection. It is identified by a slightly enlarged portion of the liquid-line tubing. A restricted strainer is easily discovered when it feels cool to the touch. It should feel about 6°F to 10°F above room temperature.

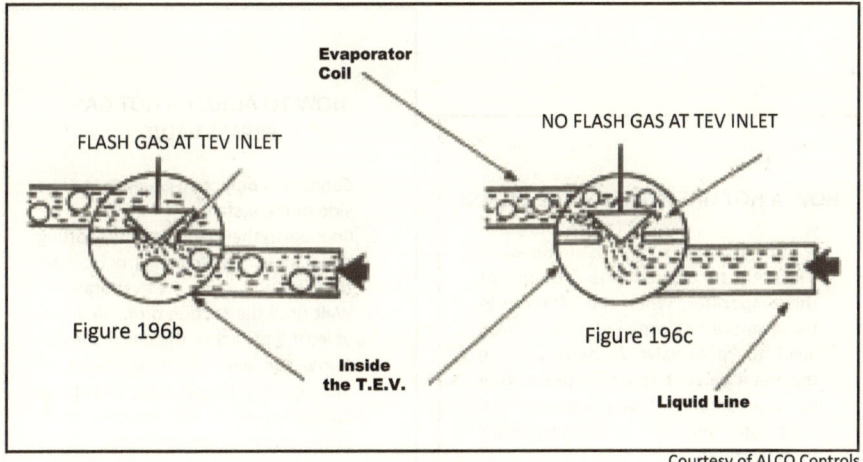

Courtesy of ALCO Controls

A field-installed service valve at the liquid receiver might be too small or not fully open, a pipe joint in the liquid line might be partially filled with solder, or a kink in the liquid line can cause a restriction. Copper tubing is soft and relatively easy to twist, kink, or flatten. The liquid line must be thoroughly inspected and any defect corrected. Also, the connections on the liquid receiver must be as large as the liquid line; be sure there are no reducer fittings or couplings installed that would create a partial restriction in the system. (See the chart on page 365 for the recommended refrigerant line sizes.) Also, replace any solenoid valve not working properly.

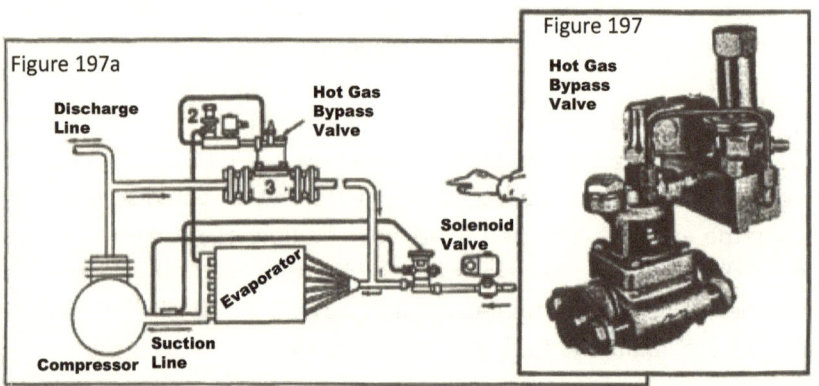

### HOW A HOT GAS BYPASS VALVE WORKS

A hot gas bypass valve is installed in a system to prevent the frosting of the evaporator. When the pressure in the evaporator drops below a desired level, the pilot valve 2 opens, causing the main valve 3 to open, permitting sufficient amount of hot gas to enter the evaporator and prevent the evaporator from frosting.

### HOW TO ADJUST A HOT GAS BYPASS VALVE

1. Connect a compound gauge to the low side of the system.
2. Cool down the evaporator by shutting down the evaporator fan(s), or block off the airflow through the evaporator.
3. Wait until the suction drops down to at least 5 psi below the desired level.
4. Allow the evaporator pressure to increase by the bypass gas. The spring load can be varied until the valve closes at the exact desired pressure.

### I-3. Improper Piping Design and High Superheat

Excessive vertical lift of the liquid line could be a probable cause. Generally, for every foot of vertical lift using R-22, there is approximately 1/2 psi drop. Flash gas forms in a system that has excessive vertical lift. Make sure before the refrigerant flows up the riser, it is subcooled by going through a heat exchanger, enough to prevent it from changing to gas by the time it reaches to the top of the riser where pressure is decreased. Usually, subcooling 10°F is sufficient for elevations up to twenty-five feet. Be sure to check the manufacturer's data.

Also, the liquid line may be too long, too small, or have too many fittings. All three cause excessive liquid-line pressure drop. Replace piping sections with the correct line size.

### I-4. Inadequate Subcooling and High Superheat

If the system is designed to provide a certain degree of subcooling to compensate for system pressure loses, and the liquid refrigerant is inadequately subscooled, flash gas at the TEV will occur. You will need to increase the subcooling for proper system performance. A heat exchanger with an increased cooling capacity or an increase in condenser fan speed may be needed. Clean linted condenser fins. (See fig. 198.)

### I-5. Low Head Pressure and High Superheat

All air-cooled condensing units are sized for a specific ambient air temperature (usually 90°F-95°F). When ambient temperature drops below 50°F, the condenser becomes oversized, and some type of head pressure control is needed. A variable-speed fan is no longer effective due to low ambient temperatures and/or prevailing winds. You will need to install a flooding-type head pressure control such as an electric, solenoid, or AEV, and base the fan cycling on the ambient temperature with a pressure override. (A pressure control connected to the high

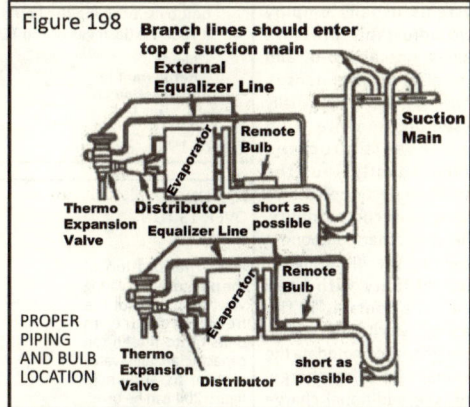

10°F of subcooling is enough for up to twenty-five feet of elevation.

Figure 198  Proper piping and bulb location.

side set to cut off power to a solenoid valve at the condenser outlet, or the condenser fan, until pressure in the condenser rises to a preset level.)

### I-6. Restricted Capillary Tube or TEV Distributor and High Superheat

Two probable causes of high superheat are a restriction of the liquid refrigerant flow into the evaporator due to contamination in the system or the use of an undersized valve or capillary tube. If the correct size valve or capillary tube is installed, replace the filter-driers.

### I-7. Excessive Load on Evaporator above Design Conditions and High Superheat

A TEV is restricted to the amount of flow it can pass by the size of its orifice. Therefore, if the system is required to absorb more heat than it is designed to do (an excessive load), high superheat will occur. If this happens, a balance-ported TEV sized for the maximum load conditions must be installed.

### I-8. System Contamination and High Superheat

The probable causes of high superheat are moisture in the system caused by condensation from moist air penetrating the system by using internally wet charging hoses or gauges, or a plugged filter-drier causing excessive pressure drop, resulting in flash gas.

Generally, only a single valve is needed for head pressure control in systems up to about fifteen-ton capacity because the factory-assembled units have components sized and preset for those specific systems. Larger systems usually employ two adjustable pressure valves, one ahead of and one after the condenser. Those systems are normally assembled on-site from components procured independently, thus the need for adjusting pressures for system compatibility. The only other component requiring specific sizing for compatibility with head pressure controls is the receiver, which must be large enough to accommodate the normal operating charge, plus the additional charge that would be necessary to totally flood the condenser.

Installation of a heat exchanger in the liquid and suction line provides for a heat transfer from the warm liquid line to the cool vapor leaving the evaporator, causing the following:

1. It helps reduce frost on the suction line.
2. It reduces the amount of liquid refrigerant in the suction line.
3. It increases the operating efficiency of the unit by subcooling the liquid refrigerant.
4. It reduces flash gas at the TEV.

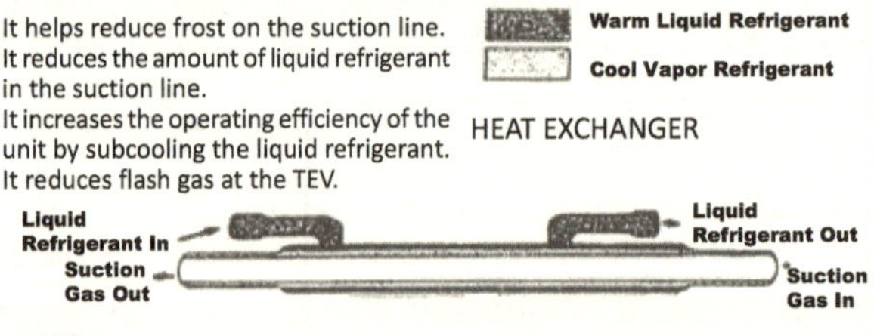

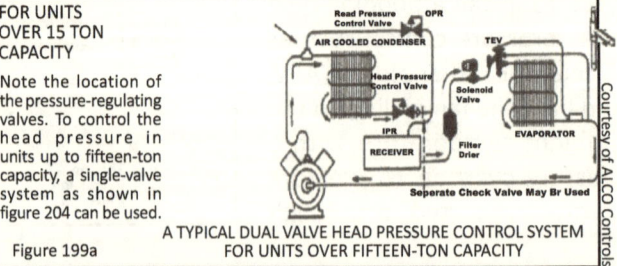

Figure 199

FOR UNITS OVER 15 TON CAPACITY

Note the location of the pressure-regulating valves. To control the head pressure in units up to fifteen-ton capacity, a single-valve system as shown in figure 204 can be used.

Figure 199a

A TYPICAL DUAL VALVE HEAD PRESSURE CONTROL SYSTEM FOR UNITS OVER FIFTEEN-TON CAPACITY

Another cause of restriction is the formation of wax in the sealed system restricting the flow of refrigerant. This may be indicated by a sudden rise in suction pressure when the system warms up after a shutdown. To check while the system is running, use a heat gun to warm the TEV (to melt the ice) and note the rise in the suction pressure. The heat gun usually does not melt the wax, making wax difficult to discover. Wax in the system may indicate that the wrong type of oil is being used.

Waxing can occur in low-temperature units such as those with evaporator temperatures of -25°F. Wax usually liquefies and flows again near 0°F or higher.

**RECOMMENDED SIZES FOR REFRIGERANT LINES**

| Btu/h | LENGTH OF RUN | | | | |
|---|---|---|---|---|---|
| | 15 ft OD Inches | 25 ft OD Inches | 36 ft OD Inches | 50 ft OD Inches | 100 ft OD Inches |
| | Suction Line | | | | |
| 18,500-22,000 | 5/8 | 5/8 | 5/8 | | |
| 22,000-24,000 | 5/8 | 5/8 | 5/8 | 3/4 | 3/4 |
| 24,000-34,000 | 5/8 | 5/8 | 3/4 | 3/4 | 3/4 |
| 38,000-40,000 | 3/4 | 3/4 | 7/8 | 7/8 | 7/8 |
| 40,000-44,000 | 3/4 | 7/8 | 7/8 | 7/8 | 7/8 |
| 44,000-51,000 | 7/8 | 7/8 | 7/8 | 7/8 | 7/8 |
| 53,000-66,000 | 7/8 | 7/8 | 7/8 | 1 1/8 | 1 1/8 |
| | Liquid Line | | | | |
| 18,500-22,000 | 5/16 | 5/16 | 5/16 | | |
| 22,000-40,000 | 5/16 | 3/8 | 3/8 | 3/8 | 3/8 |
| 40,000-51,000 | 3/8 | 3/8 | 3/8 | 3/8 | 3/8 |
| 53,000-66,000 | 1/2 | 1/2 | 1/2 | 1/2 | 1/2 |
| | Discharge Line | | | | |
| 18,500-20,000 | 5/16 | 3/8 | 3/8 | | |
| 20,000-22,000 | 3/8 | 3/8 | 3/8 | | |
| 22,000-24,000 | 3/8 | 3/8 | 3/8 | 1/2 | 1/2 |
| 24,000-34,000 | 3/8 | 3/8 | 1/2 | 1/2 | 1/2 |
| 38,000-44,000 | 3/8 | 1/2 | 1/2 | 1/2 | 1/2 |
| 44,000-51,000 | 3/8 | 1/2 | 1/2 | 1/2 | 5/8 |
| 53,000-66,000 | 1/2 | 1/2 | 5/8 | 5/8 | 3/4 |

The above figures assume standard refrigeration tubing with a wall thickness of 0.028 inches or 0.032 inches, and do not include any consideration for additional pressure drops due to reduced joint sizes, elbows, or valves.

For each ten feet of tubing over thirty-five feet, it is necessary to add three fluid ounces of refrigerant.

**ACCEPTABLE LIQUID RECEIVER VOLUMES**

Refrigerant Weight (lb)

| HP | Volume (in³) | R-12 | R-22 | R-500 | R-502 |
|---|---|---|---|---|---|
| 1/2 | 150 | 6.8 | 6.2 | 5.9 | 6.3 |
| 3/4 | 225 | 10.3 | 9.3 | 8.9 | 9.4 |
| 1 | 300 | 13.7 | 12.4 | 11.9 | 12.9 |
| 1 ½ | 450 | 20.5 | 18.6 | 17.9 | 19.3 |
| 2 | 600 | 27.4 | 24.8 | 23.8 | 25.8 |
| 3 | 750 | 35.0 | 32.0 | 31.8 | 33.0 |
| 5 | 900 | 41.0 | 37.0 | 35.5 | 38.5 |
| 7 ½ | 1500 | 70.0 | 64.0 | 61.6 | 66.0 |

If there are too many restrictions in the liquid line, or if the liquid line is too small, pressure drop will cause insufficient liquid refrigerant flow through the refrigerant control, reducing the cooling capacity of the unit. As a general rule, it is a good practice to choose liquid-line tubing about 15% oversized.

Refer to pages 285 to 287 for correct piping method.

*Courtesy of Tecumseh Product Co.*

Figure 199b

**CONTROLLING THE HEAD PRESSURE BY CONTROLLING THE CONDENSER FAN SPEED**

The high-side pressure can also be regulated by controlling the condenser fan operation. When the fan operates, air is blown through the condenser; the condenser is cooled causing the head pressure to decrease. Fans with variable speeds are also used to maintain the head pressure at a desired level by the amount of air blown through the condenser fins. The more the air velocity is increased (more rpm), the more the head pressure drops.

The condenser fan is wired to the compressor run and common terminals through a pressure-control switch. The pressure control is connected to the compressor discharge side. The switch within the control closes the electrical circuit to the fan when the head pressure, transmitted to the pressure control through an access valve connection, rises to a predetermined point. Variations in fan speed may be obtained by using a solid-state control which governs the voltage to the fan motor. The more voltage, the higher the fan speed and vice versa.

The use of positive head pressure-regulating valves becomes necessary where the condenser is exposed to severely low temperatures.

Waxing can be remedied by evacuating and recharging the system with clean, dry refrigerant and the proper refrigerant oil recommended by the manufacturers suitable for the evaporator temperature requirements of the unit.

### I-9. Undersized TEV and High Superheat

The valve orifice is too small, starving the evaporator. Use the following information to properly size the TEV:

a. Refrigerant type.
b. Evaporating temperature.
c. Pressure drop across the valve.
d. Desired load.
e. Liquid temperature at the TEV inlet.

Figure 199c

A SOLID-STATE FAN CONTROL

### I-10. Internally Equalized TEV and High Superheat

An internally equalized TEV must not be used on evaporators through which too much pressure drop occurs. (Check the rule of thumb for selecting an externally equalized TEV.) Use an externally equalized TEV and make sure that the external equalizer line is connected properly. (See section II-5, concerning poor location for bulb and equalizer in this section.)

Generally, internally equalized valves are neither used in systems using a refrigerant distributor nor in units above a two-ton capacity.

### I-11. Gas-Charged TEV and High Superheat

If the gas-charged TEV loses control, the valve head and tubing must be kept warmer than the remote valve. Heat tape can be used to wrap the valve-sensing bulb and tubing. The valve body can be insulated or installed outside the refrigerated space.

### I-12. Sensing Bulb Failure or Loss of Charge and High Superheat

If this occurs due to a puncture or a sharp bend, replace the sensing bulb assembly or replace the TEV.

### I-13. Wrong Thermostatic Charge and High Superheat

The TEV sensing element is charged with liquid refrigerant. Some sensing elements are charged with the same type of refrigerant being used in the unit, and some use a different type or a mixture of different fluids to provide the desired operating results. If the TEV installed uses, or is suspected of using the wrong type of charge, check the specifications of the valve and compare them with the operating specifications of the unit and select the proper charge based on the system refrigerant and the desired operating performance.

## I-14. Measuring and Adjusting Operating Superheat

### Measuring

1. Determine the suction pressure or saturation temperature at the evaporator outlet with an accurate gauge.
2. Determine the saturation temperature at the observed suction pressure using the temperature-pressure chart.
3. Measure the temperature of the suction gas at the remote bulb location as follows:

    a. Clean area of the suction line at the bulb location and tape the thermocouple on the cleaned area.
    b. Insulate the thermocouple and read the temperature with an electronic thermometer.

4. Subtract the saturation temperature determined in step 2 from the sensible temperature measured in step 3. The difference is the superheat of the suction gas at the evaporator outlet.

### Adjusting

1. Requires operational head pressure and a solid column of liquid refrigerant at the TEV inlet.
2. Refrigerated space must be under a full-load condition.
3. Evaporator pressure regulators (EPRs) must be in a fully open position.
4. On a multicompressor system, extra precautions are required: ensure a constant suction pressure, a constant discharge pressure, and a constant receiver outlet pressure are all at proper levels. Control compressor cycling to match design suction pressures by isolating the automatic/electronic low-pressure controls. Switch off the necessary compressors to maintain the design suction pressures. Isolate the defrost control, especially on gas-defrost systems to maintain discharge/receiver pressure, and isolate the heat-reclaim valve (fig. 98e) to ensure it doesn't operate during superheat adjustments.
5. Set TEV superheat. Remove the seal cap from the valve to expose the adjusting stem. Rotate the stem clockwise to decrease the refrigerant

flow through the valve and increase superheat. Rotate the stem counterclockwise to increase refrigerant flow and decrease superheat.
6. Recheck superheat under low-load condition.
7. Readjust EPR pressure settings.
8. Remove all evaporator/condenser false loads and reset all isolated controls.

**I-15. Oversized Evaporator or Undersized Compressor and High Superheat**

Resize evaporator or compressor to match load requirements. Compare the information on the nameplate with the compressor, condenser, and valve specifications.

**I-16. Superheat Adjustment Too High**

Reduce superheat setting.

## CAUSES OF LOW SUPERHEAT
### (Evaporator Outlet)

**II-1. Overcharge of Refrigerant and/or Oil**

1. Refrigerant added beyond the proper amount of charge reduces the cooling capacity of the unit causing the evaporator temperature to rise (on systems without a receiver). An overcharged system is far more likely to damage the compressor than one that is undercharged. This overcharge of refrigerant is returned to the compressor crankcase as a constant floodback during operation, reducing compressor life and the cooling capacity of the unit.

Charge the unit to the proper level, for a capillary tube system, charge by the superheat determined on charts available from valve manufacturers. For TEV systems without a liquid receiver, charge by subcooling the liquid refrigerant to an optimum of 10°F less than condensing temperatures (at full load, if possible). For systems using TEVs and receivers, charge by sight glass located at the TEV inlet. (Charge until the bubbles in the sight glass disappear.)

2. Oil overcharging should be avoided as this creates the possibility of oil slugs, which can damage the compressor, and it also hinders the performance of refrigerant in the evaporator. If there is excess oil in circulation, the evaporation rate of the refrigerant is slowed down because oil acts as an insulator.

Remove the oil and maintain levels according to the manufacturers' recommendations.

### II-2. Oversized Compressor and High Superheat

When an oversized compressor is used, extra suction power is applied to the system creating a drop in pressure in the evaporator. The direct drive motor compressor either runs too fast or it has a wrong-size pulley, creating lower-than-normal suction pressures.

Reselect the compressor, replace the direct drive motor for proper speed, or reduce the compressor speed by installing the proper-size pulleys. If the evaporator design capacity is less than the actual load, match the evaporator to load requirements.

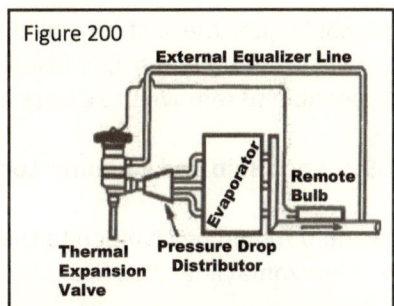

Figure 200

### II-3. Uneven or Inadequate Evaporator Loading and Low Superheat

This is caused by poor air circulation through the evaporator coil as one or more of the evaporator fan motors may be inoperative. You will need to replace any inoperative fans, increase the evaporator fan speed

> Since the TEV function is to control superheat at point C, it only makes sense that we obtain the variable opening and closing forces from that location. This can be achieved by using an externally equalized valve (see fig. 201). The externally equalized TEV senses the true closing pressure P2 at the outlet of the coil instead of pressure A at the inlet. Now pressure drop through the coil no longer affects the operation of the TEV.

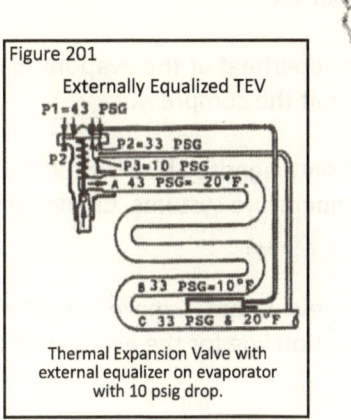

Figure 201
Externally Equalized TEV
Thermal Expansion Valve with external equalizer on evaporator with 10 psig drop.

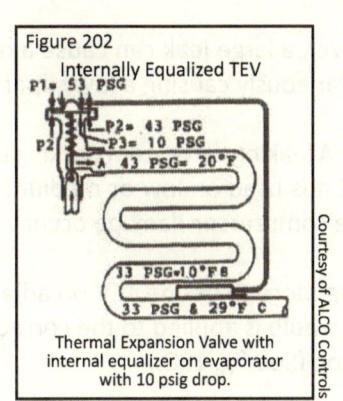

Figure 202
Internally Equalized TEV
Thermal Expansion Valve with internal equalizer on evaporator with 10 psig drop.

by replacing them with ones with higher rpm, or provide for proper air distribution (air circulation over the entire evaporator coil without any restriction in the airflow).

### II-4. Excessive Accumulation of Oil in Evaporator and Low Superheat

Too much oil in the evaporator slows the liquid refrigerant evaporation rate because oil acts as an insulator. (See sec. II-1.)

You could alter the suction piping to increase liquid refrigerant velocity to provide proper oil return, or install an oil separator, if required. It may also be possible to remove the excess amount of oil from the crankcase.

### II-5. Poor Bulb and Equalizer Location and Low Superheat

The bulb should be clamped to the suction line near the evaporator outlet on a horizontal line. (See figs. 110e and 203.)

Clean a place on a free-draining suction line before clamping the bulb in place. Insulate the remote bulb from the ambient air.

Locate the equalizer line as close to the bulb as possible on the end away from the evaporator (on the downstream side), as shown in figure 201. If the equalizer line is installed upstream of the bulb and the packing leaks, the refrigerant passing through the equalizer line will keep the sensing bulb artificially cold, thus forcing the TEV to remain closed, i.e., high superheat. If the above occurs with the equalizer line piped downstream, refrigeration will still continue.

A small leak has very little effect on superheat.

However, a large leak can cause a high superheat at the evaporator, while simultaneously causing a low superheat at the compressor.

NOTE: A leaking valve stem packing is easily diagnosed by a frosted equalizer line, if it is used on low or medium temperature systems. Change the TEV before compressor damage occurs.

If using more than one TEV on adjacent evaporators, make sure that each remote bulb is applied to the correct suction line for the evaporator it is to monitor. (See fig. 198.)

## II-6. External Equalizer Line Plugged or Crimped and Low Superheat

The external equalizer line carries the evaporator outlet pressure to the underside of the TEV diaphragm to close the valve. If that line is plugged or capped, the valve will most likely be wide open causing a flooding condition. In rare occasions, internal valve leakage could cause a constant pressure on the underside of the diaphragm and consequently create high superheat.

If the external equalizer is plugged (kinked) or capped, repair or replace it to provide an unobstructed pressure flow.

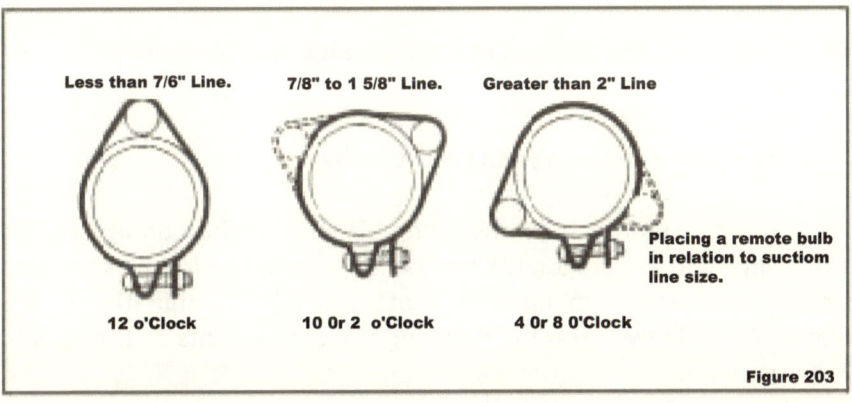

Figure 203

## II-7. Cracked Diaphragm or Bellows (Automatic Expansion Valves) and Low Superheat

A cracked diaphragm or bellows can cause floodback (valve stays open) if the adjustment cap is on and secure. (Replace the valve.)

## II-8. System Contamination and Low Superheat (See sec. I-8.)

Two probable causes of low superheat are (1) the pin and seat of the expansion valve wire eroded or held open by foreign material (resulting in liquid floodback) or (2) moisture freezing the valve pin in an open position. (If suspected, inspect the moisture indicator [sight glass] for moisture contamination reading.)

Heat the valve with a heat gun to melt the ice. Clean or replace any damaged parts in the AEV, or replace the entire valve. Install a new filter-drier.

## II-9. Evaporator Fan Blades on Backward (or Motor Running Backward) and Low Superheat

Install fan blades correctly and/or check motor rotation.

## II-10. Oversized Condenser and Low Superheat

An oversized condenser on a system without a receiver can produce excessive subcooling of the liquid refrigerant entering the valve, leading to a higher refrigerating effect, causing the TEV to become oversized (assuming the subcooling effect wasn't considered in the initial valve sizing) causing low superheat or flooding of the evaporator.

To remedy this condition, correctly balance the components in the system.

## II-11. Excessive Subcooling and Low Superheat

Probable causes are (1) subcooling circuits in the condenser, (2) liquid-line heat exchanger, (3) mechanical subcooling (such as through valves, fans, water-cooling towers, etc.), or (4) a combination of the above. Remedy by correctly matching the components to the system requirements.

## II-12. Poor Distribution through Evaporator Circuits and Low Superheat

This can be caused by incorrect sizing or distributor and/or valve orifice. The distributor tubes must be of equal dimension and length (see fig. 200). Liquid can short-circuit through unequally loaded passages and throttle the valve (cause it to react prematurely) before all passages have received sufficient refrigerant. Liquid traps should be avoided when mounting the distributor tubes.

NOTE: At proper load, there should be no more than 5°F difference between any two circuits before they enter the header.

## II-13. Faulty TEV or Wrong Charge in Sensing Bulb and Low Superheat

Depending on the type of charge, the TEV could starve or flood. Replace the valve with the proper TEV and correct charge in accordance with the evaporator requirements. Also see section II-5 on poor bulb and equalizer location.

## II-14. Interrupted Pumpdown and Low Superheat

An interrupted pumpdown leaves refrigerant in the low side in the off cycle, causing possible flooding when the system restarts. Find the cause of the interruption and adjust the control to a lower suction pressure if necessary.

## II-15. Oversized TEV and Low Superheat

An oversized TEV results in flooding to the suction line and low or negative superheat. You need to replace the TEV with one correctly sized per the information in Section I-9.

## II-16. Cold Suction Line or Compressor Location Encourages Liquid Migration to the Low Side during the Off Cycle, Causing Low Superheat When the System Starts

The remedy is to insulate the suction line and/or equip the compressor with a crankcase heater.

## III-17. TEV Seat Leaks, Liquid Line Solenoid Leaks, or the Compressor Discharge Valve Leaks during the Off Cycle, Causing Low Superheat When the System Starts

All of the above cause the liquid to make its way to the low side of the system during the off cycle. The low side fills up with liquid refrigerant and as soon as the system starts up again, the liquid floods the compressor, causing low superheat.

*Remedy*: You will need to install a heating element around the compressor crankcase.

## II-18. Excessive Coil Frosting

| Probable Causes | Remedy |
|---|---|
| a. Blocked evaporator coil (debris in the evaporator fins). | a. Clear and clean evaporator. |
| b. Incorrect fan rotation. | b. See section II-9. |
| c. Excessive room or relative | c. Provide sufficient dehumidification equipment, relocate the unit, or check usage. |
| d. Case temperature too low. | d. Adjust thermostat or regulator. |
| e. Wrong evaporator pressure regulator setting (set below 28°F saturation). | e. Repair or reset EPR setting |
| f. Faulty hot gas defrost solenoid. | f. Repair or replace the hot gas defrost solenoid. |
| g. Not enough defrost time. | g. Increase defrost time. |
| h. Evaporator fan speed set low. | h. Increase evaporator fan speed. |
| i. Inadequate automatic defrost. | i. If there is no defrost cycle and room or space temperature is 36°F to 40°F, chances are that the saturated suction temperature of the coil is 30°F or less. Air defrost should be utilized. Install defrost timers. For a room or case below 36°F, the saturated suction temperature is approximately 26°F or lower; therefore, rapid coil frosting occurs. A positive defrost system is required, either gas or electric. |

II-19. A Long, Free-Draining Suction Line to the Compressor and Low Superheat

This may lead to or add to flooding complications.

Follow good piping practice, use an accumulator.

II-20. TEV Superheat Setting Too Low

Perhaps the proper superheat instruments were not utilized for the initial superheat seating.

You will need to adjust the superheat to the desired level and take a superheat measurement using the pressure/temperature method (see sec. I-14).

III-1. High Discharge Pressure

| Probable Cause | Remedy |
|---|---|
| a. Air or debris in condenser. | a. Purge the noncondensibles. |
| b. Dirty condenser. | b. Clean the condenser. |
| c. Poor or no water flow.* | c. Check water supply, pump, and strainers. |
| d. Water supply to condenser too warm.* | d. Check, repair, adjust, condenser water-cooling facilities. |
| e. Condenser airflow restricted. | e. Clean the condenser. |
| f. Wrong airflow direction through condenser. | f. Correct airflow direction. |
| g. Wrong size condenser fan motor and/or blade. | g. Use right size components. |
| h. Condenser fan motor not running or wrong rotation. | h. Find problem and correct. |

| Probable Cause | Remedy |
|---|---|
| i. Liquid refrigerant backed up in condenser. | i. Check for faulty flooded-type head pressure control valves. |
| j. Refrigerant overcharge. | j. Charge properly. (Sec. II-1.) |
| k. Receiver too small for flooded head pressure control system in summer operation. | k. Replace with correct-size liquid receiver. |
| l. Fan pressure switch out of calibration. | l. Reset or replace as needed. |
| m. Higher rate of infiltration of ambient air inside cabinet than designed. | m. Correct air infiltration. (Panel missing or exhaust fans from elsewhere discharging under or close to condenser air intake). |
| n. Condenser air recirculation. | n. Change airflow direction. |
| o. Condenser tubing restricted internally or by physical damage. | o. Replace condenser. |
| p. Ambient conditions exceed design limitations. | p. Reevaluate suitability of unit for the application. |

Failure of Flooded Head Pressure Control System (Constant Bypass)

| Probable Cause | Remedy |
|---|---|
| a. Pressure drop through condenser exceeds 20 psi, forcing bypass port partially open. | a. Change condenser to reduce pressure drop below 20 psi |
| b. Bypass port wedged open due to foreign material between seat and seat disc. | b. Artificially reduce head pressure and tap valve body with system running to dislodge foreign material. |
| c. Bypass port seat damaged or worn. | c. Change valve. |
| d. Wrong pressure dome in valve for system refrigerant. | d. Change valve to match system refrigerant and desired pressure. |
| e. Bypass check valve in backward. | e. Install in proper flow direction. |
| f. Pressure of condenser bypass regulator (OPR) set too high. | f. Adjust condenser bypass regulator setting to the appropriate level. Normally set 20 psi below the condenser pressure regulator (IPR) setting. |

\* Applies to water-cooled condensers.

## FLOODED HEAD PRESSURE CONTROL FOR AIR-COOLED CONDENSER SYSTEMS

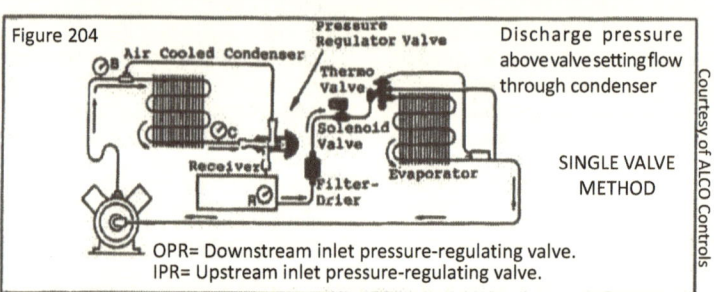

Figure 204

Discharge pressure above valve setting flow through condenser

SINGLE VALVE METHOD

OPR= Downstream inlet pressure-regulating valve.
IPR= Upstream inlet pressure-regulating valve.

*Courtesy of ALCO Controls*

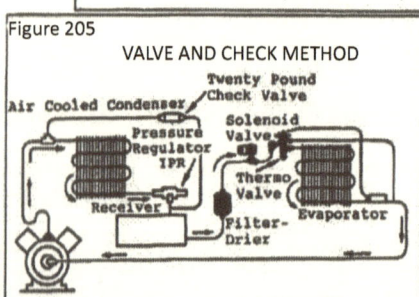

Figure 205
VALVE AND CHECK METHOD

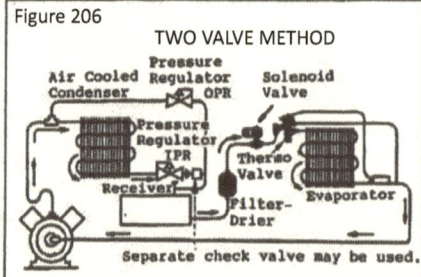

Figure 206
TWO VALVE METHOD

Separate check valve may be used.

III-2.  Low Discharge Pressure

**Probable Cause**

a. Ambient air too cold.
b. Water supply to condenser too cold.*
c. Water quantity not being regulated properly through the condenser.*
d. Refrigerant level low. (Winter charge lacking.)
e. Condenser fan and water switches improperly set.*
f. Fan cycling/variable speed not operating.
g. Uninsulated receiver in cold ambient acting as condenser.

**Remedy**

a. Install head pressure control system.
b. Check, repair, and adjust condenser water-cooling facilities.
c. Install or repair water-regulating valve.
d. Add necessary refrigerant per proper charging method.
e. Reset condenser controls.
f. Check motor and pressure setting.
g. Insulate and/or heat receiver.

Failure of Flooded Head Pressure Control System (No Bypass)

a. Foreign material wedged between condenser port and seat.
b. Wrong pressure dome on valve for system.
c. Hot gas bypass line restricted or shut off.
d. Sensing bulb lost its charge.
e. Oversized condenser or condensing unit.
f. Pressure of condenser bypass regulator (OPR) set too low.

a. Artificially raise head pressure and tap the valve body to dislodge foreign material.
b. Change dome or valve.
c. Clear obstruction.
d. Change the valve.
e. Replace with properly sized condenser.
f. Adjust condenser bypass regulator setting to appropriate level. Normally set 20 psi below condenser pressure regulator (IPR) setting.

III-3. Fluctuating Discharge Pressure.

**Probable Cause**

a. Faulty condensing water-regulating valve.*
b. Insufficient charge, usually accompanied by corresponding fluctuation in suction pressure.
c. Inadequate and fluctuating supply of cooling water to condenser.*
d. Cooling fan for condenser cycling.
e. Fluctuating discharge pressure controls on low-ambient, air-cooled condenser.
f. Fan cycling caused by pressure switches.

   * Applies to water-cooled condensers.

**Remedy**

a. Replace condensing water-regulating valve.
b. Add charge to the system.
c. Check water-regulating valve and repair or replace if defective. Check water circuit for restrictions.
d. Find cause and correct.
e. Adjust, repair, or replace controls.
f. Normal for fan-cycling operation. To eliminate, base fan cycling on ambient air temperature. Use variable speed fans or flooded-type head pressure control system (but not both together).

## CAUSES OF SUCTION PRESSURE PROBLEMS

IV-1. High Suction Pressure—High Superheat (Evaporator Outlet)

**Probable Cause**

a. Unbalanced system, load in excess of design conditions.
b. Compressor discharge valve leaking.
c. Leaking hot gas defrost solenoid / hot gas bypass valve.
d. Hot gas bypass regulator piped direct to suction, without a liquid injection TEV.

**Remedy**

a. Balance system components for the appropriate load requirements if necessary.
b. Check valve per page 47. Replace valve if necessary.
c. Check and replace hot gas bypass valves if required.
d. Install properly sized liquid injection thermo valve.

IV-2. High Suction Pressure—Low Superheat (Evaporator Outlet)

**Probable Cause**

a. Oversized expansion valve.
b. Pin and seat of expansion valve drawn, eroded, or held open by foreign material resulting in liquid floodback.
c. Ruptured diaphragm in an AEV resulting in liquid floodback.
d. External equalizer line plugged or equalizer connection capped without providing a new valve body with an internal equalizer.
e. Moisture-freezing valve in open position.

f. Leaking valve stem packing.
g. Valve superheat setting too low.

h. Leaking liquid line in suction heat exchanger.

i. Unit coming out of defrost.

**Remedy**

a. Replace with correct-size expansion valve.
b. Replace valve. Install filter-drier to remove foreign material from system.

c. Replace valve-sensing assembly.

d. If equalizer is plugged, repair or replace. Otherwise, replace with a valve having the correct equalizer.

e. Apply heat gun to valve to melt ice and install a new filter-drier to ensure a moisture-free system.
f. Replace TEV.
g. Increase superheat setting (turn valve stem clockwise).
h. Liquid to suction heat exchanger defective. (May not be needed; review requirements.)
i. Normal. Let the system balance. See section I-14 on measuring and adjusting superheat.

IV-3. Low Suction Pressure—High Superheat (Evaporator Outlet)

The probable cause is a starving evaporator. See section I on high superheat.

IV-4. Low Suction Pressure—Low Superheat.          (Evaporator Outlet)

**Probable Cause**

a. Light load condition.

b. Poor distribution through evaporator causing liquid to short-circuit through favored passes. At full load, there should be no more than 5°F difference in superheat between any two circuits as they enter the header.
c. Compressor oversized/evaporator undersized.
d. Uneven or inadequate evaporator loading due to poor air distribution.
e. Dirty evaporator filters.
f. Coil icing.
g. Frozen or slushed chiller.
h. Low water flow through chiller.*
i. Excessive accumulation of oil in evaporator.

**Remedy**

a. Shut off some compressors, install hot gas bypass. Slow down compressor rpm. Check process flows.

b. Clamp TEV sensing bulb to free draining suction line. Clean suction line thoroughly before clamping bulb in place. Install proper-size refrigerant distributor. Balance evaporator load distribution (i.e. check airflow over entire coil surface.)

c. Balance the components to load requirements.
d. Balance evaporator load distribution by providing adequate air distribution.
e. Clean evaporator filters.
f. Check, reset, or replace defrost controls.
g. Check control valves for proper setting.
h. Clean strainers, balance water flow, check pump.
i. Alter suction piping to increase gas velocity to provide proper oil return, or install an oil separator if required.

IV-5. Fluctuating Suction Pressure

**Probable Cause**

a. Incorrect superheat adjustment.
b. Improper remote bulb location or installation.

c. Floodback of liquid refrigerant caused by poorly designed liquid distribution device, or uneven evaporator loading.

d. External equalizer lines tapped at a common point although there is more than one expansion valve on the same system.

e. Faulty condensing-water regulator.*
f. Restricted external equalizer line.
g. Condenser fan cycling based on power switch settings.


h. Oversized TEV.

**Remedy**

a. See section I-14.
b. Clamp remote bulb to free draining suction line. Clean suction thoroughly before installing bulb. (See sec. II-5.)

c. Replace faulty distributor (must be properly sized). If evaporator loading is uneven, install proper load distribution fans to balance air evenly over evaporator coil.

d. Each valve must have its own separate equalizer line going directly to its own separate evaporator outlet to ensure proper operational response of each one. (See illustration in sec. II-5.)

e. Replace condensing-water regulator.
f. Clean blockage or replace equalizer line.
g. Normal for fan-cycling operation. In order to eliminate, base fan cycling on ambient air temperature with pressure override, or use variable fan speed or flooded head pressure control system (but not both together).

h. See section I-9.

| Probable Cause | Remedy |
|---|---|
| i. Faulty or oversized evaporator pressure regulator (EPR) valve. | i. Repair or replace and install EPR valve. |
| j. Compressors cycling in multicompressor system. | j. This is normal. |

*Applies to water-cooled condensers.

## MISCELLANEOUS PROBLEMS

V-1. Compressor Starts, but Motor Cycles Off on Overload Protector

| Probable Cause | Remedy |
|---|---|
| a. Excessive suction pressure. | a. See section IV-1. Unload compressors when starting. Use interval unloaders if present. |
| b. Excessive discharge pressure. | b. See section III-1. Check for mechanical damage, check motor and compressor bearings for temperature, lubricate motor bearings. |
| c. Tight bearings or mechanical damage in the compressor. | c. Determine reason and correct. |
| d. Low line voltage. | d. Find source and correct. |
| e. Improperly wired. | e. Rewire correctly. |
| f. Defective run capacitor or relays. | f. Replace faulty equipment. |
| g. Defective overload protector. | g. Replace overload protector, if internal, replace compressor. |
| h. Shorted or grounded motor winding. | h. Replace compressor and install new filter-drier(s) as previously instructed. |

V-2. Unit Starts, but Short-Cycles

| Probable Cause | Remedy |
|---|---|
| a. High-pressure lock-out.<br>  1. Dirty condenser.<br>  2. Excessive refrigerant charge on nonreceiver system.<br>  3. High suction pressure. (See sec. IV-1.)<br>  4. High/low voltage, high amperage on three-phase supply, voltage out of balance. (Voltage should not vary more than 3% between lines). | a. Find source of problem and take corrective action. |
| b. Low-pressure lock-out.<br>  1. Low refrigerant flow. (See section I, high superheat)<br>  2. Low airflow through evaporator.<br>  3. Low outside temperature on air-cooled condenser.<br>  4. Evaporator discharge air recirculating.<br>  5. Liquid-line solenoid leaking during the off cycle.<br>  6. Undercharged system.<br>  7. TEV problems. | b. Find source of problem and take corrective action. |

c. Uninsulated receiver exposed to low ambient air.

c. Insulate or artificially heat receiver.

### V-3. Unit Runs Continuously—Adequately Sized, Inadequate Cooling

**Probable Cause**

a. Load too high. (Has there been a recent addition to the design load, product, people, equipment, etc.?)
b. Low refrigerant charge.

**Probable Cause**

c. Low refrigerant flow due to liquid-line restrictions. (See sec. I-1 to I-3.)
d. Evaporator coil iced or dirty. (See sec. II-19)
e. Refrigerated or air-conditioned space has excessive load or poor insulation.
f. Dirty condenser.
g. Blocked filters.
h. Hot gas bypass valve stuck.
i. Compressor valve leaking.

j. Undersized TEV.
k. Faulty EPR or pressure setting too high.

**Remedy**

a. Choose a condensing unit with increased capacity to match load requirement, or reduce load.
b. Repair leak and recharge.

**Remedy**

c. Check and repair liquid feed problem.
d. Check and repair defrost system, clean evaporator and/or drains.
e. Determine fault and correct problem.
f. Clean condenser.
g. Clean or replace filters.
h. Check, repair, or replace.
i. Check high—and low-side pressures (leaky valve will not be able to develop proper suction or discharge pressures). See pages 46 and 47.
j. See section I-9 for TEV selection.
k. Replace EPR or readjust pressure setting where needed.

### V-4. Unit Will Not Start

Check pressure controls, loads, relays, and capacitors. Check power source. Check circuit breaker. Check fuses (right size?); if blown, find the cause. Check thermostat (contact made, or loose wire?).

### V-5. Faulty Controls

Check the control circuit transformer for size. Check contacts. Are they stuck open? Check connections of terminals and contactors.

### V-5. Causes of High Amperage Draw

**Probable Cause**

a. Excessive system load.
b. Defective capacitor.
c. High or low voltage.
d. Incorrect wiring.
e. Tight compressor.
f. Burned contacts.
g. Too small of wire gauge used in hookup.
h. Voltage out of balance on three-phase supply.

**Remedy**

a. Reduce load.
b. Replace capacitor.
c. Correct voltage problem.
d. Rewire properly.
e. Replace compressor.
f. Replace contactor.
g. Rewire with proper size.
h. Voltage should not vary more than 3% between lines, or call power company.

V-7.  Loss of Oil, Loss of Oil Pressure, or Oil Pressure Control Trips Out

**Probable Cause**

a. Insufficient oil in system.
b. Suction riser too large.
c. Insufficient traps in suction risers.
d. Suction superheat too high.
e. Lack of oil separator on systems operating below—30°F suction.
f. Low refrigerant charge.
g. Refrigerant floodback.
h. Ruptured suction line to the liquid heat exchanger.
i. Crankcase heater burnt out.
j. Iced-up evaporator coil. (See sec. II-18.)
k. Distributor and/or expansion valve too large (oil dilution, creating crankcase "foaming," or loss of oil pressure.
l. Liquid refrigerant feeding through oil separator.
m. Evaporator fan motor(s) not operating.
n. Defective oil pump.
o. Plugged oil pump inlet screen.
p. Worn bearings, pump, or compressor.
q. Oil-pressure control trip-out due to compressor short-cycling.
r. Control voltage too high, causing premature trip out.
s. Defective oil-pressure control.
t. Defective compressor overload.
u. Oil-pressure control wired incorrectly.

**Remedy**

a. Add oil per specifications.
b. Check line sizing at design conditions and change tubing if correct.
c. Install a suction trap in the vertical riser per the manufacturer's specs.
d. Adjust superheat.
e. Add oil separator.
f. Add refrigerant.
g. Adjust superheat to 20°F minimum at the compressor. (Check Manufacturer's specifications.)
h. Replace heat exchanger.
i. Replace crankcase heater.
j. Clean evaporator coil.
k. Check for proper size and change if incorrect.
l. Check oil separator heater.
m. Check and replace if necessary.
n. Replace oil pump.
o. Clean oil pump screen.
p. Replace bearings, pump, or compressor.
q. Check high—and low-pressure control settings. Check refrigerant charge. Check for dirty condenser. Replace burnt out condenser motors.
r. Adjust to correct voltage.
s. Replace control.
t. Replace overload protector.
u. Check wiring and correct.

Different sizes of motors come with different ampere ratings. The table on this page indicates the average expected ampere rating according to the horsepower of the unit.

Since all electrical components are selected by their voltage and amperage ratings, the replacement part must have a rating that is equal to, or greater than, the actual *full-load amperage* and *locked-rotor amperage* of the motor. Therefore, when the compressor motor ratings cannot be determined by labeling or the wiring diagram, this table can be used as a guide to determine the condition of a compressor motor.

Tight compressors and other factors (see p. 380) cause a higher than average amperage draw. These factors should be carefully considered before condemning a compressor.

| Approximate Horsepower | 120 Volts AC | | 240 Volts AC | | |
|---|---|---|---|---|---|
| | Full Load | Locked Rotor | Full Load | Locked Rotor | |
| 1/10 | 3.0 | 18.0 | 1.5 | 9.0 | Use a clamp-on ammeter to determine the amperes drawn while the motor compressor operates. |
| 1/8 | 3.8 | 22.8 | 1.9 | 11.4 | |
| 1/8 | 4.4 | 26.4 | 2.2 | 13.2 | |
| ¼ | 5.8 | 34.8 | 2.9 | 17.4 | |
| 1/3 | 7.2 | 43.2 | 3.6 | 21.6 | |
| ½ | 9.8 | 58.8 | 4.9 | 29.4 | |
| ¾ | 13.8 | 82.8 | 6.9 | 41.4 | |
| 1 | 16.0 | 96.0 | 8.0 | 48.0 | |
| 1½ | 20.0 | 120.0 | 10.0 | 60.0 | |
| 2 | 24.0 | 144.0 | 12.0 | 72.0 | |
| 3 | 34.0 | 204.0 | 17.0 | 102.0 | |

*Courtesy of White-Rodgers, Division of Emerson Electric Company*

| Watts | Voltage (AC—Single Phase) | | | | Watts | Voltage (AC—Single Phase) | | | | Watts | Voltage (AC—Single Phase) | | | | Conversion Table for Watts Amperes Volts |
|---|---|---|---|---|---|---|---|---|---|---|---|---|---|---|---|
| | 120 | 208 | 240 | 277 | | 120 | 208 | 240 | 277 | | 120 | 208 | 240 | 277 | |
| | Amperes | | | | | Amperes | | | | | Amperes | | | | |
| 500 | 4.2 | 2.4 | 2.1 | 1.8 | 2500 | 20.9 | 12.0 | 10.4 | 9.0 | 4500 | 37.5 | 21.6 | 18.8 | 16.3 | |
| 1000 | 8.3 | 4.8 | 4.2 | 3.6 | 3000 | 25.0 | 14.4 | 12.5 | 10.8 | 5000 | 41.7 | 24.0 | 20.8 | 18.0 | |
| 1500 | 12.5 | 7.2 | 6.3 | 5.4 | 3500 | 29.2 | 16.8 | 14.6 | 12.6 | | | | | | |
| 2000 | 16.7 | 9.6 | 8.3 | 7.2 | 4000 | 33.3 | 19.2 | 16.7 | 14.4 | | | | | | |

# CHECKING OUT THE SOLID-STATE TEV (THERMAL ELECTRIC VALVE)

As a general rule of thumb, when a thermal electric valve (TEV) is incorporated in a refrigeration system, that system becomes more energy efficient.

Sooner or later every service technician will be faced with troubleshooting and servicing an electric valve system. This section will help you to successfully meet that challenge.

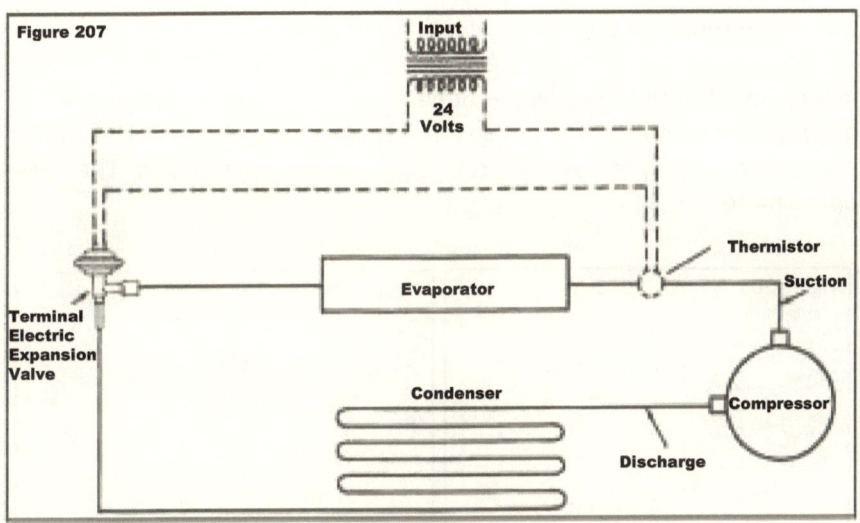

Actually, the job is easy, even though it is entirely different from working with a conventional pressure-controlled system.

Being electrically operated, the thermal electric valve does not depend on system pressure or temperature, nor does it need a charged element or capillary line. It can be used in any type of system with any noncorrosive refrigerant. And it automatically compensates for head pressure, evaporator load, and ambient conditions. The electric valve (figs. 107a and 207) is operated by low-voltage current. It has two working parts: a wire-wound bimetal heater and a spring-loaded needle.

At zero voltage, the valve is closed. As voltage is applied, the heater deflects the bimetal upward. The spring-loaded needle rises as it follows the movement of the bimetal and opens the valve; the more voltage applied, the greater the opening. (See figs. 208 and 209.)

Voltage to the valve is regulated by a liquid-sensing thermistor. This thermistor is installed in the suction line at the exact point where a complete change of refrigerant from liquid to vapor is desired. The thermistor reacts to the amount of liquid refrigerant as it leaves the evaporator.

Figures 210 and 211 illustrate the operation of the thermistor and the electric valve when they are wired in series. (This is a system in which the valve and thermistor replace a thermostatic expansion valve.)

Voltage input to the valve heater depends on thermistor resistance. When it is exposed to hot refrigerant gas, the thermistor self-heats. This lowers its resistance and increases current input to the valve heater. The valve opens more.

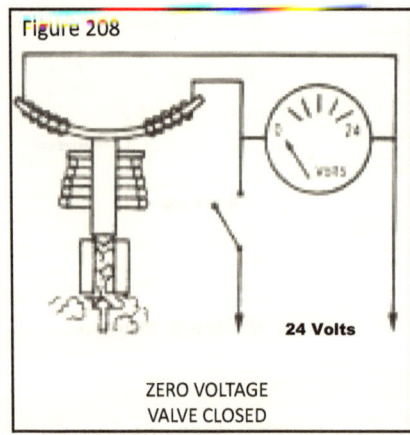

Figure 208

24 Volts

ZERO VOLTAGE
VALVE CLOSED

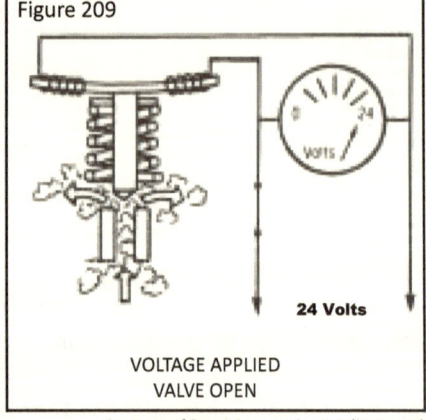

Figure 209

24 Volts

VOLTAGE APPLIED
VALVE OPEN

Courtesy of Eaton Corporation Appliance and Specialty Controls Division

This continues until enough liquid refrigerant is fed into the evaporator to reduce superheat.

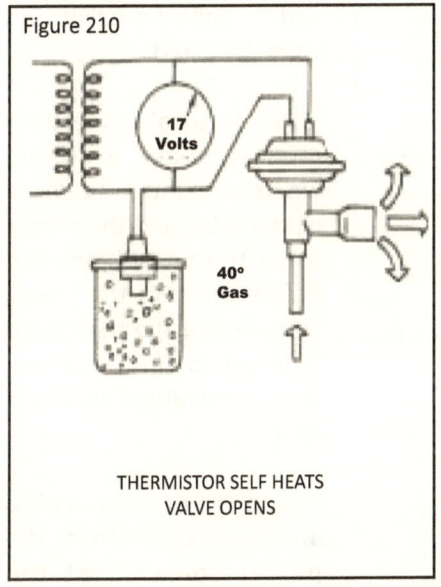

Figure 210 — THERMISTOR SELF HEATS VALVE OPENS (17 Volts, 40° Gas)

Figure 211 — THERMISTOR COOLS VALVE BEGINS TO CLOSE (5 Volts, 40° Liquid)

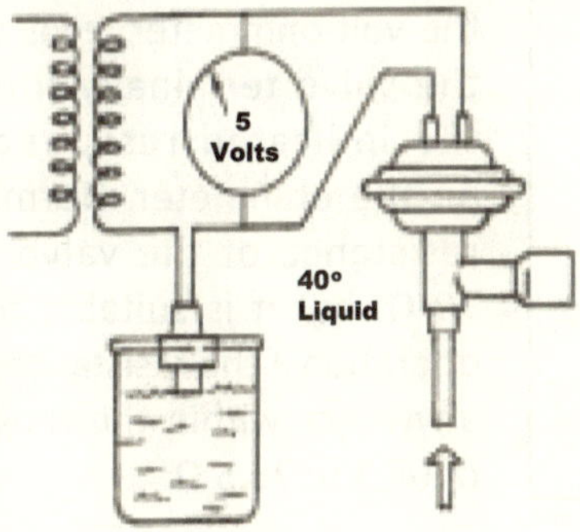

Courtesy of Eaton Corporation, Appliance and Specialty Controls Division

The wet refrigerant gas cools the thermistor, increasing its resistance. Less voltage is sent to the valve heater, and the valve moves toward closing.

Troubleshooting and servicing thermal electric valves is relatively easy. All you need is a standard volt-ohmmeter, a service gauge manifold, and a few minutes of your time.

By using the volt-ohmmeter and normal service procedures, you can quickly identify system conditions and check valve operation. Be sure to observe these precautions:

1. Never put more than the output of a 24-volt transformer into the valve circuit.
2. Do not put 24-volt current directly to the thermistor unless the valve is also wired into the circuit.
3. Do not touch both terminals of the valve with one of the volt-ohmmeter leads. You will short out the valve and damage the thermistor.

If you determine that the system is not working properly, here are the steps to follow:

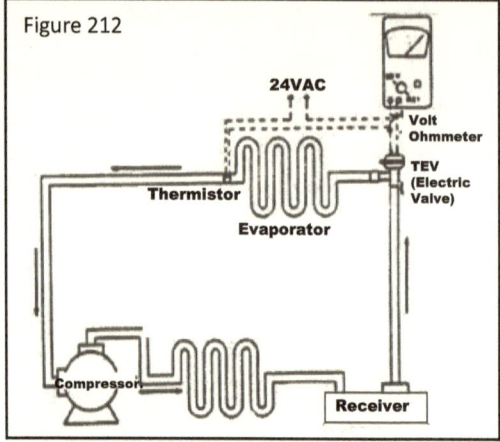

Figure 212

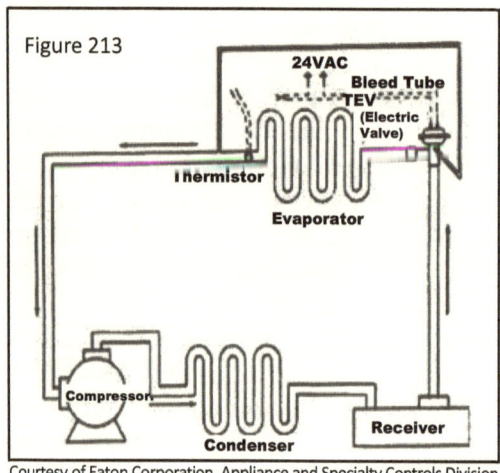

Figure 213

Courtesy of Eaton Corporation, Appliance and Specialty Controls Division

Make initial electrical check. Be sure that the electrical connections to the valve and thermistor are tight and correctly wired.

If the electrical connections are OK, make a simple electrical continuity check of the valve and thermistor using the volt-ohmmeter. Before you start this check, be sure that the thermistor is at room temperature. Thermistor ratings are based on 77°F.

Disconnect the valve and thermistor from the power source. Set the volt-ohmmeter to the RX1 scale.

Check the valve by connecting the volt-ohmmeter leads to the valve terminals. Note the indicated resistance on the ohmmeter. Normal resistance of the valve is 70 Ω, but it is suitable for operation if the resistance is anywhere within the range of 66.5 to 73.5 Ω.

Check the thermistor by connecting the volt-ohmmeter to the thermistor leads. The thermistor is OK if the volt-ohmmeter reading is within the tolerance range.

Defective thermistors generally have higher than normal readings. No reading indicates an open circuit.

If the valve and thermistor pass these checks, you're ready for the next step.

Check system performance. Reconnect the valve and thermistor to the 24-volt power source.

Switch the volt-ohmmeter to the 50-volt scale and connect its leads to the valve terminals (see fig. 212). If the valve operates correctly, the volt-ohmmeter reading will fluctuate as the valve modulates refrigerant flow. The reading will drop each time droplets of liquid refrigerant reach the thermistor. It will rise when dry suction gas is at the thermistor.

During normal operation, the volt-ohmmeter reading may be as high as 20 volts and as low as 5 volts.

At peak loads, the ohmmeter should register a reading within the 15 VAC to 24 VAC range. It may hover at some point between 15 VAC and 20 VAC. If the reading stays between 15 VAC and 24 VAC for more than 5 minutes and the system does not become satisfied, the valve is probably not opening properly. It may be restricted by solder or other foreign particles.

At low loads, the readings will fall between 5 VAC and 14 VAC and may hover at some point. If it stays in this range for more than 5 minutes and the system is not at minimum load conditions, the valve may be overfeeding the evaporator.

To pinpoint either of these problems, attach a gauge manifold to the system and check the high-side and low-side pressures

If, after conducting the volt-ohmmeter tests, you still haven't found a problem, you're ready for the next step.

Conduct gauge manifold tests. Connect a service gauge manifold to the system. Observe high-side and low-side pressures as you perform the following procedures:

To see if the valve is closing, remove voltage to the valve by disconnecting any wire in the circuit. With the circuit open, low-side pressure should begin to drop. Wait three or four minutes. If pressure does drop, the valve is closing freely—it is not clogged or sticking.

To see if the valve is opening, remove the thermistor from the circuit by disconnecting the thermistor wires. Connect the open wire from the 24-volt source to the open wire from the valve (fig. 213). This will send 24-volt current directly to the valve, and low-side pressure should begin to increase. If it does, the valve is opening properly.

This type of testing cannot be done with a capillary tube system. With a thermostatic expansion valve, you will have to remove the thermal bulb and heat it to see if the valve opens properly.

Replace components, if necessary. If you find the thermal electric valve or thermistor to be defective, replace those components with exact duplicates. Replace only the part that has failed; it is not necessary to replace any valve or thermistor that is operating properly.

When replacing the valve, mount it with the head up, or within 30° of upright, if possible. This prevents any liquid from migrating to the bimetal chamber where it might affect valve operation.

If you mount the valve head down, use a valve with a bleed tube connection. The bleed connection is the third connection on solder-type valves. It prevents migration of liquid refrigerant to the bimetal chamber. On heat pumps, always use a valve with a bleed tube. (Bleed tubes are not available on flare-type electric valves.)

Connect the bleed tube downstream of the thermistor in the system suction line (see fig. 213a). On heat pumps, connect the bleed tube to the common suction line. Remember to use a chill block or wet cloths to protect the valve during soldering.

When replacing the thermistor, simply unscrew the defective one and install the new one. Even though the threads are of the dry-seal type, use a good thread sealer. Turn the replacement thermistor tightly into the saddle to guarantee a leakproof seal.

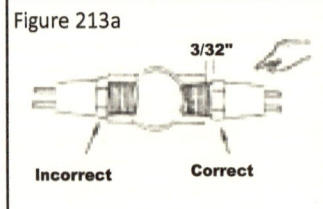

Figure 213a

Incorrect / Correct / 3/32"

The threads on the thermistor are the dry-seal type. However, use a good thread sealer (like Teflon tape). Turn the thermistor tightly into the saddle to guarantee a leakproof seal. The 3/32" clearance (one or two threads) between the thermistor and the saddle top shown in this sketch will properly locate the thermistor, which should be flush with the inside of the tube wall. If a new location is necessary, the suction-line adapter saddle should be mounted over a 27/64" (0.422") line opening, or a Z-size drill bit hole (0.412"). When a Z-size bit is used, retap the thermistor with a 1/2" OD to 1 5/8" OD are available.

Courtesy of ALCO Controls

## WATER-COOLED CONDENSERS

Circulating water removes heat from water-cooled condensers. When the flow of this cooling water becomes too slow, or if it stops, the pressure in the condenser rises very rapidly, causing the high-pressure control to shut off the unit.

Once the compressor stops running, the head pressure drops sharply, and very soon it reaches the high-pressure control cut-in point, causing the power to the compressor to be restored. The unit resumes operation, and the short-cycling continues.

The interruption or reduction of this circulating flow of water can be due to a defective (or out of adjustment) valve, a clogged-up screen, a decomposed rubber condenser hose through which the exhaust water runs, or a water line in the condenser choked with sediment.

---

**REASONS FOR A RESTRICTED WATER FLOW**

a. Leaky valve bellows. (Valve replacement is required.)
b. Leaky valve. (Valve replacement is required.)
c. Valve adjusted for too slow a flow. (Readjustment of valve is required.)
d. Partial or no flow of water due to obstructed screen. (Cleaning is required.)
e. Chattering valve. (Valve replacement is required.)
f. Valve obstructed by an excessive amount of sediment. (Valve cleaning or replacement required.)

---

## EXCESSIVE WATER FLOW

Excessive water flow results in the use of more water than actually needed to properly cool the condenser. Refrigeration is unaffected, but a typical complaint is a higher than normal water bill.

---

**THREE POSSIBLE CAUSES OF EXCESSIVE WATER FLOW**

a. Water valve adjusted for a higher flow. (Readjust the valve.)
b. Water supply pressure too high. (Rarely happens.)
c. Leaky water valve, as evidenced by a continuous flow of water when the unit is turned off. (Replace the valve.)

### Figure 214a

See recommended valve opening pressures (psi) on page 364.

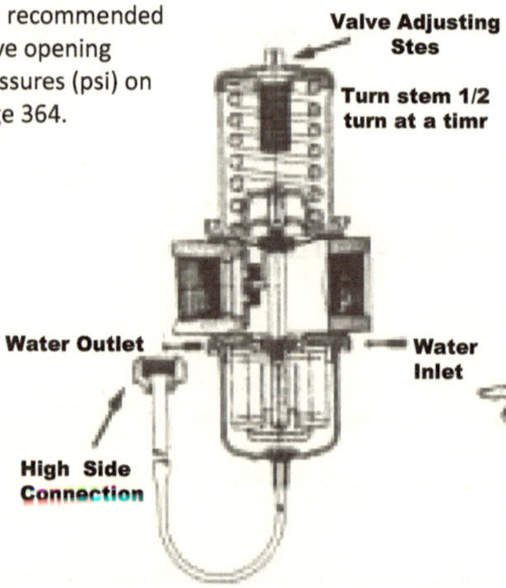

Valve Adjusting Stes
Turn stem 1/2 turn at a timr

Water Outlet
Water Inlet
High Side Connection

CROSS SECTION OF A WATER-REGULATING VALVE

See fig. 214b.

If the maximum desired water temperature is 65°F, the valve must be adjusted to open at 74 psi for systems using R-12, 123 psi for systems using R-22, etc.

### HOW A WATER-REGULATING VALVE WORKS

High-side pressure actuates bellows which move the valve slide across the seat. A rise in high-side pressure opens the valve, while a drop in high-side pressure will close the valve. Adjustment is simple because the adjusting stem is completely accessible.

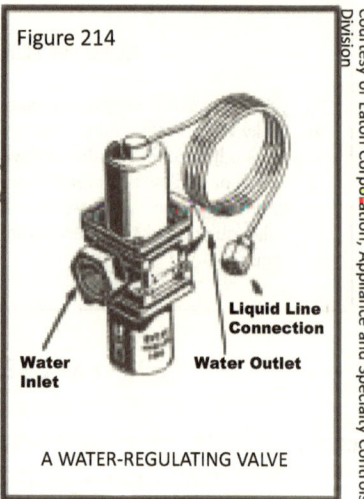

Figure 214

Liquid Line Connection
Water Inlet
Water Outlet

A WATER-REGULATING VALVE

Courtesy of Eaton Corporation, Appliance and Specialty Controls Division

## WATER-REGULATING VALVE INSTALLATION TIPS

A water-regulating valve is installed in the condenser water inlet or outlet line, with the head pressure connection attached directly to the high-pressure side of the system. The valve may be installed in any position.

Care should be exercised that any rubber hose connections are always in good condition. Condenser outlet hose connection tends to deteriorate due to the blistering of its inner wall by warm water, thus restricting or even stopping the flow of water through the condenser.

Failure of valve to close off when the condensing unit stops may be due to improper valve-setting adjustment. For proper setting, see the table, figure 214b.

Another common cause for failure of valve to close during the unit shutdown period, is air which has been trapped in the unit high side. This can be corrected by purging the noncondensible gases from the condensing unit high side. Air should always be purged both from the condenser water line and from the high-pressure side of the refrigeration system.

The water-regulating valve can be manually opened to aid in draining water from the system. Two openings are provided in the spring housing through which a screwdriver, or similar tool, can be inserted. Place the tip of the screwdriver under the spring pad and exert a force to oppose the spring force. This will allow system pressure on the bellows side to open the valve and drain the water (compressor does not have to be running), thus the valve is manually opened without disturbing the setting previously established with the adjusting screw.

Normally, no filtering is required. When the water source contains hard particles, such as sand, the water must be filtered prior to entry to the valve to prevent internal damage.

---

**VALVE OPERATING PRESSURE ADJUSTMENT**

This pressure adjustment is easily and quickly made by turning the adjusting stem with a standard service wrench. To increase flow (decrease head pressure), turn stem counterclockwise. To decrease flow (increase head pressure), turn stem clockwise. Actual operating condenser pressure is as much as 25-30 psi above the opening pressure. Low suction pressures or high water pressures tend to lower condenser pressure. High suction pressures or low water pressures tend to raise condenser pressure. Generally, it is better to determine the maximum summer temperature of the water to be used and adjust the valve accordingly, rather than adjusting it for each season. See minimum recommended valve opening pressures (psi) on page 392.

# WATER VALVE ADJUSTMENT

Referring to the chart, you will find the corresponding head pressures expected for different water temperatures.

Adjust the valve-adjusting screw to the desired head pressure and let it run for some time. It should automatically when the head pressure reaches the point indicated on the chart.

Use a thermometer to check the water temperature and a compound gauge to check the head pressure.

| REFRIGERANT | WATER TEMPERATURE (°F) | | | | | | | | | | | |
|---|---|---|---|---|---|---|---|---|---|---|---|---|
| | 50 | 55 | 60 | 65 | 70 | 75 | 80 | 85 | 90 | 95 | 0 | HEAD PRESSURE TABLE FOR WATER TEMPERATURE USED FOR VARIOUS SYSTEMS AND REFRIGERANT TYPES |
| | HEAD PRESSURE (psi) | | | | | | | | | | | |
| R-12 | 56 | 62 | 68 | 74 | 80 | 87 | 93 | 101 | 108 | 117 | 125 | |
| R-22 | 95 | 104 | 113 | 123 | 133 | 144 | 155 | 158 | 180 | 194 | 208 | |
| R-717 | 98 | 108 | 119 | 130 | 140 | 152 | 164 | 177 | 191 | 205 | 220 | |
| R-500 | 71 | 78 | 86 | 94 | | 112 | 121 | 131 | 142 | 153 | 165 | Figure 214b |
| R-502 | 116 | 126 | 137 | 147 | 160 | 173 | 186 | 200 | 214 | 230 | 246 | |

### HOW TO REMOVE A WATER VALVE

Figure 214c

1. Disconnect power to the unit.
2. Disconnect the water valve wires from the unit.
3. Close the manual water shut off valve.
4. Remove this valve from the system. If the valve pressure tube is connected into the cylinder head of the compressor, proceed as follows:

    a. After connecting the gauge manifold to the system, turn the suction-line service valve all the way clockwise.
    b. Turn on the compressor until pressure in the crankcase drops to 0 psi.
    c. Using a heat lamp, heat the valve and line for about five minutes to move the liquid refrigerant that has condensed in the valve and this tube, back into the condensing unit.
    d. Turn the discharge service valve all the way clockwise.
    e. Open both manifold valves. The high pressure in the water valve refrigerant line and the manifold bypasses to the low side. Heat the water valve again being aware that no one is standing in front of the line, as any remaining refrigerant is released with considerable force.

# SERVICING A COOLING (WATER) TOWER

The accumulation of water deposits on evaporative water-cooled condensers acts as insulation and reduces the efficiency of the unit if it is not periodically removed from the system.

Water-softening chemicals are used for water treatment to reduce water deposits.

The pH factor is a means of measuring the impurities in water. The pH scale runs from one to fourteen; one through seven is the degree of acid in solution and eight through fourteen, the degree of alkalinity. An ideal pH factor is seven to eight, indicating no excess of acid or alkaline. The water should be tested when its temperature ranges from 70°F to 80°F. Certain chemicals can be added as needed to keep the water close to a neutral pH.

When servicing a water tower, check the following:

a. All lines exposed to possible freezing in cold weather should be wrapped with insulation and protected with heating tapes.
b. Clean strainer.
c. Check and clean air inlet screens if necessary.
d. Check pump. Clean and flush.
e. Check tension of belts and tighten if necessary.
f. Check water for foreign objects such as leaves, debris, and algae. Clean as necessary.
g. Check water level and adjust float if required.
h. Check spray nozzles and clean if necessary.
i. Check water-overflow drain. Clean if required.

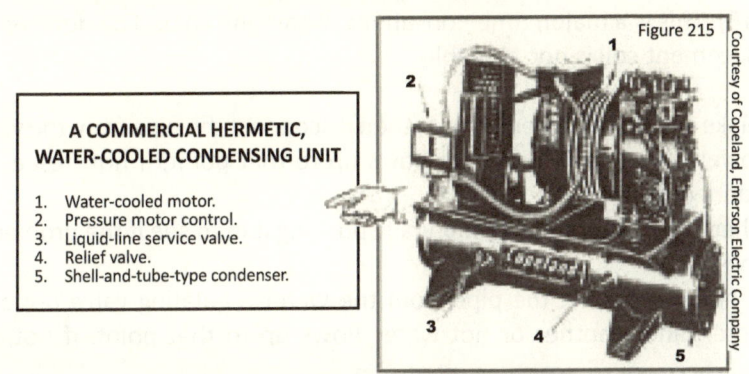

Figure 215

A COMMERCIAL HERMETIC, WATER-COOLED CONDENSING UNIT

1. Water-cooled motor.
2. Pressure motor control.
3. Liquid-line service valve.
4. Relief valve.
5. Shell-and-tube-type condenser.

Courtesy of Copeland, Emerson Electric Company

# HOW TO TRACE WATER CIRCUIT TROUBLES

To pinpoint the source of a blockage, disconnect one by one the water inlet and outlet connected to each component and see where the water flow stops.

1. Disconnect the water outlet hose from the condenser to see if the water flows up to that point.
2. If not, reseal the connection and disconnect the condenser water inlet hose to see if water flows up to that point. If water flows to the condenser inlet, but does not flow out of the condenser, or if it flows too slowly, the problem is a restricted condenser (caused by the formation of deposits on the inside of the tubing walls).

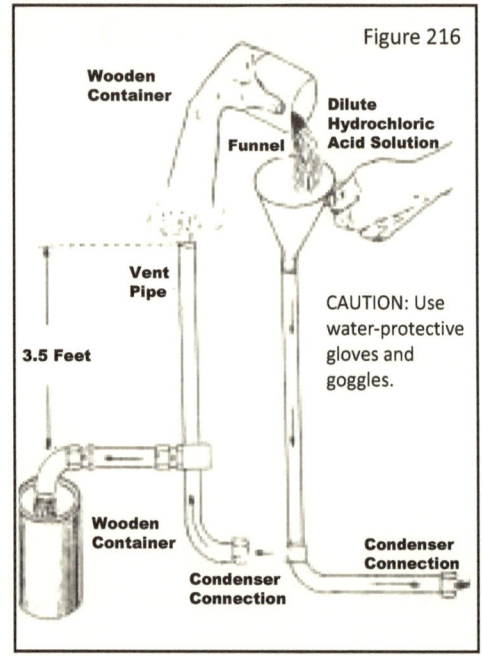

Figure 216

CAUTION: Use water-protective gloves and goggles.

Use containers made of acid-resistant material. Never use aluminum or galvanized vessels.

One way to clean a water-cooling tube is to connect a vent pipe to the upper condenser connector and pour the dilute hydrochloric acid solution through a funnel as shown.

Either the water-cooled condenser is replaced or the water lines within the condenser must be cleaned with a power-driven wire brush after removing the condenser end plates, or by using a dilute hydrochloric acid solution (see fig. 216). This is a major, time-consuming repair and should be done only if a replacement coil is not available.

3. Make sure the strainer is clean, then disconnect the pipe from the water tower outlet to see if water flows up to that point; if not, reseal the connection.
4. Check the water float by manually pushing it in to see if the problem is the float.
5. If not, disconnect the pipe from the water-regulating valve outlet to determine whether or not water flows up to that point. If not, the

source of the blockage must be in the water valve. Generally, the water-cooled condenser is connected to the cooling system by a plastic or rubber hose, and a blockage can occur in that short length of hose due to its decomposition. Refer to figure 214c for the removal of the water-regulating valve.

A water-hammering problem can be remedied by installing a short length of vertical pipe just ahead of the water-circulating valve. This absorbs the shock of the water after a sudden disruption in the flow.

As a rule of thumb, the water temperature should rise about 10°F as it flows through the condenser and absorbs heat. To determine the expected unit head pressure, add 10°F to 15°F to the temperature of the water, leaving the condenser from its discharge line. Using the chart on page 124, convert the sum of these temperatures to the pressure shown under the appropriate type of refrigerant.

Connect the high-pressure gauge to the service valve and compare the reading with the pressure calculated from the chart. If the reading exceeds the expected pressure by more than 5 psi, shut down the unit. The problem may be due either to the presence of air in the system or the system being overcharged.

Open the gauge connected to the discharge service valve and purge the system for about twenty seconds. Turn on the compressor again. Purge the system for another twenty seconds and watch the gauge. A head pressure drop (even a small one) is an indication of air in the system. If there is no pressure drop, the problem can be due to an overcharged system. Resume purging until the last condenser coil and the bottom of the liquid receiver feel cold to the touch.

Another common water-cooled-condenser problem is the settlement of waterborne impurities on the tubing walls. This is easily identified by the higher than normal head pressures and a hot liquid line, provided, of course, there is no air in the system and it is not overcharged.

The most common means of controlling waterborne deposits is through the addition of a weak sulfuric acid/chromate solution to the water in the cooling towers. Commercially prepared scale-prevention chemicals are the safest cleaners. These can be purchased from refrigeration dealers. Follow the directions carefully.

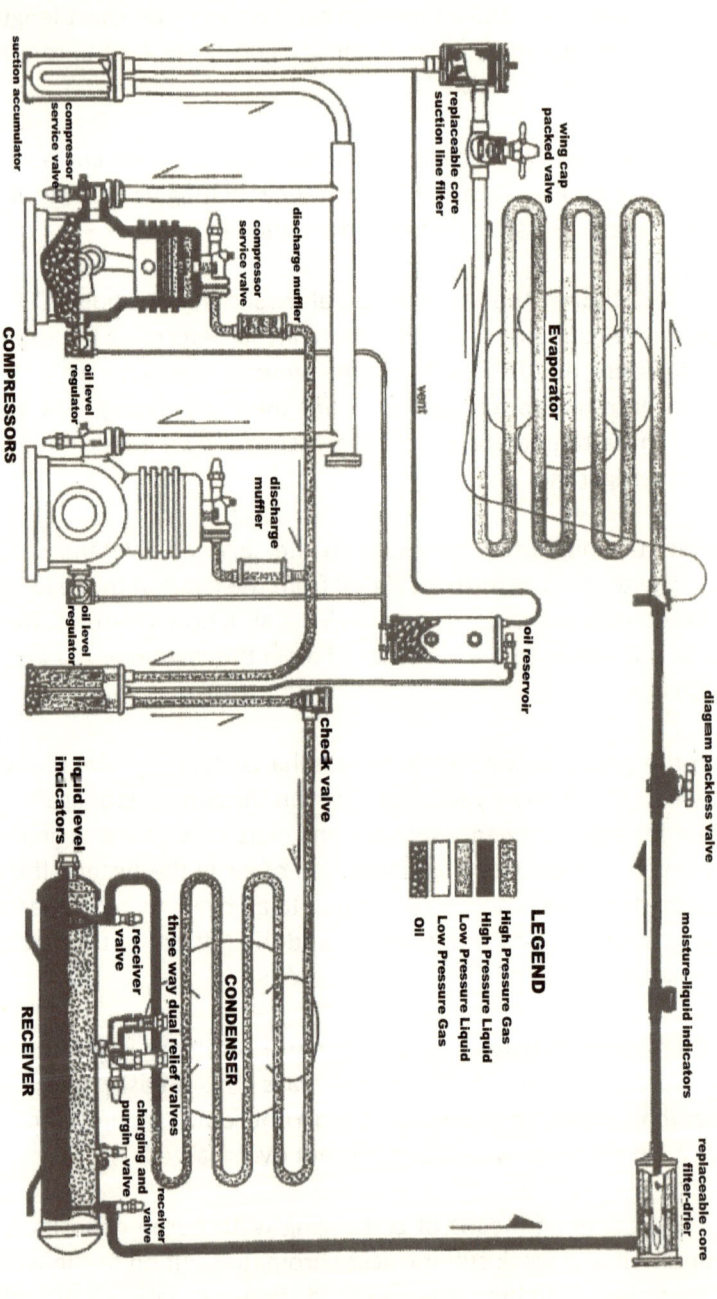

## A TEST OF KNOWLEDGE ON TROUBLESHOOTING REFRIGERANT FLOW CONTROLS

1. What are the possible causes that generally lead to relay burnout? (pp. 354, 355)
2. Explain flash gas. (p. 360)
3. How is a flash gas problem remedied? (p. 360)
4. In which direction should the stem on the water-regulating valve be turned to increase the flow of water? (p. 391)
5. What regulates voltage to an electric (solid-state expansion) valve? (p. 384)
6. As a general rule, what locked rotor amperage reading should be expected on a three-ton compressor motor that operates on 240 VAC? (p. 382)
7. In what position should a TEV remote sensing bulb be installed on a seven-eighth-inch line? (p. 371)
8. How does inadequate subcooling affect the superheat? (p. 363)
9. What are the three possible causes of excessive water flow in a water-cooled condenser system? (p. 390)
10. How can a water-cooled condenser that is filled with mineral deposits be repaired? (p. 394)
11. Explain how an evaporator pressure-regulating valve operates. (p. 362)
12. What scale should be chosen to set an ohmmeter when checking a solid-state expansion valve? (p. 386)
13. Why should you not touch both of the terminals of an electric valve with one of the voltmeter leads? (p. 385)
14. When testing an electric valve with a voltmeter, what range of reading should you expect at peak loads? (p. 387)
15. What are the symptoms indicating that the tubing walls in a water-cooled condenser is covered with sediment? (p. 395)
16. What are the possible causes of high amperage draw in a system? (p. 382)
17. What are the possible causes for a fluctuating discharge pressure? (pp. 376, 377 and 378)
18. Can flash gas be detected by a sight glass? (p. 360)
19. Can a solenoid valve be used to increase the head pressure in cold seasons? (p. 363)
20. Explain how a crimped external equalizer line reduces the superheat. (p. 371)

# Suva

## SUBSTITUTING SUVA REFRIGERANTS FOR CFCS

### Introduction

Chlorofluorocarbons (CFCs) have been used as refrigerants for over sixty years. They contain all the desirable properties by being nonflammable, low in toxicity, noncorrosive, noncarcinogenic, nonpoisonous, and have a compatibility with other materials.

However, depletion of the ozone layer has been linked to the stability of CFC compounds coupled with their chlorine content. Consequently, CFCs are being phased out, and SUVA refrigerant blends substituted.

This chapter deals with the introduction and handling of those refrigerants. It explains step-by-step how to retrofit the existing units from CFCs to the SUVA refrigerant blends.

This chapter reflects the latest data from DuPont Fluorochemicals.

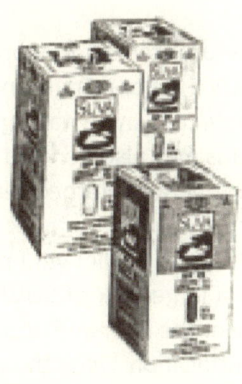

## SAFETY RULES FOR HANDLING SUVA REFRIGERANT CONTAINERS

1. Use safety glasses, steel-toe safety shoes, and gloves when handling containers.
2. Disposable containers should never be refilled with anything.
3. Containers must be protected from objects that could cause cuts or abrasions to the metal surface.
4. Don't force connections that do not fit. The threads on the regulators or any other equipment must mate with those on the container valve outlet.
5. Containers must be stored under a roof and protected from extremes in temperature.
6. Do not allow any liquid refrigerant to come into contact with skin as it will cause frostbite.
7. SUVA containers must never be used as supports, rollers, etc. Containers must be used only for storage and transportation of SUVA refrigerants.
8. Live steam or a direct flame must never be applied to a valve or container.
9. Do not attempt to alter or repair containers or valves.
10. Containers must never be heated to a temperature exceeding 125°F (52°C).
11. Never tamper with the safety devices in the valves or containers.
12. Returnable cylinders must never be refilled without approval from the manufacturers.
13. A rope, chain, or lifting magnet must be used when handling containers. A crane can be used when a safe platform or cradle is available to hold the container.
14. When containers are not in use, the valves must be kept tightly closed, and valve caps and hoods securely in place.

SUVA refrigerant blends cause no harm if they are handled in accordance to the manufacturer's safety recommendations.

Before starting to work, make sure there is no dangerous concentration of SUVA vapor refrigerant in the air caused by a leak or spill. This can be easily detected by installing an air monitor in the area. If needed, blowers or fans may also be used to prevent dangerous concentrations of SUVA vapor refrigerants in the area.

Purge and relief vent piping should be routed outdoors far from air intakes. If a large leak of SUVA refrigerant occurs, it may accumulate in low spots near the floor and occupy the space of available oxygen (vital for life) causing suffocation. Open the windows, turn on an exhaust fan(s), and leave the area, allowing the vapors to escape. Portable fans may also be used to circulate air near the floor. Don't return to the room unless an air monitor indicates there is no concentration of vapors in the area or unless self-contained breathing apparatus (or an airline mask) is used.

Since SUVA refrigerants are virtually odorless and hard to smell, a permanently installed air monitor may become necessary for confined areas. Inhalation of high concentrations of SUVA refrigerant vapors may cause dizziness, loss of coordination, and confusion. Overexposure to such high concentrations may even cause cardiac irregularities, unconsciousness, or even death without warning. Any person exposed to high concentrations of SUVA refrigerant vapors who experiences any of the above symptoms must immediately move to fresh air and seek medical attention.

Usually, SUVA refrigerant vapors have little or ignorable effect on eyes or skin. However, overexposure to some SUVA refrigerant blends such as SUVA 123 vapor refrigerant can cause a mild to moderate eye irritation, blurring of vision, or tearing. If liquid SUVA refrigerant blends come in contact with eyes or skin, they can cause frostbite. (They can freeze eyes or skin on contact). In which case, immediately soak the exposed areas in lukewarm water for about ten to fifteen minutes and seek medical attention.

## LEAK DETECTION

There are two categories of leak detectors: area monitors and leak pinpointers. Area monitors are installed to signal the presence of a concentration of any targeted compound in an entire room on a continual basis. Leak pinpointers are used when servicing equipment (see fig. 82a and 218).

There are three types of leak detectors:

1. Nonselective
2. Halogen-selective
3. Compound-selective

### I. NONSELECTIVE LEAK DETECTORS

These leak detectors are very durable, simple to use, and inexpensive. They detect any type of vapor or emission indiscriminately. Due to their insensitivity, incapability to be calibrated, and lack of selectivity, their use for area monitoring is very limited. Many nonselective leak detectors on the market today are not sufficiently sensitive for use with SUVA MP refrigerants. In fact, where SUVA MP components are concerned (such as HCFC-22), even leak detectors made especially for these blends may not be sensitive to them.

## II. HALOGEN-SELECTIVE LEAK DETECTORS

A special sensor is attached to these leak detectors to enable the monitor to sense and detect only compounds containing iodine, chlorine, fluorine, and bromine with no interference from any compound other than the targeted one. This gives the halogen-selective detector an advantage in the number of nuisance (false) alarms due to the presence of other nontargeted compounds.

These leak detectors are more sensitive than the nonselective type. They are able to detect leaks as little as 0.05 oz per year when used as a leak pinpointer, and a 5 ppm detection limit when used as an area monitor. They are easy to handle and calibrate.

## III. COMPOUND-SPECIFIC LEAK DETECTORS

They are the most expensive type of leak detector with the capability of detecting the presence of a single compound without any nuisance alarms.

## FLUORESCENT ADDITIVES

In addition to electronic leak pinpointers, the use of fluorescent additives has been practiced by refrigeration repair personnel for a long time. At the time of servicing the equipment, the fluorescents are added to the lubricant in the sealed system. When a leak occurs, the additive escapes with the refrigerant, generally leaving a bright yellow or green ring around the leak hole which is visible only under ultraviolet (UV) light. (You can find battery-operated UV lights on the market to serve this purpose.) These additives are capable of detecting leaks a low as 0.25 oz per year.

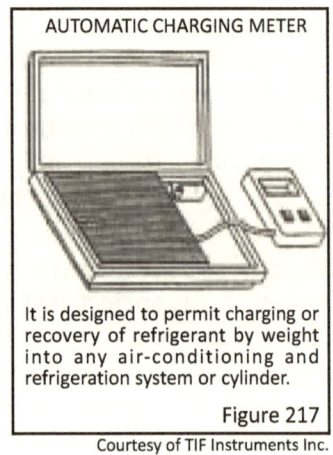

AUTOMATIC CHARGING METER

It is designed to permit charging or recovery of refrigerant by weight into any air-conditioning and refrigeration system or cylinder.

Figure 217

Courtesy of TIF Instruments Inc.

It is important to use only the type of fluorescent additive compatible with the lubricant and refrigerant being used in the system. If you have any doubt about the use of any specific additive, contact the fluorescent additive manufacturers for detailed information. The only disadvantage with the use of additives is that certain parts of a sealed system may be hard to reach and unobservable.

## SUVA MP 66 REFRIGERANT

This is the recommended, environmentally acceptable alternative for refrigerant R-12 used in low-temperature systems with evaporator temperatures below -10°F (-23°) (such as commercial and residential freezers, transport refrigeration equipment, ice machines, etc.). It can also be used as an acceptable alternative for existing R-500 systems. SUVA MP 66 and 39 will decompose if exposed to high temperature, producing elements such as electrical resistant heaters, open flames, etc. The pungent odors produced by decomposition may irritate the nose and throat.

SUVA MP 66 comes in 30 lb, 12 lb, and 1700 lb containers, color-coded light gray green. Nylon hose is recommended for use with this refrigerant. Existing low temperature R-12 and R-500 refrigeration systems can be retrofitted to use MP 66.

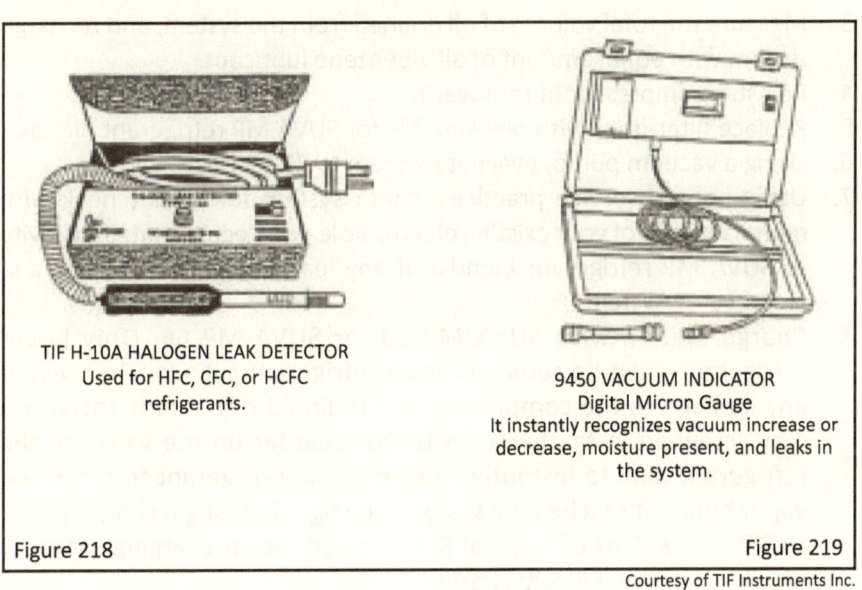

TIF H-10A HALOGEN LEAK DETECTOR
Used for HFC, CFC, or HCFC refrigerants.

9450 VACUUM INDICATOR
Digital Micron Gauge
It instantly recognizes vacuum increase or decrease, moisture present, and leaks in the system.

Figure 218

Figure 219

Courtesy of TIF Instruments Inc.

## SUVA MP 39 REFRIGERANT

This is the recommended, environmentally acceptable alternative refrigerant for medium-temperature R-12 system evaporators operating at -10°F (-23°C) or higher, such as food and dairy display cases, beverage dispensers, walk-in coolers, residential and commercial refrigerators, and beverage vending machines. This refrigerant comes in 30 lb, 120 lb, and 1700 lb containers, color-coded coral red. Nylon hose is recommended for use with SUVA MP 39 refrigerant. Existing R-12 refrigeration systems can be retrofitted to use this refrigerant.

## STEP-BY-STEP RETROFIT INSTRUCTIONS FOR SUVA MP 66 AND MP 39

1. Remove the R-12 or R-500 charge from the unit (15 inHg vacuum is required to remove the charge).
2. Drain the existing oil from the compressor (unless alkylbenzene lubricant is already in the system).
    Small hermetic compressors which have no oil drain, must be removed from the system in order to drain the existing oil through their suction line. In larger systems, oil must also be removed from low spots around the evaporator. All the existing lubricant in the oil separator must be drained too.
3. Measure the total volume of oil drained from the system, and recharge system with equal amount of alkylbenzene lubricant.
4. Reinstall compressor (if removed).
5. Replace filter-drier with one suitable for SUVA MP refrigerant blends.
6. Using a vacuum pump, evacuate system to 29.9 inHg vacuum.
7. Using normal service practices, check system for leaks (check with manufacturers of your existing electronic leak detector for its sensitivity to SUVA MP refrigerant blends). If any leak detected, seal leak and re-evacuate system.
8. Charge system with SUVA MP 39, or SUVA MP 66. Only liquid refrigerant must be removed from refrigerant cylinder. To prevent any damage to the compressor due to liquid refrigerant entry, you may screw an insta-charge restrictor adapter on the valve on the refrigerant tank to instantly convert liquid refrigerant to saturated vapor before it reaches the system (see fig. 220). Begin charging 70% to 75% by weight of original R-12 charge. Begin charging 100% by weight of original R-500 charge.

9. Run equipment and adjust charge until optimum operating conditions are reached (only liquid refrigerant must be removed from refrigerant cylinder). If more charge is required, add refrigerant to the system in increments of 3% to 5% of original R-12 or R-500 charge.

## SUVA 134A REFRIGERANT

It is the environmentally acceptable alternative replacement for R-12 used in medium-temperature refrigeration units such as supermarket cases, residential and commercial refrigerators and display cases, industrial and commercial chillers, and automotive air conditioners.

It is nonflammable with low toxicity. It has a boiling point of -15°F (-26°C). It is carried in 30 lb and 123 lb containers, color-coded light blue.

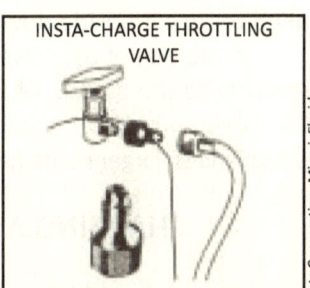

The Insta-Charge fitting is a fast and safe method for charging *liquid* refrigerant through the low-pressure side of an air-conditioning or refrigeration system without damaging the compressor. It converts liquid refrigerant to saturated vapor for fast entry. It connects directly to the refrigerant cylinder valve.

Figure 220

It is sold under different trade names such as:

    Hydrofluorocarbon-134A
    HFC 134A
    HFA 134A
    SUVA 134A
    SUVA Trans A/C (automotive market)
    SUVA Cold MP (stationary refrigeration/air-conditioning market)
    Formacel 24
    Dymel 134A (general aerosol market)
    Dymel 134P (aerosol pharmaceutical market)

At ambient temperatures and atmospheric pressure, HFC-134A is nonflammable. It becomes combustible at pressures 5.5 psig and lower at 350°F (177°C) when mixed with 60% or more volume of air. Equipment should never be leak-tested with a pressurized mixture of HFC-134A and air.

HFC-134A can be safely pressured with dry nitrogen. The 30-pound cylinders come with the same outlet fittings as CFC-12 (R-12) cylinders when used for stationery refrigeration equipment. HFC-134A 30-pound cylinders used for automotive air conditioners have a CGA-167 outlet valve to avoid mixing R-12 and HFC-134A at the time of servicing automotive air conditioners. The 123-pound containers are furnished with a two-way valve to remove SUVA 134A refrigerant either as liquid or as vapor from the cylinder without the need to turn it upside down.

On the side of the valve, the liquid handwheel is connected to a dip tube reaching to the bottom of the cylinder. The vapor wheel is placed on top of the valve. Both wheels are distinctly marked as liquid or vapor. For information pertaining to larger containers, contact DuPont Chemicals at 1-800-441-9442.

## CHARGING AND FILLING SAFETY MEASURES

1. Remove all of the remaining refrigerant from the system (or cylinder, if you are filling one).
2. Make sure that vacuum lines are free from restrictions that might produce discharge pressures above 15 psig (205 kPa) and cause the formation of combustible mixtures.
3. Never fill refrigeration equipment (or cylinders) when under positive air pressure. They must be evacuated before filling.
4. Stop charging equipment (or filling cylinders) before the pressure goes above 300 psig (2170 kPa).
5. Cylinders filled with SUVA 134A refrigerant should be periodically analyzed for air content.

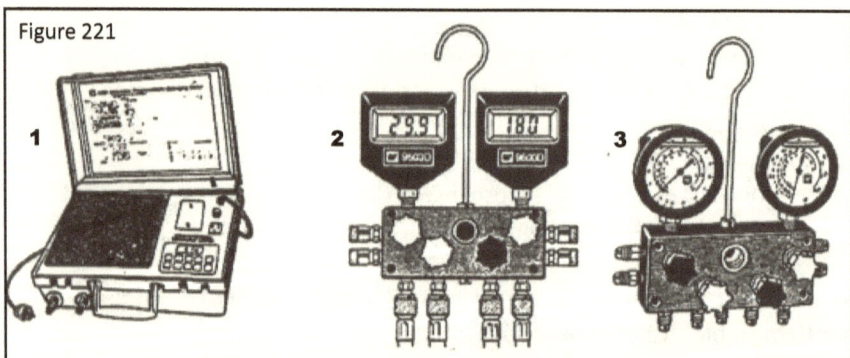

Figure 221

1. An automatic, programmable charging meter designed to permit charging or recovery of refrigerant by weight, into any air-conditioning and refrigeration system or cylinder.
2. A four-way digital gauge set with sight glass. The combination low-side gauge reads from -30 inHg to 99.9 psi and displays both vacuum and pressure readings.
3. A standard manifold gauge set manufactured by the same company.

Courtesy of TIF Instruments Inc.

## STEP-BY-STEP RETROFIT INSTRUCTIONS FOR SUVA 134a

1. Remove R-12 from the system (20 inHg vacuum is needed to remove charge).
2. Drain the existing oil from the compressor. Small hermetic compressors which have no oil drain must be removed from the system to drain the oil out through the compressor suction line. In larger systems, oil must be removed from low spots around the evaporator. All existing lubricant must be drained from the oil separator, too.
3. Measure the total volume of oil drained from the system. Recharge the system with an equal amount of Polyol Ester Lubricant.
4. Reinstall compressor (if necessary).
5. Replace filter-drier with one compatible with SUVA 134a.
6. Evacuate system to 30 inHg (use a vacuum pump).
7. Using normal service practices, check system for leaks. (Check with the manufacturer of your leak detector for its sensitivity to SUVA 134a). If any leak is detected, seal leak and re-evacuate system.
8. Charge system with SUVA 143a. (Charge 90% by weight of the amount of the original R-12 charge.)
9. Run equipment and adjust charge until optimum operating conditions are reached.

If a flushing technique is being used, repeat steps 1, 2, 3, 5, 6, and 7 three times. Run the system with R-12 refrigerant twenty-four to forty-eight hours after each time. Filter-drier replacement is not necessary during the flushing procedure.

---

#### TYPES OF OIL USED WITH SUVA BLENDS

When performing any retrofit, you should always remove as much of the old oil as possible before adding the new lubricant. The most important thing is to use only the product or type recommended by the compressor manufacturer. The type of lubricant is determined by the type of refrigerant. HFC refrigerants (134a and HP62) will require a polyol ester lubricant for stationary units. But cars with R-134a systems use polyalkylene glycol. HCFCs like SUVA MP series or HP80 or 81 require alkylbenzine.

## SUVA HP REFRIGERANTS

SUVA HP refrigerants (SUVA HP 62, 80, and 81) are odorless and nonflammable at atmospheric pressure in temperatures up to 176°F (80°C). They should not be mixed with air when testing for leaks. They can be used for retrofitting of existing nonflooded systems using R-502. They're suitable for both medium and low temperature equipment such as transport refrigeration, ice machines, and supermarket display cases. SUVA HP62 is intended primarily for use in new equipment. Although we have provided retrofit guidelines for this product, HP80 will usually provide a lower cost retrofit of equipment now using R-502. SUVA HP refrigerant blends are carried in 15 lb and 30 lb disposable cans, 123 lb or 1682 lb cylinders, 5000 gal. truck tank, and 170,000 lb rail containers. Disposable cans used for stationary refrigeration applications, are furnished with the same outlet fittings as R-502 cylinders.

A 123 lb SUVA HP (nonrefillable) cylinder comes with a liquid/vapor two-way valve for dispensing refrigerant either as a liquid or vapor without inverting it. On the side of the valve, a liquid handwheel is connected to a dip tube reaching to the bottom of the cylinder. A vapor wheel is placed on top of the valve. Both wheels are distinctly marked for liquid or vapor.

SUVA HP refrigerants become combustible if they are permitted to come in contact with high concentrations of air above atmospheric pressure. Never leak-test equipment with a pressurized mixture of air and SUVA HP refrigerants.

When filling refrigerant tanks, make sure they are not pressurized over one and a half times above the normal SUVA HP refrigerant operating pressure. Relief valves on either the refrigerant supply system or the tanks, must be set below this point.

## PREPARATION FOR CHARGING AND FILLING

- ★ Refrigeration units or cylinders must never be under positive air pressure. Normally, they should be evacuated before charging begins.
- ★ Remaining refrigerant should be removed from refrigeration units or cylinders before evacuation begins.
- ★ Make certain vacuum pump discharge lines are free of restrictions. Any restriction may cause the formation of a combustible mixture due to the discharge pressure increase.
- ★ Be sure filled cylinders are regularly checked for air content.

## CHARGING A SYSTEM WITH SUVA HP 80 OR HP 81

SUVA HP 80 and HP 81 replace R-502 in the existing systems equipped with expansion valve(s) or capillary tube(s), such as food service and warehousing, refrigerated transport units, freezers and display cases used in supermarkets, walk-in freezers, etc.

Systems using HP 80 or 81 should be charged with alkylbenzine or Polyol Ester lubricant. Furthermore, filter-driers used in existing systems must be replaced with ones compatible with the new generation refrigerant blends. Systems equipped with reciprocating, screw, scroll, or rotary compressors can be cost effectively retrofitted with HP 80 or HP 81.

SUVA HP 80 offers a slightly higher capacity as compared with R-502. It is suitable for use in frozen food and dairy display cases, ice cream dispensers, walk-in coolers/freezers, beverage and vending machines.

Due to the difference in liquid density of R-502 and HP 80 or 81, the system will require less weight of the HP 80 or 81 than the R-502.
     The amount of charge will vary according to the size of the evaporator, condenser, length of tubing runs, and the size of the receiver, if so equipped. Generally, an optimum charge will be 90% to 95% of the original weight of the R-502 charge. It is recommended that you start with a 90% charge and increase charge as necessary to obtain optimum operation.

Since the liquid composition in the cylinder is different from the vapor composition, it is important that only liquid be charged into the system to ensure the correct refrigerant composition. Liquid is withdrawn from the bottom of the cylinder through the dip tube only when the cylinder is in the upright position.

The initial charge should be added to the high-pressure side when the unit is not running. When the cylinder pressure is equal to the pressure in the system, the remainder of the charge may be added to the suction side with the compressor turned on.

Since you should remove only liquid from the charging cylinder, some compressors may be damaged if liquid HP refrigerant enters, so care should be taken to introduce SUVA slowly to allow time for it to vaporize. You may find it necessary to install a throttling valve to ensure that only vapor refrigerant enters the compressor.

## STEP-BY-STEP RETROFIT INSTRUCTIONS FOR SUVA HP 80 AND HP 81 REFRIGERANTS

1. Remove existing R-502 charge from the system. (If the amount of charge in the unit is unknown, weigh the R-502 removed).
2. Drain the existing oil charge from the compressor and measure the amount drained. Small hermetic compressors may be physically removed from the system to drain oil through their suction (or discharge) line. Open drive and semihermetic compressors are equipped with an oil drain hole. In large commercial systems, lubricant present in the accumulator(s) and oil separator(s) must also be drained.
3. Charge compressor with an equal volume of lubricant compatible with SUVA HP 80 or HP 81.
4. Reinstall compressor (if removed).
5. Change filter-drier to one suitable for use with HP 80 or 81.
6. Evacuate system with a vacuum pump. (Vacuum to 29.9 inHg.)
7. Using normal service practices, check system for leaks. (Check with the manufacturer of your detector for its sensitivity to HP 80 and HP 81).
8. Charge system to 90% by weight of the original charge with SUVA HP 80 or 81 with the refrigerant cylinder in an upright position. Only liquid refrigerant should be removed from the cylinder. The initial charge should be added to the system with the compressor turned off.
9. Run equipment and adjust charge to achieve optimum operating conditions. (If more charge is needed, add more SUVA HP 80 or 81 in small amounts from the cylinder.)

SUVA HP 80 and SUVA HP 81 come in 15 lb, 30 lb, and 123 lb cylinders that are color coded light brown.

## SUVA HP 62 REFRIGERANT

It is an environmentally acceptable alternative for R-502 and formulated for use in new commercial refrigeration equipment producing medium and low temperatures, such as vending machines, ice machines, food service, and transport.

SUVA HP 62 refrigerant comes in medium blue containers. Compared to R-502, it produces as much as 14°F (9°C) lower condenser discharge temperature and better lubricant stability. Due to this property, it can prolong compressor life. It can be used in virtually all R-502 applications.

## STEP-BY-STEP RETROFIT INSTRUCTIONS FOR SUVA HP 62 REFRIGERANT

★ 1. Remove the existing oil charge from the compressor (if Polyol Ester oil is not already in the system).
2. Measure the total volume of oil removed from the system and record it.
3. Charge the system with an equal amount of Polyol Ester lubricant. Run system for at least seventy-two hours.
4. Repeat steps 1, 2 and 3 twice.
5. Remove 502 charge from the unit (10-20 inHg vacuum is needed to remove all of the charge).
6. Reinstall compressor (if removed).
7. Change filter-drier with one suitable for use with SUVA HP 62 refrigerant.
8. Evacuate system with a vacuum pump (vacuum to 29.9 inHg).
9. Using normal service practices, check the system for leaks. (Check with the manufacturer of your leak detector for its sensitivity to SUVA HP 62). If any leak(s) detected, seal and re-evacuate the system.
10. Charge the system with SUVA HP 62 (90% by weight of the original 502 charge). Only liquid refrigerant should be removed from the cylinder.
11. Run equipment and adjust charge to achieve optimum operating conditions. (If more charge is needed, add in increment of 2% to 3% of original R-502 charge.)

★ If system is equipped with a small hermetic compressor which has no oil drain:

   a) R-502 must be removed from the system (10-20 inHg vacuum needed to remove charge).
   b) Drain the existing oil out the compressor suction line.
   c) Reinstall compressor.
   d) Evacuate system to 29.9 inHg vacuum.
   e) Leak check system.
   f) Charge system with SUVA HP 62.

## SUVA-123 (HCFC-123) REFRIGERANT

SUVA-123 (HCFC-123), a nonflammable liquid refrigerant, is the environmentally acceptable replacement for R-11. With a boiling point of 81.7°F (28°C), it is used in centrifugal chillers and the brine system. It is sold under different trade names such as SUVA Centri-LP, SUVA 123, HCFC 123, or Hydrochlorofluorocarbon 123. It must never be used as an uncontained flushing agent in refrigeration repairs. Drums used for this refrigerant are color coded light gray and must be stored in an upright position and placed out of direct sunlight at temperatures below 125°F (52°C).

If HCFC-123 drums are to be transported or refilled indoors, an air monitor will be required.

When charging or vacuuming a chiller, HCFC-123 drums must be connected to the chiller with both a vent line and a liquid transfer line. This is done to minimize venting the gas into the work area while liquid HCFC-123 is being transferred and also to avoid an overpressure or vacuum condition in the drum.

If HCFC-123 vapor comes into contact with high temperature producing elements such as electric resistance heaters, flames, etc., it will decompose producing toxic and irritating compounds such as hydrogen chloride and hydrogen fluoride with pungent odors irritating the nose and throat.

Do not store HCFC-123 in locations containing alkali or materials such as powdered aluminum, beryllium, or zinc if there is no air monitor in the room to indicate that HCFC-123 concentration is below 30 ppm AEL in the working space.

Also as a safety measure, be certain to wear proper respiratory protection at the time you are breaking into the sealed system. It produces almost the same operating temperatures and pressures in a chiller with a relatively lower capacity compared with R-11.

Solid core driers used with R-11, R-12, and R-22 can be used with HCFC-123 refrigerant. Lubricants used with R-11 may be used with HCFC-123.

For indoor applications, use an HCFC specific air monitor (0 to 150 ppm range). For more efficiency, install air sensors in a location where the average concentration of HCFC-123 vapor is more likely to occur. According to ASHRAE (American Society of Heating, Refrigeration and Air-Conditioning Engineers), proper respiratory measures should be available for immediate use under circumstances such as a large leak or release.

## NOTES ON RECOVERY, RECLAIMING, and RECYCLING

With the exception of a different filter-drier and/or moisture indicator, most of the machinery and recovery equipment used for R-12 and R-502 can be used to retrofit SUVA MP 39, MP66, HP62, HP80, and HP81. In some cases, even the same compressor oil can be used. Check with the manufacturer of your equipment for specific recommendations. Care must be taken not to cross-contaminate when switching from CFC refrigerants to the new SUVA blends.

Be certain to use the new pressure/temperature charts instead of the ones designed for the CFC refrigerants. Several experiments have proven that only a small percentage of the expansion devices now in place (TEVs, AEVs, capillary tubes, etc.) will have to be replaced after retrofitting. Sometimes superheat may require some readjustment. Liquid-line sight glasses can be used to determine the proper charge. But in some cases, bubbles will show even when the system is properly charged with SUVA refrigerant. In which case, other methods may be adopted to determine the proper amount of the new refrigerant charge.

Several compressor manufacturers provide retrofit instructions. For instance Bristol has approved SUVA HP81. Copeland has approved and will warrant their compressors retrofitted with MP39, MP66, and HP80 if their retrofit procedures are followed. Manitowoc Ice Machines are charged in the factory with HP81. Tecumseh uses alkylbenzene lubricant in its R-12 compressor and has approved HP81, so you needn't change lubricants when retrofitting with MP39 or 66.

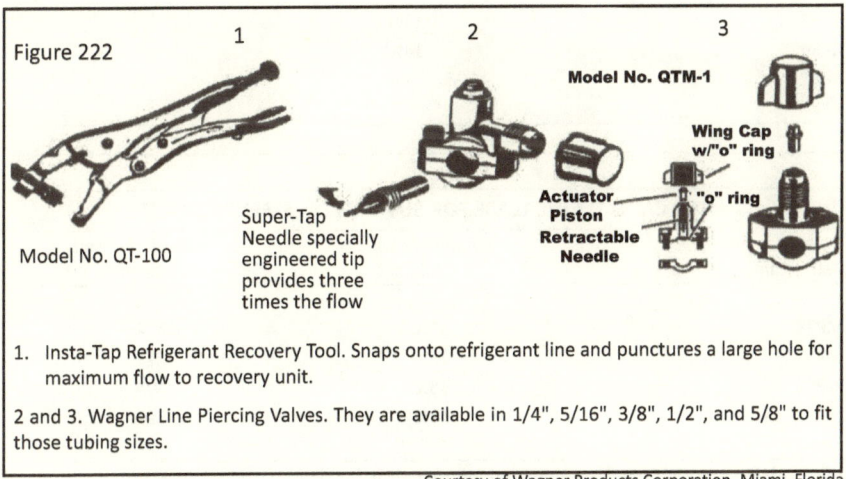

Figure 222

Model No. QT-100

1. Super-Tap Needle specially engineered tip provides three times the flow

2. Model No. QTM-1 — Actuator Piston, Retractable Needle

3. Wing Cap w/"o" ring, "o" ring

1. Insta-Tap Refrigerant Recovery Tool. Snaps onto refrigerant line and punctures a large hole for maximum flow to recovery unit.

2 and 3. Wagner Line Piercing Valves. They are available in 1/4", 5/16", 3/8", 1/2", and 5/8" to fit those tubing sizes.

Courtesy of Wagner Products Corporation, Miami, Florida

| TEMP. (°F) | HCFC 123 | HCFC 124 | HFC 125 | HFC 134a | TEMP. * | HCFC 123 | HCFC 124 | HFC 125 | HFC 134a |
|---|---|---|---|---|---|---|---|---|---|
| \multicolumn{10}{c}{SUVA VAPOR PRESSURE/TEMPERATURE RELATIONSHIP CHART} |
| -100 | 29.9* | 29.2* | 24.4* | 27.8* | | | | | |
| -90 | 29.8* | 28.8* | 21.7* | 26.9* | 110 | 10.5 | 79.7 | 291.6 | 146.8 |
| -80 | 29.7* | 28.2* | 18.1* | 25.6* | 120 | 15.4 | 94.9 | 334.3 | 171.9 |
| -70 | 29.6* | 27.4* | 13.3* | 23.8* | 130 | 21.0 | 111.7 | 380.3 | 199.8 |
| -60 | 29.5* | 26.3* | 7.1* | 21.5* | 140 | 27.3 | 130.4 | 430.2 | 230.5 |
| -50 | 29.2* | 24.8* | 0.3 | 18.5* | 150 | 34.5 | 151.0 | 482.1 | 264.4 |
| -40 | 28.9* | 22.8* | 4.9 | 14.7* | 160 | 42.5 | 173.6 | | 301.5 |
| -30 | 28.5* | 20.2* | 10.6 | 9.8* | 170 | 51.5 | 198.4 | | 342.0 |
| -20 | 27.8* | 16.9* | 17.4 | 3.8* | 180 | 61.4 | 225.6 | | 385.9 |
| -10 | 27.0* | 12.7* | 25.6 | 1.8 | 190 | 72.5 | 255.1 | | 433.6 |
| 0 | 26.0* | 7.6* | 35.1 | 6.3 | 200 | 84.7 | 287.3 | | 485.0 |
| 10 | 24.7* | 1.4* | 46.3 | 11.6 | 210 | 98.1 | 322.1 | | 540.3 |
| 20 | 23.0* | 3.0 | 59.2 | 18.0 | 220 | 112.8 | 359.9 | | |
| 30 | 20.8* | 7.5 | 74.1 | 25.6 | 230 | 128.9 | 400.6 | | |
| 40 | 18.2* | 12.7 | 91.2 | 34.5 | 240 | 146.3 | 444.5 | | |
| 50 | 15.0* | 18.8 | 110.6 | 44.9 | 250 | 165.3 | 491.8 | | |
| 60 | 11.2* | 25.9 | 132.8 | 56.9 | 260 | 185.8 | | | |
| 70 | 6.6* | 34.1 | 157.8 | 70.7 | 270 | 207.9 | | | |
| 80 | 1.1* | 43.5 | 186.0 | 86.4 | 280 | 231.8 | | | |
| 90 | 2.6 | 54.1 | 217.5 | 104.2 | 290 | 257.5 | | | |
| 100 | 6.3 | 66.2 | 252.7 | 124.3 | 300 | 285.0 | | | |

Vapor pressures are shown as psig. *Indicates inches of mercury vacuum.

| \multicolumn{2}{c}{TEV CAPACITY INCREASE FOR SUVA REPLACEMENTS TO R502} |
|---|---|
| Suva Refrigerant | Equipment used in buildings |
| HP62 | 11% |
| HP80 | 14% |
| HP81 | 22% |

| \multicolumn{3}{c}{TEV CAPACITY INCREASE FOR SUVA REPLACEMENTS TO 412} |
|---|---|---|
| Suva Refrigerant | Equipment used in buildings | Equipment used in vehicles |
| MP39 | 35% | N/A |
| MP52 | 28% | 27% |
| MP66 | 38% | N/A |
| HFC-134a | 38% | 30% |

NOTE: MP39 and MP66 are recommended for use only in buildings.

## SATURATED VAPOR TEMPERATURES OF R12 AND ITS SUVA REPLACEMENTS

| psiA (psiG) | Temperature, °F | | | | | kPA (Bar) | Temperature, °C | | | | |
|---|---|---|---|---|---|---|---|---|---|---|---|
| | R-12 | 134a | MP39 | MP52 | MP66 | | R-12 | 134a | MP39 | MP52 | MP66 |
| 12.2(5*) | -29.3 | -22.1 | -23.2 | -14.8 | -26.6 | 80(0.80) | -35.2 | -31.1 | -31.7 | -27.1 | -33.6 |
| 14.7( 0) | -21.6 | -14.9 | -16.0 | -7.4 | -19.4 | 100(1.00) | -30.1 | -26.3 | -26.9 | -22.2 | -28.9 |
| 16.7( 2) | -16.1 | -9.8 | -10.8 | -2.1 | -14.3 | 110(1.10) | -27.8 | -24.2 | -24.8 | -20.0 | -26.8 |
| 18.7( 4) | -11.1 | -5.2 | -6.1 | 2.6 | -9.7 | 120(1.20) | -25.7 | -22.3 | -22.8 | -18.0 | -24.8 |
| 20.7( 6) | -6.5 | -0.9 | -1.8 | 7.0 | -5.4 | 130(1.30) | -23.8 | -20.5 | -21.0 | -16.1 | -23.0 |
| 22.7( 8) | -2.9 | 3.0 | 2.2 | 11.1 | -1.5 | 140(1.40) | -21.9 | -18.8 | -19.2 | -14.3 | -21.3 |
| 24.7(10) | 1.6 | 6.7 | 5.9 | 14.9 | 2.2 | 150(1.50) | -20.1 | -17.1 | -17.6 | -12.7 | -19.6 |
| 26.7(12) | 5.4 | 10.1 | 9.4 | 18.5 | 5.7 | 160(1.60) | -18.5 | -15.6 | -16.0 | -11.1 | -18.1 |
| 27.7(13) | 9.0 | 13.4 | 12.7 | 21.8 | 8.9 | 170(1.70) | -16.9 | -14.1 | -14.5 | -9.5 | -16.6 |
| 30.7(16) | 12.3 | 16.5 | 15.8 | 25.0 | 12.0 | 180(1.80) | -15.4 | -12.7 | -13.1 | -8.1 | -15.2 |
| 32.7(18) | 15.5 | 19.4 | 18.8 | 28.1 | 14.9 | 190(1.90) | -13.9 | -11.4 | -11.8 | -6.7 | -13.8 |
| 34.7(20) | 18.5 | 22.2 | 21.6 | 31.0 | 17.8 | 200(2.00) | -12.5 | -10.1 | -10.5 | -5.4 | -12.6 |
| 36.7(22) | 21.4 | 24.9 | 24.3 | 33.7 | 20.4 | 210(2.10) | -11.2 | -8.8 | -9.2 | -4.1 | -11.3 |
| 38.7(24) | 24.2 | 27.4 | 26.9 | 36.4 | 23.0 | 220(2.20) | -9.9 | -7.6 | -8.0 | -2.8 | -10.1 |
| 40.7(26) | 26.9 | 29.9 | 29.4 | 38.9 | 25.5 | 230(2.30) | -8.6 | -6.5 | -6.8 | -1.6 | -9.0 |
| 42.7(28) | 29.5 | 32.2 | 31.8 | 41.4 | 27.9 | 240(2.40) | -7.4 | -5.4 | -5.7 | -0.5 | -7.8 |
| 44.7(30) | 32.0 | 34.5 | 34.1 | 43.8 | 30.2 | 250(2.50) | -6.2 | -4.3 | -4.6 | 0.6 | -6.8 |
| 46.7(32) | 34.4 | 36.7 | 36.4 | 46.0 | 32.4 | 260(2.60) | -5.1 | -3.2 | -3.5 | 1.7 | -5.7 |
| 48.7(34) | 36.7 | 38.9 | 38.5 | 48.2 | 34.5 | 270(2.70) | -4.0 | -2.2 | -2.5 | 2.8 | -4.7 |
| 50.7(36) | 39.0 | 41.0 | 40.6 | 50.4 | 36.6 | 280(2.80) | -2.9 | -1.2 | -1.5 | 3.8 | -3.7 |
| 52.7(38) | 41.2 | 43.0 | 42.7 | 52.5 | 38.6 | 290(2.90) | -1.9 | -0.3 | -0.5 | 4.8 | -2.7 |
| 54.7(40) | 43.3 | 44.9 | 44.6 | 54.5 | 40.6 | 300(3.00) | -0.8 | 0.7 | 0.4 | 5.8 | -1.8 |
| 56.7(42) | 45.4 | 46.8 | 46.6 | 56.5 | 42.5 | 310(3.10) | 0.2 | 1.6 | 1.3 | 6.7 | -0.9 |
| 58.7(44) | 47.4 | 48.7 | 48.4 | 58.4 | 44.4 | 320(3.20) | 1.1 | 2.5 | 2.2 | 7.6 | 0.1 |
| 60.7(46) | 49.4 | 50.5 | 50.3 | 60.3 | 46.2 | 330(3.30) | 2.1 | 3.3 | 3.1 | 8.5 | 0.9 |
| 62.7(48) | 51.3 | 52.2 | 52.0 | 62.1 | 47.9 | 340(3.40) | 3.0 | 4.2 | 4.0 | 9.4 | 1.8 |
| 64.7(50) | 53.2 | 53.9 | 53.8 | 63.9 | 49.7 | 350(3.50) | 3.9 | 5.0 | 4.8 | 10.3 | 2.6 |
| 66.7(52) | 55.0 | 55.6 | 55.5 | 65.6 | 51.3 | 375(3.75) | 6.1 | 7.0 | 6.9 | 12.3 | 4.6 |
| 68.7(54) | 56.8 | 57.2 | 57.1 | 67.3 | 53.0 | 400(4.00) | 8.2 | 8.9 | 8.8 | 14.3 | 6.5 |
| 70.7(56) | 58.5 | 58.9 | 58.8 | 69.0 | 54.6 | 425(4.25) | 10.2 | 10.7 | 10.6 | 16.2 | 8.3 |
| 72.7(58) | 60.3 | 60.4 | 60.4 | 70.6 | 56.2 | 450(4.50) | 12.0 | 12.4 | 12.4 | 18.0 | 10.1 |
| 74.7(60) | 61.9 | 62.0 | 61.9 | 72.2 | 57.7 | 475(4.75) | 13.9 | 14.1 | 14.0 | 19.7 | 11.8 |
| 76.7(62) | 63.6 | 63.4 | 63.4 | 73.7 | 59.2 | 500(5.00) | 15.6 | 15.7 | 15.7 | 21.4 | 13.4 |
| 78.7(64) | 65.2 | 64.9 | 64.9 | 75.3 | 60.7 | 525(5.25) | 17.3 | 17.2 | 17.2 | 23.0 | 14.9 |
| 80.7(66) | 66.8 | 66.4 | 66.4 | 76.8 | 62.2 | 550(5.50) | 18.9 | 18.7 | 18.7 | 24.5 | 16.4 |
| 82.7(68) | 68.4 | 67.8 | 67.8 | 78.2 | 63.6 | 575(5.75) | 20.5 | 20.2 | 20.2 | 26.0 | 17.8 |
| 84.7(70) | 69.9 | 69.2 | 69.2 | 79.7 | 65.0 | 600(6.00) | 22.0 | 21.5 | 21.6 | 27.4 | 19.2 |
| 86.7(72) | 71.4 | 70.6 | 70.6 | 81.1 | 66.4 | 625(6.25) | 23.5 | 22.9 | 22.9 | 28.8 | 20.6 |
| 88.7(74) | 72.9 | 71.9 | 72.0 | 82.5 | 67.7 | 650(6.50) | 24.9 | 24.2 | 24.2 | 30.2 | 21.9 |
| 90.7(76) | 74.3 | 73.2 | 73.3 | 83.9 | 69.1 | 675(6.75) | 26.3 | 25.5 | 25.5 | 31.5 | 23.2 |
| | | | | | | 700(7.00) | 27.7 | 26.7 | 26.8 | 32.7 | 24.4 |
| | | | | | | 725(7.25) | 29.0 | 27.9 | 28.0 | 34.0 | 25.6 |

* Inches of mercury

NOTE: kPa and Bar are absolute pressure

Used by permission of DuPont Fluoroproducts

(psiG) (The numbers in parentheses = the reading on the manifold gauge).

## SATURATED VAPOR TEMPERATURES OF R502 AND ITS SUVA REPLACEMENTS

| psiA (psiG) | Temperature, °F | | | | kPA (Bar) | Temperature, °C | | | |
|---|---|---|---|---|---|---|---|---|---|
| | R502 | HP80 | HP81 | HP62 | | R502 | HP80 | HP81 | HP62 |
| 12.2(5*) | -56.8 | -59.7 | -55.8 | -57.7 | 80(0.80) | -50.4 | -51.9 | -49.8 | -50.8 |
| 14.7( 0) | -49.7 | -52.9 | -49.0 | -50.8 | 100(1.00) | -45.7 | -47.4 | -45.3 | -46.3 |
| 16.7( 2) | -44.8 | -48.1 | -44.1 | -45.9 | 110(1.10) | -43.6 | -45.5 | -43.3 | -44.2 |
| 18.7( 4) | -40.2 | -43.8 | -39.7 | -41.4 | 120(1.20) | -41.7 | -43.6 | -41.4 | -42.4 |
| 20.7( 6) | -36.0 | -39.7 | -35.7 | -37.3 | 130(1.30) | -39.9 | -41.9 | -39.7 | -40.6 |
| 22.7( 8) | -32.2 | -36.0 | -31.9 | -33.5 | 140(1.40) | -38.2 | -40.3 | -38.0 | -38.9 |
| 24.7(10) | -28.7 | -32.7 | -28.4 | -30.0 | 150(1.50) | -36.6 | -38.8 | -36.5 | -37.4 |
| 26.7(12) | -25.2 | -29.3 | -25.1 | -26.6 | 160(1.60) | -35.1 | -37.3 | -35.0 | -35.9 |
| 27.7(14) | -22.0 | -26.2 | -22.0 | -23.5 | 170(1.70) | -33.7 | -35.9 | -33.6 | -34.5 |
| 30.7(16) | -19.0 | -23.3 | -19.1 | -20.5 | 180(1.80) | -32.3 | -34.6 | -32.3 | -33.1 |
| 32.7(18) | -16.1 | -20.5 | -16.3 | -17.7 | 190(1.90) | -31.0 | -33.3 | -31.0 | -31.8 |
| 34.7(20) | -13.3 | -17.9 | -13.6 | -15.0 | 200(2.00) | -29.7 | -32.1 | -29.8 | -30.6 |
| 36.7(22) | -10.7 | -15.4 | -11.1 | -12.4 | 210(2.10) | -28.5 | -30.9 | -28.6 | -29.4 |
| 38.7(24) | -8.2 | -12.9 | -8.6 | -10.0 | 220(2.20) | -27.3 | -29.8 | -27.4 | -28.2 |
| 40.7(26) | -5.8 | -10.6 | -6.2 | -7.6 | 230(2.30) | -26.2 | -28.7 | -26.3 | -27.1 |
| 42.7(28) | -3.4 | -8.4 | -4.0 | -5.3 | 240(2.40) | -25.1 | -27.6 | -25.3 | -26.0 |
| 44.7(30) | -1.2 | -6.2 | -1.8 | -3.1 | 250(2.50) | -24.0 | -26.6 | -24.2 | -25.0 |
| 46.7(32) | 1.0 | -4.1 | 0.3 | -1.0 | 260(2.60) | -23.0 | -25.6 | -23.2 | -24.0 |
| 48.7(34) | 3.1 | -2.1 | 2.4 | 1.1 | 270(2.70) | -22.0 | -24.7 | -22.3 | -23.0 |
| 50.7(36) | 5.2 | -0.1 | 4.4 | 3.1 | 280(2.80) | -21. | -23.7 | -21.3 | -22.1 |
| 52.7(38) | 7.1 | 1.8 | 6.3 | 5.0 | 290(2.90) | -20.1 | -22.8 | -20.4 | -21.1 |
| 54.7(40) | 9.1 | 3.7 | 8.2 | 6.9 | 300(3.00) | -19.2 | -21.9 | -19.5 | -20.2 |
| 56.7(42) | 10.9 | 5.5 | 10.0 | 8.7 | 310(3.10) | -18.3 | -21.1 | -18.6 | -19.4 |
| 58.7(44) | 12.8 | 7.2 | 11.8 | 10.5 | 320(3.20) | -17.4 | -20.2 | -17.8 | -18.5 |
| 60.7(46) | 14.5 | 8.9 | 13.5 | 12.2 | 330(3.30) | -16.5 | -19.4 | -16.9 | -17.7 |
| 62.7(48) | 16.3 | 10.6 | 15.2 | 13.9 | 340(3.40) | -15.7 | -18.6 | -16.1 | -16.8 |
| 64.7(50) | 18.0 | 12.2 | 16.8 | 15.6 | 350(3.50) | -14.9 | -17.8 | -15.3 | -16.0 |
| 66.7(52) | 19.6 | 13.8 | 18.4 | 17.2 | 360(3.60) | -14.1 | -17.0 | -14.5 | -15.3 |
| 68.7(54) | 21.2 | 15.4 | 20.0 | 18.7 | 375(3.75) | -12.9 | -15.9 | -13.4 | -14.1 |
| 70.7(56) | 22.8 | 16.9 | 21.6 | 20.3 | 400(4.00) | -11.0 | -14.1 | -11.6 | -12.3 |
| 72.7(58) | 24.4 | 18.4 | 23.0 | 21.8 | 425(4.25) | -9.2 | -12.4 | -9.8 | -10.6 |
| 74.7(60) | 25.9 | 19.8 | 24.5 | 23.3 | 450(4.50) | -7.5 | -10.7 | -8.2 | -8.9 |
| 76.7(62) | 27.4 | 21.3 | 26.0 | 24.7 | 475(4.75) | -5.9 | -9.2 | -6.6 | -7.3 |
| 78.7(64) | 28.9 | 22.7 | 27.4 | 26.1 | 500(5.00) | -4.3 | -7.6 | -5.0 | -5.7 |
| 80.7(66) | 30.3 | 24.0 | 28.8 | 27.5 | 525(5.25) | -2.8 | -6.2 | -3.6 | -4.3 |
| 82.7(68) | 31.7 | 25.4 | 30.1 | 28.9 | 550(5.50) | -1.3 | -4.8 | -2.2 | -2.8 |
| 84.7(70) | 33.1 | 26.7 | 31.5 | 30.2 | 575(5.75) | 0.1 | -3.4 | -0.8 | -1.5 |
| 86.7(72) | 34.4 | 28.0 | 32.8 | 31.6 | 600(6.00) | 1.5 | -2.1 | 0.6 | 0.1 |
| 88.7(74) | 35.8 | 29.3 | 34.1 | 32.8 | 625(6.25) | 2.8 | -0.8 | 1.8 | 1.2 |
| 90.7(76) | 37.1 | 30.6 | 35.4 | 34.1 | 650(6.50) | 4.1 | 0.4 | 3.1 | 2.4 |
| 92.7(78) | 38.4 | 31.8 | 36.8 | 35.4 | 675(6.75) | 5.4 | 1.6 | 4.3 | 3.6 |
| 94.7(80) | 39.7 | 33.0 | 37.8 | 36.6 | 700(7.00) | 6.6 | 2.8 | 5.5 | 4.8 |
| 99.7(85) | 42.8 | 36.0 | 40.8 | 39.6 | 725(7.25) | 7.8 | 3.9 | 6.7 | 6.0 |
| 104.7(90) | 45.8 | 38.8 | 43.7 | 42.5 | 750(7.50) | 9.0 | 5.1 | 7.8 | 7.1 |
| 109.7(95) | 48.6 | 41.6 | 46.5 | 45.2 | 775(7.75) | 10.1 | 6.1 | 8.9 | 8.2 |
| 114.7(100) | 51.4 | 44.3 | 49.2 | 47.9 | 800(8.00) | 11.2 | 7.2 | 9.9 | 9.2 |
| 119.7(105) | 54.1 | 46.8 | 51.8 | 50.5 | 825(8.25) | 12.3 | 8.2 | 11.0 | 10.3 |
| 124.7(110) | 56.8 | 49.3 | 54.3 | 53.0 | 850(8.50) | 13.3 | 9.2 | 12.0 | 11.3 |
| 129.7(115) | 59.3 | 51.8 | 56.8 | 55.5 | 900(9.00) | 15.4 | 11.2 | 14.0 | 13.3 |
| 134.7(120) | 61.8 | 54.1 | 59.2 | 57.9 | 950(9.50) | 17.4 | 13.1 | 15.9 | 15.2 |
| 139.7(125) | 64.2 | 56.4 | 61.5 | 60.2 | 1000(10.0) | 19.3 | 14.9 | 17.7 | 16.7 |
| 144.7(130) | 66.5 | 58.6 | 63.7 | 62.4 | 1050(10.5) | 21.1 | 16.6 | 19.5 | 18.7 |

| | | | | | | | | | |
|---|---|---|---|---|---|---|---|---|---|
| 147.9(135) | 68.8 | 60.8 | 65.9 | 64.6 | 1100(11.0) | 22.9 | 18.3 | 21.2 | 20.4 |
| 154.7(140) | 71.0 | 62.9 | 68.1 | 66.8 | 1150(11.5) | 24.6 | 19.9 | 22.8 | 22.1 |
| 159.7(145) | 73.2 | 65.0 | 70.2 | 68.8 | 1200(12.0) | 26.2 | 21.5 | 24.4 | 23.6 |
| 164.7(150) | 75.4 | 67.0 | 72.2 | 70.9 | 1250(12.5) | 27.9 | 23.0 | 25.9 | 25.2 |
| 169.7(155) | 77.5 | 69.0 | 74.2 | 72.9 | 1300(13.0) | 29.4 | 24.5 | 27.4 | 26.7 |
| * Inches of mercury | | | | | NOTE: kPa and Bar are absolute pressure | | | | |

Used by permission of DuPont Fluoroproducts

(psig) (The numbers in parentheses = the reading on the manifold gauge)

# GLOSSARY OF TERMS

A-coil: An A-shaped evaporator normally installed in the plenum chamber of a central air-conditioning unit

accumulator: A small storage cylinder placed anywhere in the suction line. It collects liquid refrigerant and allows it to vaporize before getting into the compressor.

air infiltration: Penetration of the outside air into a refrigerated or an air-conditioned area

ammeter: A device to measure the flow of electrical current in a circuit

ampere: The standard of measurement for the amount of current flowing past any given point in an electrical circuit

atmospheric pressure: 14.7 lb/in$^2$ pressure exerted from the whole mass of air surrounding the earth upon different objects at sea level

automatic expansion valve (AEV): A pressure controlled valve that controls the flow of refrigerant entering the evaporator

axial flow fan: A regular fan (with blades perpendicular to its shaft, moving air along the direction of its shaft)

boiling point: The temperature at which any liquid boils and changes to vapor. Boiling point of water at sea level is 212°F.

bending spring: A tube in the form of a coiled spring used in bending tubing

back pressure: Another term for low-side pressure

bimetal strip: A flat spring composed of two different metals soldered together which is flexed by temperature changes

bin thermostat: A thermostat or lever installed in the storage compartment of an ice-making machine to stop the production of ice when the bin is full

capacitor: A device for holding or storing an electric charge, used with compressor motors to boost starting and/or running efficiency.

capillary tube: A slender tube having a very small bore. Its small diameter maintains a pressure difference in the sealed system as well as controlling the flow of refrigerant into the evaporator when the compressor runs.

check valve: A device installed in the sealed system that allows refrigerant to flow in only one direction

charge: The amount of refrigerant put into a sealed system

circuit (electrical): The path of electric current

closed circuit: An uninterrupted path for electrical current

compressor: A motor-driven pump that draws vapor refrigerant from the evaporator, compresses it, and forces it out under high pressure to the condenser.

cooling coil: Another name for an evaporator coil

cold: The absence of heat

condensation tray: A drip man

condensing unit: An assemblage of refrigerating components that change vapor refrigerant to liquid refrigerant usually consisting of the compressor, condenser, condenser fan, and a receiver tank. It could be roof or ground mounted.

condenser fan: A fan located near the condenser to move the air rapidly over the surface of the condenser coil to aid in the transfer of heat

cold control: Another name for a thermostat

compound gauge (refrigeration): A device capable of measuring pressures that are above or below atmospheric pressure

condensation: The process by which moisture in the air forms droplets of water when coming in contact with a cold surface

conductor: A body through which heat or current is transmitted

convection: As air is cooled, it becomes heavier and tends to seek the lower levels. As it becomes warmer it gets lighter and tends to rise; thus, creating air circulation.

capacitor: A device for storing an electrical charge

de-energize: To stop current flow through a circuit

defrost bimetal: A temperature-sensing device installed on the evaporator coil to stop current flow to the defrost heater when frost on the coil has melted. (Most commercial units employ a sensing bulb for this purpose.)

defrost cycle: One of the refrigerating cycles in which the accumulation of ice on the evaporator coil is defrosted. In this cycle, sometimes the compressor is turned off and the defrost heaters turned on; and sometimes a solenoid valve bypasses not gas to the evaporator with the compressor running.

dehumidifier: A device to remove moisture from the air

distributor (refrigeration): An apparatus having a single inlet port and more than one outlet for directing the flow of refrigerant into multiple ports of a condenser or an evaporator(s)

drier (filter-drier) (dehydrator): A device to remove moisture from refrigerant circulating in a sealed system

defrost timer: An apparatus that causes refrigeration cycles to change. (See defrost cycle above.)

defrost heater: The heating element placed around or adjacent to the evaporator coils. During the defrost cycle, it melts the ice built up during the cooling cycle.

duct work: Channel(s) through which air flows

**differential:** The pressure difference between the cut-in and cut-out pressures in a pressure control switch

**energize:** To connect power to an electrical component

**electrical circuit:** The complete path of electrical current

**electromagnet:** A device creating a magnetic field when energized

**electromotive force:** Electrical energy that induces current flow in an electrical circuit which is measured in volts

**epoxy:** A substance characterized by great adhesiveness and strength. Can be used to seal small evaporator leaks.

**equalizer line:** Auxiliary tubing linking the evaporator outlet to the TEV to exert an average true pressure on the valve diaphragm when there is a significant pressure drop in the evaporator

**evacuation:** Removing moist air from a sealed system by creating a vacuum with the help of a vacuum pump

**evaporator pressure drop:** The difference in pressure between the inlet and the outlet of the evaporator. (Pressure drop also occurs in a filter-drier.)

**expansion valve:** A pressure-operated device in a refrigerating system to create and maintain the pressure difference between the high and low sides

**evaporator:** Part of the sealed system in a refrigeration unit that converts liquid refrigerant into a vapor to absorb heat

**Fahrenheit:** The scale used by GD Fahrenheit on his thermometer, which, at sea level, water boils at 212°F and freezes at 32°F.

**farad:** The unit of electrical capacity; the capacity of a capacitor (called capacitance). Because this unit of measure is too large, capacity is usually shown in microfarads (one millionth of a farad). Symbol f or mf. After Michael Faraday, English physicist.

**fittings:** Male and female connectors or adapters used in joining together the ends of tubing

flare: A small funnel-like enlargement in the ends of tubing enabling the flare nut to make an airtight seal

flare nut: One type of fitting which joins tubing after the ends have been flared, and its companion nut has been fitted to the other tube to be joined.

flash gas: That portion of the liquid refrigerant which evaporates instantly (flashes) and turns into vapor as it passes through the orifice of a refrigerant control

float valve: A valve to control the level of liquid in a container, operated by a device floating on the surface of the liquid.

floodback: The migration of liquid refrigerant into a part of a system during the off cycle.

flooding: Permitting liquid refrigerant to flow into certain parts of the refrigerating system.

flooded system: Refrigerating units employing a high or low side float system. Uncommon in modern refrigeration.

flux: The chemical contained in some solders, or applied separately to surfaces prior to brazing to prevent oxidation. (Electrical): A circular magnetic field created around a conductor carrying a current.

four-way valve: Also referred to as a reversing valve. A device used to reverse the flow of refrigerant between the evaporator and the condenser without passing through the compressor.

freezing point: The temperature at which a liquid solidifies

Freon: See refrigerant

frost-free refrigerator: A refrigerator using a defrost timer

fuse: A protective device in an electrical circuit. A conductor with less electrical tolerance than other elements in the circuit. An overload or electrical surge causes it to melt and open the circuit.

gas: The state of a substance which is neither solid nor liquid gasket: Rubber or other pliable material placed around the inner edge of a refrigerator door to seal the cabinet from outside air

gauge (high or low pressure): Measures pressure in pounds per square inch (psi). A refrigeration gauge scale reads up to 500 lb/in$^2$; the low-pressure gauge is capable of measuring vacuum in inches of mercury (inHg).

ground: The connection to transfer excess electrical current to earth; used to protect personnel from accidental electrical shock

ground wire: A safety device. Usually a third wire from the unit to a ground source to carry off excess electrical current.

guardette: Another name for an overload protector

head pressure: Another name for high side or discharge pressure

heat: A form of energy which causes a body to rise in temperature

heat exchanger (residential units): That part of the sealed system where the capillary tube and suction line are joined (side by side) for the purpose of transferring heat as is passes through

heat pump: An air conditioner capable of heating as well as cooling by the use of a four-way valve which reverses the direction of refrigerant flow

hermetic compressor: A compressor and its motor sealed within a metal housing. (Hermetically sealed.)

hermetic system: Refers to the hermetic compressor commonly found in residential and small commercial units

high-pressure gauge: An instrument used for measuring pressures above atmospheric pressure in the high side of the sealed system

high side: The parts of the sealed system that are under high pressure (condensing pressure); including the side of the compressor that discharges hot gas, the condenser, the filter-drier, and the receiver

high-side float control: A control mechanism that maintains a constant level of liquid refrigerant in the high-pressure side of a refrigerating unit

high-side pressure: Pressure in the high side of a sealed system

high-side-pressure control: (high-pressure cut out) A switch that limits the high-side pressure of a refrigerating unit

high vacuum pump: The apparatus used to create a high vacuum in a sealed system. The pump causes a drop in pressure to vaporize and draw out moisture from the system.

horsepower: The unit of power equal to 746 watts or 550 ft/lbs of work per second

hot gas bypass: The tubing system to transfer hot vapor refrigerant from the condenser to the low temperature side of a refrigerating unit

insulation (electrical): Encloses electrical conductors (like wires) with a nonconducting material to prevent unintentional grounding

insulation (thermal): Materials that are poor heat conductors. The liners of cold compartments and freezers.

latent heat: Heat not measurable on a thermometer which causes the change of state of a substance

lead (leed): That portion of electrical wiring connected to a component

leak: Escape of refrigerant from a sealed system

leak detector: (electronic type) A battery-operated instrument that emits a high-pitched sound when its probe is near a source of a refrigerant leak

liquid line: The tube connecting the condenser (or receiver) to the capillary tube or any other refrigerant control (TEV or AEV)

load: The work required of the system. The amount of cooling necessary to accomplish the desired results.

loading: The period in which the TEV permits the refrigerant to flow into the evaporator

low-side float control: A control mechanism that maintains a constant level of liquid refrigerant in the low-pressure side of a refrigerating unit low-side pressure: Pressure in the compressor suction line and the evaporator

low-side-pressure control (low-pressure cut-out switch): A switch connected to the compressor that limits the low-side pressure of a refrigerating unit

mercury-type thermostat: An electrical switch that controls the temperature by the expansion or contraction of mercury as it is affected by heat or cold

moisture: Water in the system caused by leakage of outside air

motor compressor: The mechanism creating the pressure difference in the sealed system

off cycle: That part of the refrigeration cycle in which the compressor does not run

open circuit: Interrupted electrical path preventing current flow

overload: Current above the capacity of the mechanism

overload protector: A device placed in series with the common terminal of the compressor. It shuts off power to the compressor upon sensing above-normal amperage. It reconnects power to the compressor when the temperature drops.

parallel circuit: The connection of separate parts in a circuit so that all positive poles are connected to one conductor and all negative poles connected to another

plenum chamber: In a central air-conditioning system, a housing that holds heating and cooling elements, and a system of mechanical ventilation from which temperature (and sometimes moisture) controlled air is forced into the conditioned area(s).

pressure control: A safety device that connects and disconnects the power supply to the compressor (and certain other components depending on the unit) as pressures change in the sealed system

pressure drop: Sudden decrease in pressure

purging: The process of removing moisture from a sealed system. This process allows a less-than-full-charge to circulate and "flush out" the unwanted vapor and moisture.

**receiver:** A cylindrical container in a sealed system for storing refrigerant

**refrigerant:** Any of a class of nonflammable hydrocarbons containing fluorine used in refrigeration systems

**refrigeration:** The cooling of a space or object below ambient temperature

**relay:** An electromagnetic device that normally functions as a switch to energize or de-energize specific circuits

**resistance:** To oppose, restrict, or govern the flow of current. It is measured in ohms.

**restriction (partial):** Occurs when moisture gets into the sealed system and turns to ice in the filter-drier, capillary tube, or any other refrigerant control, blocking the flow of refrigerant. Also caused by wax or other small particles in the sealed system.

**retainer strip:** A metal or plastic strap used to hold another object in place

**refrigeration with cycle defrost:** In these units, each time the thermostat is satisfied, the evaporator coil is defrosted by an electric heating element (or a solenoid is energized shunting hot refrigerant through tubing adjacent to the evaporator coil).

**reversing valve:** See four-way valve

**saddle valve:** A type of piercing valve silver-brazed to tubing

**saturated vapor pressure:** A certain pressure imposed upon vapor refrigerant at which no more refrigerant can vaporize

**schematic diagram:** A line drawing of an electrical circuit in which the wiring and components are simplified and depicted as commonly recognized symbols

**sealed system:** An airtight passage for refrigerant circulating through the compressor, filter-drier, capillary tube, evaporator, and condenser.

**self-contained unit:** A refrigerating or air-conditioning system in which all of the components of the entire sealed system are contained within the same console

sensing bulb: A tubular extension of a temperature control device that is filled with refrigerant and attached to the evaporator. As temperature changes occur, the expansion and contraction of the refrigerant in the tube causes the contacts in the temperature control to open or close the circuit to the compressor.

series circuit: The connection of separate parts end-to-end in an electrical circuit to form a single path for current

series-parallel circuit: Having the structure of both series and parallel circuits

service valve: An access valve, such as a saddle valve or piercing valve, that allows entry into the sealed system for pressure testing, recharging, etc.

shaded-pole motor: A type of motor having a run winding but no start winding used only for light loads

short: A circuit with no load

short circuit: Unintentional contact between wires or components resulting in lowered resistance and excessive flow of current. (This may cause an overload.) A circuit with no load.

sight glass: A device installed in a sealed system with a clear glass insert in its top for viewing the refrigerant to detect a deficiency or moisture in the system

silver brazing: The process of joining metals with a nonferrous substance with some percentage of silver using enough heat to bring it its melting point.

slant-type coil: A flat evaporator coil placed at an angle to the flow of air.

solenoid: A coil of insulated wire that produces a strong magnetic field in its center when current passes through it

solenoid valve: A valve operated by a moving armature activated by an electromagnetic field, used to control the flow of refrigerant or other liquid.

split-phase motor: A motor with two separate stator windings. Both windings are energized during start-up. When motor gains speed, one winding is de-energized by the starting relay and the motor continues running on the remaining winding.

split system: A refrigerating or air-conditioning system in which the evaporator is installed in separate location from the rest of the unit.

squirrel cage fan: A fan with blades parallel to its shaft that moves air perpendicular to its shaft

starting relay: An electrical device that alternatingly connects and disconnects electric current to the start winding of a compressor motor

strainer: A filter or screen placed in the sealed system to retain any solid particles that may be circulating with the refrigerant

subcooling: Lowering the temperature of the refrigerant to compensate for the liquid-line pressure losses and to prevent flash gas at the TEV

suction line: The tube connecting the evaporator and compressor

suction side: The low-pressure side that begins at the inlet of the capillary tube and extends to the inlet of the compressor

superheat: The temperature difference between the vaporizing refrigerant in the evaporator and the evaporator outlet

superheated refrigerant: The temperature of vaporized refrigerant

surge tank: A container installed in the low-pressure side of a refrigerating system to prevent short cycling by decreasing the pressure fluctuation in the low side

swaging: The enlargement of the end of one tube to allow another tube of the same size to be inserted into it

switch: A device to connect or disconnect an electrical circuit

temperature: The degree of hotness or coldness measured on a specific scale such as a thermometer

terminal: A device attached to the end of a wire or to an apparatus for convenience in making electrical connections termination switch: See defrost bimetal.

test light: A bulb with a pair of leads used in testing a specific range of voltages

thermistor: A semiconductor in which the resistance varies with its temperature

thermostat: An automatic heat-sensing device for regulating temperature by opening and closing an electrical circuit

thermostatic expansion valve (TEV): A valve operated by the temperature and pressure in the evaporator to control the flow of refrigerant

throttling: Expansion of gas through a controlled opening

transformer: A device with two sets of windings that will generate the desired voltage(s) from its secondary winding

tubing: Pipes that carry the refrigerant. Usually made of copper, aluminum, or steel.

two-temperature valve: This valve is located in the suction line at the outlet of the warmer evaporator in multiple evaporator systems. When a different temperature is required in each evaporator, this valve maintains a predetermined pressure (and consequently temperature) in the warmer evaporator.

vacuum: A space exhausted to a high degree (by an air pump) to a point well below atmospheric pressure

vaporization: The conversion of liquid into steam by increasing temperature or reducing pressure; or a combination of both

volt: The measure of energy (electromotive force) which pushes electrons through a circuit

voltmeter: An instrument for measuring voltage in a circuit

watt: The rate of work represented by a current of one ampere under the pressure of one volt; a volt-ampere. (After James Watt.)

water-cooling tower: A unit in which heated water is rapidly cooled by bringing it into contact with fan-forced circulating air

water-cooled condenser: A type of condenser which is rapidly cooled by the action of water flowing over or adjacent to its coil

wax: An ingredient in compressor oil. When cooled enough, it settles in a refrigerant metering device (such as the capillary tube) causing a restriction in the sealed system.

winding: A coil of wire that produces a magnetic field when current is applied

# INDEX

Brazing and flaring, 68
Capacitors, 89, 90, 91
Capillary Tube:
    How it works, 31, 64, 65
    Test for restriction, 178
    Size and length chart, 66
    How to unclog, 64
Central Air Conditioners, 247 through 289
Charging Methods:
    By using Ammeter, 110, 112
    By using a charging Cylinder, 111, 113
    By observing evaporator frost pattern, 114, 115
    By gauge pressure, 115, 116
    By using a charging chart, 283
    By high side, 406, 409
    By name plate information, 113
Causes of high head pressure, 180, 181, 182
Compressor:
    Adding oil, 68
    Burnout, 174, 175
    Hard start kit, 63
    Hermetic, 61, 62
    How to identify unmarked terminals, 58, 59
    Short Cycling, 171, 172, 173
    Starting a stuck compressor, 60
    Supply voltage checking, 53
    Testing by using an Ohmmeter, 57
    Testing by using a test cord, 56
    Testing compressor power supply voltage, 53
    Testing by using Wattmeter, 50, 52
    Three phase compressor installation, 273
    Types of compressor, 48, 49, 61
Condenser:
    Cleaning, 181
    Fan motors, 33, 82, 83, 84, 85
    High head pressure, 180
    Water condenser, 212, 213, 393, 394
    Water condenser valve adjustment, 390, 391, 392
Defrost systems, 32, 33
Evaporator, 30, 249, 251, 252, 253
    Fan, 32, 233, 236, 239
Fans, 236, 237, 238, 239
Filter-drier, 27, 108, 109
Heats pumps, 265 through 271
Humidifiers and de-humidifiers, 228, 229
Ice machines, 186 through 202
Installation of:
    Window air conditioning, 230 through 234
    Central air conditioning, fig. 249, 272, 253, 255
    Determining tonnage for a new system, 274
Learning how to read electrical diagrams, 306 through 353
Low side and high side float systems, 169
Mini pressure control, 248
Oil separator, 154
Piping method for a/c and

433

refrigeration,285,286,287,365
Pressure control commercial refrigeration,140 through 144
Questions and answers about schematic diagrams,334 through 351
Relay:
- Commercial,243,70,71,74,75
- Starting-residential,26,27,73
- Wiring,fig. 153,74

Refrigeration,7 through 10
Refrigerator Types: 14,103
- Manual Defrost 17,30
- Frost Free 19,20,22
- Cycle defrost 17,18,32

Short cycling of compressor,170,171,172,173, 176
Saturated vapor refrigerant temperature,132
Test of knowledge,222,223,284,397, 34,38,11,105,128,129
Testing:
- Fan motor,82,83,85
- Relays,74,75,71,72
- Defrost limit switch 86,87,88
- Capacitors 89,91
- Compressor 50,51,52,54,56,57,63,67
- Defrost heaters 96,97,98
- Expansion valves(automatic),371,388
- Thermostats,92,93
- Timers,76,77,78,79,80
- Supply voltage,52,53
- Contractors,245,246,247

Troubleshooting charts:
- Residential refrigerators and freezers,118,121,127
- Commercial ice-machines,198 through 201
- Water fountain,204
- Refrigeration system by touch,207,208
- Compressor(commercial and residential),216 through 221

Central air conditioning system (in cooling mode),264
Central air conditioner,227 through 282
Compressor,54,354,355
Refrigerant flow controls,357 through 373
Excessive oil frosting,374 through 381

Valves:
- Hot bypass valve (how to adjust),362
- Pressure regulating (reducing) valve,145,146,147,359
- Check valves,144,145
- Solenoid valves,148 through 153
- Reversing valves,156,157
- Thermostatic expansion (TEV) valves,369,371,273,158,160, 164
- Automatic expansion (AEV) valves,158,383,384,386

Wiring Method for:
- Central air conditioners,fig. 154,153,163
- Residential/commercial refrigerators and freezers,fig. 7

Made in United States
Troutdale, OR
10/15/2024

23791481R00268